Simulation Foundations, Methods and Applications

Series Editor

Louis G. Birta, University of Ottawa, Ottawa, ON, Canada

Advisory Editors

Roy E. Crosbie, California State University, Chico, CA, USA

Tony Jakeman, Australian National University, Canberra, ACT, Australia

Axel Lehmann, Universität der Bundeswehr München, Neubiberg, Germany

Stewart Robinson, Loughborough University, Loughborough, Leicestershire, UK

Andreas Tolk, Old Dominion University, Norfolk, VA, USA

Bernard P. Zeigler, University of Arizona, Tucson, AZ, USA

More information about this series at http://www.springer.com/series/10128

José L. Risco Martín · Saurabh Mittal ·
Tuncer Ören

Editors

Simulation for Cyber-Physical Systems Engineering

A Cloud-Based Context

Springer

Editors
José L. Risco Martín
Department of Computer Architecture
and Automation
Universidad Complutense de Madrid
Madrid, Spain

Saurabh Mittal
The MITRE Corporation
Fairborn, OH, USA

Tuncer Ören
School of Electrical Engineering
and Computer Science
University of Ottawa
Ottawa, ON, Canada

ISSN 2195-2817   ISSN 2195-2825   (electronic)
Simulation Foundations, Methods and Applications
ISBN 978-3-030-51911-7   ISBN 978-3-030-51909-4   (eBook)
https://doi.org/10.1007/978-3-030-51909-4

This Springer imprint is published by the registered company Springer Nature Switzerland AG
The registered company address is: Gewerbestrasse 11, 6330 Cham, Switzerland

To Esther, my amazing wife, and to our two children Jaime and Gabriel, who are indeed a treasure for us.

José L. Risco Martín

To my wife and kids who create my world and everything in it.

To the Almighty who breathes life around me.

Saurabh Mittal

To my life-long friend and wife Füsun

Tuncer Ören

Preface

Cyber-Physical Systems (CPS) are physical systems augmented with software to allow them the ability to process knowledge, communicate, and control. They can interact between themselves, as well as with humans, through a variety of modalities running in complex environments. CPS are increasingly applied across various sectors of the society with the advent of Internet of Things (IoT)-enabled devices. Modeling and simulation (M&S) is appearing as a viable way to perform CPS engineering but there exist many problems in performing CPS M&S. Many different modeling paradigms, methods, and solutions are in use, which lead to challenges known to the M&S community from hybrid simulation approaches. While the technical problem to be solved in a lab-setting is within the purview of current engineering methods, the deployment of CPS in a net-enabled society introduces complexity that changes the risk profile of the engineered solution. CPS constituent elements may also manifest autonomy in its entire spectrum, i.e., from a passive computational element to a highly sophisticated adaptive learning system that contextualizes its behavior to the dynamic environment. The problem is further exacerbated when such CPS have constituent elements that adapt and learn from a novel situation, and coordinate with other systems that were not in the design workbench, to begin with.

This new class of complex systems such as CPS have computational elements that may either add intelligence or provide decision-making into the larger socio-technical CPS or may act as a Test and Evaluation (T&E) infrastructure for the very CPS itself. M&S is at the center-stage of such T&E and experimentation. The increased use of cloud-based technologies introduces problems of virtualizing various CPS elements. Remote access mechanisms need to be instituted to keep CPS secure. The contribution of M&S Cloud-based solutions in CPS engineering cannot be overstated. Because of the distributed nature of CPS, cloud infrastructure is an unavoidable possibility in CPS engineering. The Cloud Infrastructure is becoming the foundation for using simulation as a service in various M&S endeavors. While the M&S community has been researching M&S as-a-service solutions for more than a decade, there has not been a definitive text that illustrates various aspects of bringing the power of cloud computing to the M&S discipline.

Rapid advancements in cloud systems engineering have speeded up the adoption of Simulation as-a-service with only a handful of research groups developing infrastructure that is production ready.

This book supplies a landscape of M&S technologies and the infrastructure about the usage of cloud-based environments for CPS engineering. The book covers the engineering, design, and application of cloud-simulation technologies and infrastructure as applicable to CPS engineering. It captures the knowledge and lessons learned from developing real-time embedded and robotic systems and their deployment in a cloud-based infrastructure for their application in CPS engineering and deployment in an IoT-enabled society. Cloud-based M&S also act as a medium to facilitate CPS engineering and governance. The disciplines of cloud-based simulation and CPS engineering are evolving at a rapid rate, but are not aptly supporting each other's advancement. This book brings together these two communities that already serve multi-disciplinary applications, and provides the state-of-the-art in methods, technologies, and approaches that elaborate on the cloud-based M&S support to CPS engineering across many sectors such as Healthcare, Smart Grid, Manufacturing, Education, Defense, and Energy. The book is organized into four main parts, each making up several chapters. Part I guides the reader through the fundamental concepts of different infrastructures to perform Cloud-based CPS engineering. Part II describes methodologies to perform service composition, scheduling, and the integration of nature-inspired modeling in Cloud-based CPS engineering. Part III describes some real-world Cloud-based CPS engineering. Finally, Part IV analyzes various aspects of reliability, truth, resilience, and ethical requirements of Cloud-based CPS Engineering.

We invite the reader to explore the state-of-the-art in M&S-based CPS engineering in a cloud-enabled context.

Madrid, Spain   José L. Risco Martín, Ph.D.
Fairborn, OH, USA   Saurabh Mittal[1], Ph.D.
Ottawa, ON, Canada   Tuncer Ören Ph.D.

---

[1]The author's affiliation with The MITRE Corporation is provided for identification purposes only, and is not intended to convey or imply MITRE's concurrence with, or support for, the positions, opinions or viewpoints expressed by the author(s). Approved for Public Release. Distribution Unlimited. Case Number 19-1916-8.

Contents

Editors and Contributors

About the Editors

José L. Risco Martín is an Associate Professor of the Computer Science Faculty at Universidad Complutense de Madrid, Spain, and head of the Department of Computer Architecture and Automation at the same University. He received his M. Sc. and Ph.D. degrees in Physics from Universidad Complutense de Madrid, Spain, in 1998 and 2004, respectively. His research interests focus on Computer Simulation and Optimization, with emphasis on Discrete Event Modeling and Simulation, Parallel and Distributed Simulation, Artificial Intelligence in Modeling and Optimization, and Feature Engineering. In these fields, he has co-authored more than 150 publications in prestigious journals and conferences, several book chapters, and three Spanish patents. He has received the SCS Outstanding Service Award in 2017, and the HiPEAC Technology Transfer Award in 2018. He is an associate editor of SIMULATION: Trans. of Soc. Mod. and Sim. Int., and has organized several Modeling and Simulation conferences like DS-RT, ANSS, SCSC, SummerSim or SpringSim. He is ACM Member and SCS Senior Member. Prof. José L. Risco Martín has participated in more than 15 research projects and more than 10 contracts with industry. He has elaborated simulation and optimization models for companies like Airbus, Repsol or ENAGAS. He has been CTO of DUNIP Technologies LLC, and co-founder of BrainGuard SL.

Saurabh Mittal is the Chief Scientist for Simulation, Experimentation, and Gaming Department at The MITRE Corporation, the President for the Society for the Society for Modeling and Simulation (SCS) International. He served as the Director-at-Large for SCS from 2017–2020. He holds a Ph.D. and MS in Electrical and Computer Engineering with dual minors in Systems and Industrial Engineering, and Management and Information Systems from the University of Arizona, Tucson. He has co-authored over 100 publications as book chapters, journal articles, and conference proceedings including 4 books, covering topics in the areas of complex systems, system of systems, complex adaptive systems, emergent behavior,

Modeling and Simulation (M&S), and M&S-based systems engineering across many disciplines. He serves on many international conference program/technical committees, as a referee for prestigious scholastic journals and on the editorial boards of Transactions of SCS and Journal of Defense M&S. He is a recipient of Herculean Effort Leadership award from the University of Arizona, US DoD's highest civilian contractor recognition: Golden Eagle Award, and Outstanding Service and Professional Contribution awards from SCS.

Tuncer Ören is a professor emeritus of computer science at the School of Electrical Engineering and Computer Science of the University of Ottawa, Canada. He has been involved with simulation since 1965. His Ph.D. is in Systems Engineering from the University of Arizona, Tucson, AZ (1971). His basic education is from Galatasaray Lisesi, a high school founded in his native Istanbul in 1481, and in Mechanical Engineering at the Technical University of Istanbul (1960). His **research interests** include: advancing methodologies for modeling and simulation; agent-directed simulation; agents for cognitive and emotive simulations (including representations of human personality, understanding, misunderstanding, emotions, and anger mechanisms); computational awareness; reliability, QA, failure avoidance, ethics; as well as body of knowledge and terminology of modeling and simulation. He has over 540 **publications**, including 54 books and proceedings (+ 3 in press and in preparation). He has contributed to over 500 **conferences and seminars** held in 40 countries. Dr. Ören **has been honored** in several countries: **USA**: He is a Fellow of SCS (2016), an inductee to SCS Modeling and Simulation Hall of Fame–Lifetime Achievement Award (2011), and received SCS McLeod Founder's Award for Distinguished Service to the Profession (2017). **Canada**: Dr. Ören has been recognized, by IBM Canada (2005), as a pioneer of computing in Canada. He received the Golden Award of Excellence from the International Institute for Advanced Studies in Systems Research and Cybernetics (2018). **Turkey**: He received "Information Age Award" from the Turkish Ministry of Culture (1991), an Honor Award from the Language Association of Turkey (2012), and Lifetime service award from the Turkish Informatics Society and Turkish Association of Information Technology (2019). A book was edited by Prof. Levent Yilmaz: Concepts and Methodologies for Modeling and Simulation: A Tribute to Tuncer Ören. Springer (2015).

Contributors

José L. Ayala Department of Computer Architecture and Automation, Complutense University of Madrid, Madrid, Spain

Thomas Bitterman Wittenberg University, Springfield, USA

Xudong Chai Aerospace Cloud Science and Technology Co., Ltd, Beijing, China

Daniel M. Dubois Centre for Hyperincursion and Anticipation in Ordered Systems (CHAOS), CHAOS ASBL, Institute of Mathematics B37, University of Liège, Liège, Belgium;
HEC Liège, Université de Liège, Liège, Belgium

Douglas Flournoy The MITRE Corporation, Bedford, MA, USA

Sarada Prasad Gochhayat Virginia Modeling Analysis and Simulation Center, Old Dominion University, Suffolk, VA, USA

Hui Gong CASIC Intelligence Industry Development Co., Ltd, Beijing, China

Carlos González Department of Computer Architecture and Automation, Complutense University of Madrid, Madrid, Spain

Hemant Gupta School of Computer Science, Carleton University, Ottawa, Canada

Md. Ariful Haque Computational Modeling and Simulation Engineering, Old Dominion University, Norfolk, VA, USA

Kevin Henares Department of Computer Architecture and Automation, Complutense University of Madrid, Madrid, Spain

Román Hermida Department of Computer Architecture and Automation, Complutense University of Madrid, Madrid, Spain

Baocun Hou Aerospace Cloud Science and Technology Co., Ltd, Beijing, China

Rob Kewley simlytics.cloud LLC., Morristown, NJ, USA

Bheshaj Krishnappa Risk Analysis and Mitigation, ReliabilityFirst Corporation, Cleveland, OH, USA

Sanja Lazarova-Molnar University of Southern Denmark, Odense, Denmark

Bo Hu Li State Key Laboratory of Intelligent Manufacturing System Technology, Beijing Institute of Electronic System Engineering, Beijing, China;
School of Automatic Science and Electrical Engineering, Beihang University, Beijing, China;
Beijing Complex Product Advanced Manufacturing Engineering Research Center, Beijing Simulation Center, Beijing, China

Feng Li School of Computer Science and Engineering, Nanyang Technological University, Singapore, Singapore

Ting Yu Lin School of Automatic Science and Electrical Engineering, Beihang University, Beijing, China;
Beijing Complex Product Advanced Manufacturing Engineering Research Center, Beijing Simulation Center, Beijing, China

Yang Liu Aerospace Cloud Science and Technology Co., Ltd, Beijing, China

Margaret L. Loper Georgia Tech Research Institute, Atlanta, USA

Matthew T. McMahon The MITRE Corporation, Mclean, VA, USA

Saurabh Mittal The MITRE Corporation, Fairborn, OH, USA

Nader Mohamed California University of Pennsylvania, Harrisburg, USA

Tuncer Ören School of Electrical Engineering and Computer Science, University of Ottawa, Ottawa, Ontario, Canada

Josué Pagán Department of Electronic Engineering, Technical University of Madrid, Madrid, Spain

Ernest H. Page The MITRE Corporation, Mclean, VA, USA

José L. Risco Martín Department of Computer Architecture and Automation, Universidad Complutense de Madrid, Madrid, Spain

Sachin Shetty Virginia Modeling Analysis and Simulation Center, Old Dominion University, Suffolk, VA, USA

Robert Siegfried Aditerna GmbH, Riemerling/Munich, Germany

Mayank Singh University of KwaZulu-Natal, Durban, South Africa

Xiao Song State Key Laboratory of Intelligent Manufacturing System Technology, Beijing Institute of Electronic System Engineering, Beijing, China

Andreas Tolk The MITRE Corporation, Hampton, VA, USA

Brian M. Wickham The MITRE Corporation, Mclean, VA, USA

Chen Yang School of Computer Science and Technology, Beijing Institute of Technology, Beijing, China

Bernard Zeigler RTSync Corp, Arizona Center for Integrative Modeling and Simulation, Sierra Vista, AZ, USA

Lin Zhang School of Automation Science and Electrical Engineering, Beihang University, Beijing, China

Longfei Zhou Computer Science & Artificial Intelligence Laboratory, Massachusetts Institute of Technology, Cambridge, MA, USA

Part I
Foundations

Chapter 1
Cloud-Based M&S for Cyber-Physical Systems Engineering

José L. Risco Martín and Saurabh Mittal

Abstract Cyber-Physical Systems (CPS) are complex systems that have two essential elements: the cyber element and the physical element. These are analogous to the early hardware/software (HW/SW) systems that predated the internet. CPS are fundamentally HW/SW systems with an additional capability of being remotely controlled, which introduces a significant amount of risk in CPS operations. Today, such CPS leverage cloud computing infrastructures to provide scale and wider usage, and as such many such systems are remotely deployed as well. The use of Modeling and Simulation (M&S) in HW/SW engineering is a standard practice but incorporating M&S in the cloud environment to support CPS engineering brings forth a new set of challenges for both the M&S technology and CPS engineering methodologies. This chapter will provide an overview of M&S Cloud computing technology and its impact on CPS engineering, along with various challenges. It will also provide a brief overview of the chapters that follow.

1.1 Introduction

According to a definition provided by the National Science Foundation (NSF), Cyber-Physical Systems (CPS) are hybrid networked cyber and engineered physical elements co-designed to create adaptive and predictive systems for enhanced performance. These systems are built from and depend upon the seamless integration of computation and physical components. NSF early identified CPS as "a key area of research" [1]. Examples of CPS include autonomous automobile systems, auto-

J. L. Risco Martín (✉)
Department of Computer Architecture and Automation, Universidad Complutense de Madrid, C/Prof. José García Santesmases 9, 28040 Madrid, Spain
e-mail: jlrisco@ucm.es

S. Mittal
The MITRE Corporation, Fairborn, OH, USA
e-mail: smittal@mitre.org

© Springer Nature Switzerland AG 2020
J. L. Risco Martín et al. (eds.), *Simulation for Cyber-Physical Systems Engineering*,
Simulation Foundations, Methods and Applications,
https://doi.org/10.1007/978-3-030-51909-4_1

3

matic pilot avionics, industrial control systems, medical monitoring, smart grid, and robotics systems.

A typical CPS is comprised of the following components [2]:

- Sensors
- Actuators
- Hardware platforms (that host sensors and actuators)
- Software interfaces (that access hardware directly or remotely through a cyber environment)
- Computational software environments (that may act both as controller or service provider)
- Networked environments (that allow communication across geographical distances)
- End user autonomy (that allows CPS to be used as a passive system or an active interactive system)
- Critical infrastructures (water, power, etc., that provide the domain of operation and operational use-case)
- Ensemble behaviors
- Emergent behaviors.

Figure 1.1 shows various aspects of CPS divided into Left-Hand Side (LHS) and Right-Hand Side (RHS). LHS consists of a collection of users, systems (both hardware and software), and devices (physical platforms). Traditional systems engineering practices and end user use-cases can be developed in LHS. The RHS shows aspects related to infrastructures. Fundamentally, they can be characterized by Information Technology (IT) and Operational Technology (OT). Between the LHS and RHS is the network/cyber environment that allows information exchange between the two. With the network spanning large geographical distances, the presence of a large number of entities/agents and their concurrent interactions in the CPS result in the ensemble and emergent behaviors. The "infrastructure-in-a-box" is largely unavailable but can be brought to bear with various existing domain simulators in an integrated simulation environment.

Today more and more CPS design problems are reaching insoluble levels of complexity. System complexity continues to grow by leaps and bounds. Multi-level complexity is a fundamental nature of heterogeneous systems today. Current CPS may incorporate large- scale systems, which may be under independent operational and managerial control. These are better termed as Industrial CPS. The Internet of Things (IoT), which is also considered as a CPS, is beginning to incorporate all of these characteristics and is becoming a significant contributor to the increase in complexity. However, the IoT phenomenon is still in the formative stages of apparently exponential growth. Designing these systems is equally complex and methodologies available through traditional systems engineering practices fall short of engineering these complex systems [3]. New methods are needed to advance the engineering practices. Research must advance in two directions: (1) methodological, defining standards, languages, and protocols to handle such complexity, and (2) technological, adapting

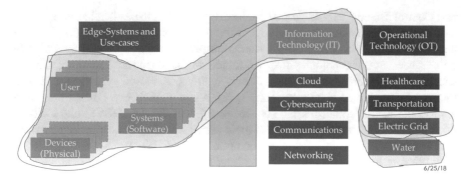

Fig. 1.1 CPS landscape (reproduced from [2])

new engineering processes to the newly available computing infrastructures, so the technology could be applied at multiple levels of CPS specifications.

A challenge in the development of CPS is large differences in the design practices between all the involved engineering disciplines, like software engineering or systems engineering. The emergence and proliferation of CPS has proved that the historical distinction between software engineering and systems engineering is becoming less relevant. Understanding physical, biological, artificial, and social systems require a well-founded, formal, yet intuitive methodology and language that is capable of modeling the complexities inherent in these systems in a coherent, straightforward manner. Considering the current scenario where rapid innovations are assumed to be essential, engineers must be able to explore and exploit software and systems designs collaboratively, analyzing trade-offs and obtaining rapid conclusions. A solid Modeling and Simulation (M&S) methodology to guide all these processes will allow disciplines to cooperate seamlessly [4]. In this regard, the realization that models should serve as foundational and design artifacts has started to gain momentum among both software and systems engineers, resulting in new model-based methodologies. For software engineers, this happened in the early 1990s. After the object-oriented paradigm, the Unified Modeling Language (UML) was adopted in 1997, under the auspices of the Object Management Group with nine different diagram types, which grew to 13 with the transition to UML 2.0 in 2005 [5]. In response to systems engineers, the System Modeling Language (SysML) was developed and adopted in 2007. Like UML 1.x, it had nine diagram types, but not exactly the same set; some were removed from UML 2.0, some modified, and new diagrams were added [6]. Later, the community developed methodologies to integrate software and systems engineering to perform software systems engineering. However, these practices fall short of doing multi-domain systems modeling and simulation.

UML and SysML based approaches were largely used to develop Information Technology (IT) systems and found limited use with formal systems engineering practices that required heavy engineering analysis. For example, an electrical system, or a control system that relied on scientific disciplines like Electrical Engineering and Control Theory. To date, formal engineering methodologies that have been used to develop hardware-software systems in the past four decades are finding limited used with UML/SysML for their systems engineering endeavors. They still rely on engineering methodologies that are rooted in the scientific discipline, such as Electrical Engineering, to engineer closed systems. A closed system is a system that can be expressed in a closed form using a scientific theory. Formal modeling approaches using formalisms such as Discrete Event System (DEVS) and Colored Petri Nets are extensively used due to their verification and validation rigor. When many such closed systems (physical components in CPS) are required to operate with other systems across the network, along with human interaction as an integral part of functioning, new architectures are needed. The nature of architecture also changed to a distributed one, infusing a new set of challenges.

To facilitate architecture development, various architecture frameworks came into existence in the next decade that facilitated interconnection between various component systems (e.g., US Department of Defense Architecture Framework (DoDAF), UK Ministry of Defense Architecture Framework (MODAF), Unified Architecture Framework (UAF), etc.). However, these architecture frameworks were described using UML/SysML and did not mandate any M&S paradigm, theory or a tool to perform simulation-based systems engineering for closed or hybrid systems. Consequently, the gap between a formal model specified through a user-friendly representation, that is implemented and executed correctly by a simulator, still remains. To fill this gap, work by Mittal S, Risco-Martín [7] on DEVS Unified Process (DUNIP) and the incorporated DEVS Modeling Language (DEVSML) Stack [8, 9] using the DEVS formalism, defined a methodology and an abstract language for conceptual modeling and complex systems architecting that integrated the conceptual, structural, and functional aspects of the modeled system.

Contemporary CPS are considered as a new trend in IoT, where the physical systems are the sensors that collect real-world information. This information is transferred to the cyber layer, i.e., the computational modules. The cyber layer analyzes and notifies the findings to the corresponding physical systems through a feedback loop [10]. In the CPS model, the integration of cloud technologies in the CPS cyber layer to ensure scalability, communication, and computation limits the model's accuracy, and adequate energy consumption is highly recommended. A cloud-based M&S architecture can facilitate the deployment of cloud-based CPS in two phases. The first one is the conceptual phase, when the system is initially conceived. Here a solid M&S architecture is fundamental to have an initial idea of the structure and behavior of the whole system at a reduced cost. Because of the inherently distributed nature of the current CPS, this M&S framework must support distributed system design as well [11]. The second phase is the production system, i.e., when the system is actually deployed in the real world.

This process can be done based on the 5 C architecture: Connection, Conversion, Cyber, Cognition, and Configuration [12]. In the *Connection* level, devices are designed to auto-connect and auto-sense its behavior. In the *Conversion* level, meaningful information has to be inferred from the data. In the *Cyber* level, information is being pushed to it from every connected machine to form the machine-to-machine

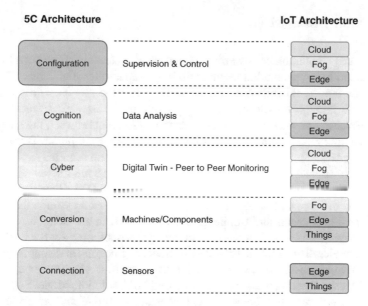

Fig. 1.2   Relation between 5 C and IoT architectures

network. An efficient methodology for managing and analyzing information at this level is creating "digital twins" to study performance for further analysis. The *Cognition* level generates knowledge of the monitored system. Proper presentation of the acquired knowledge to expert users supports the correct decision to be taken. Finally, in the *Configuration* level, the production system can be reconfigured based on the priority and risk criteria, i.e., it acts as a resilience control system. Figure 1.2 illustrates the relation of the 5 C and IoT layers. A contemporary M&S architecture can facilitate the design of twin models, in which physical operations are coupled with virtual operations using intelligent reasoning agents.

Since cloud infrastructure usage is becoming ubiquitous in our day-to-day life, an integrative Cloud-based M&S architecture for CPS engineering provides a bridge between design practices in different engineering disciplines, physical CPS layers, and application CPS layers. As a result, CPS application design, reconfiguration, and autonomy becomes features that can now be explored in an efficient manner.

This chapter provides an overview of M&S Cloud computing as a universal methodology to act as a medium to conceptualize current complex CPS and the technology to handle the design and evolution of contemporary IoT/CPS applications. The chapter is organized as follows. Section 1.2 provides current aspects and architectures of Cloud-based M&S and ongoing community efforts. Section 1.3 describes some challenges and cloud implications with Cloud-based M&S CPS engineering. Section 1.4 provides a quick overview of the book chapters. Section 1.5 concludes the chapter.

1.2   Cloud-Based M&S

Modeling and simulation are two distinct activities. Likewise, cloud-based modeling and cloud-based simulation requires different capabilities. Fundamentally, they require that the digital infrastructure be deployed in the cloud environment and is accessible through remote mechanisms. While modeling activity requires editor workbenches that may be accessible through an Internet browser, the simulation activity requires specific simulation architecture to be cloud compliant. Transitioning any existing M&S application in a cloud-based environment requires explicit M&S infrastructure engineering. Additionally, cloud-based M&S has challenges that traditional simulation systems engineering has largely solved [13]. However, as Mittal and Tolk [2] discuss, CPS are multi-modal and multi-domain systems that involve domain-specific concepts and architectures from more than one domain.

Cloud-based M&S foundation is built on offering modeling as a service and simulation as a service. This implies that both the model engineering and simulation engineering activities must be service-oriented. Further, these services must be deployed in a cloud environment for realizing a cloud-based M&S solution. Model Engineering (ME) [14] incorporates full lifecycle management of the model and credibility to the model engineering process, which includes establishing standards, theory, methods, and tools for doing modeling, and the management of the model, data, knowledge, activities, and organizations/people involved in the model engineering process in a collaborative manner. ME, to be cloud-deployed, requires the availability of a model repository and the execution of the ME process through a browser-based access mechanism. There exist many such tools (e.g., NoMagic Cameo, IBM Rhapsody, Eclipse IDEs, etc.) that facilitate ME in a cloud environment. Simulation engineering incorporates the execution of model in a computational environment by a simulator and various tools and software dependencies that are required by the simulator. Deploying a simulator in a cloud-environment is not straightforward though. Simulators may require high computational resources and the High-Performance Computing (HPC) community has been working on bringing large computational resources for simulation execution for quite a long time. Leveraging the HPC community body of work of the past few decades and masking it behind the service interface for cloud-enabled access is indeed the easiest solution when the simulation system is purely a software system. As Mittal and Tolk [15] explore in their recent book, CPS M&S takes the form of a Live, Virtual, and Constructive (LVC) system. LVC system incorporating simulators at varying levels of fidelity involve both hardware and software components. While one can make the software (purely constructive) components available in a cloud environment, bringing virtual (may include hardware) and live (hardware and hybrid) components is not practical and requires simulation engineering to rely on specific technologies, standards, and methods developed by the Distributed Simulation Engineering community engaged in LVC SoS engineering.

Simulation Interoperability Standards Organization (SISO) stood up the Cloud-Based Modeling and Simulation (CBMS) Study Group in 2016, [16] under the leadership of Col. Robert Kewley, to identify and document the existing M&S in the cloud activities, document best practices, highlight lessons learned, and identify various potential standards to facilitate adoption by other practitioners. Their focus was strictly on CBMS, and application to CPS was out-of-scope. The group identified several focus areas or themes into which the efforts and ideas could be organized to answer important questions or to identify best practices. Potential themes included (but are not limited to)

1. Developing composable services
2. Service discovery
3. Security
4. Deployment, management and governance of services
5. DEVS modeling and other alternative modeling frameworks
6. Development of a reference architecture
7. Service-oriented architectures
8. Business case analysis and return on investment
9. Application of the Distributed Simulation Engineering and Execution Process (DSEEP)
10. The emerging role of the cloud service provider
11. Impact on Validation, Verification, and Accreditation (VV&A) practices
12. Data services (including terrain).

The CMBS study group organized the group in four broad areas

- Models, simulators, and data: Analyze cloud M&S efforts from an interoperability perspective for the models, simulators, and data aspects. This includes investigating semantic model interoperability, simulation architectures, and handling of structured and unstructured data.
- Architecture: Synthesize concepts and constructs from the current M&S architectures into a coherent vision for the future optimized for modern cloud computing and Big Data environments. One of the goals is to enable interoperability with legacy architectures while providing an unconstrained path to the future.
- Cloud infrastructure: Investigate the impact of cloud computing technologies on various aspects of M&S, including system scalability, advanced visualization, scalability of data systems, and high bandwidth or low latency connections to computing and memory. Also investigate the ease of integration into the internet of things type scenarios, which is similar to embedding M&S into live military hardware.
- Services: Investigate, propose, and evaluate standards, agreements, architectures, implementations, and cost-benefit analysis of Modeling and Simulation (M&S) as a Service (MSaaS) approaches.

The CBMS literature survey is available in [13]. The CBMS SG produced a report [17], pending approval by SISO Standards Activity Committee. The CMBS SG also synchronized their effort with North Atlantic Treaty Organization (NATO) Modeling and Simulation Group (MSG) effort named NATO MSG-136 [18]. NATO MSG-136 effort developed a Reference Architecture for M&S as a Service (MSaaS) (Fig. 1.3).

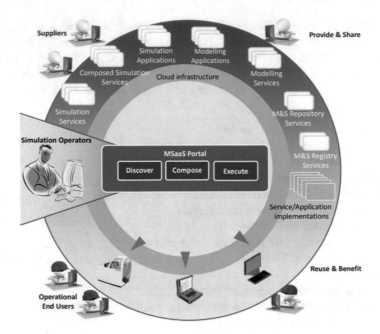

Fig. 1.3 Allied framework for MSaaS (reproduced from [18])

The NATO MSaaS Technical Reference Architecture is described in the form of architecture building blocks and architecture patterns. An architecture building block defines a capability, capturing amongst other requirements, applicable M&S and data standards, and enabling technology. An architecture pattern suggests ways of combining Architecture Building Blocks. The idea is that the Reference Architecture is not a final product but provides a structure where more content can be added over time. In principle, all capabilities for M&S in the cloud could (eventually) be captured in this Technical Reference Architecture. The NATO MSaaS Engineering Process (EP) is a process description for the development and execution of simulations within an existing MSaaS implementation (or infrastructure). An existing MSaaS implementation is assumed to have M&S Enabling Services in place (such as repository services, composition services, and management and control services) and provides capabilities to create a simulation. The process is described as an overlay to the DSEEP and addresses the MSaaS specific engineering considerations during DSEEP execution. The process description will be updated as the Technical Reference Architecture evolves.

1.3 M&S-Based CPS Engineering

M&S has been stated as a powerful design technique for CPS [19] but has inherent challenges involved in the design of contemporary CPS. Through the use of M&S, models take the center stage of the entire design process. Every single system specification (and the underlying components) are defined using models. These models are able to show the evolution of the system, and can be used since the very first stages of the design: preparation of concerns, trace generation, impact analysis, verification and validation, simulation, synthesis, etc. Models can be used for early identification of design defects prior to prototyping, significantly reducing costs. Additionally, the use of M&S can facilitate the automation of some key design processes like the synthesis in prototype devices, automatic code generation or the system deployment on complex and heterogeneous platforms [11]. However, current heterogeneity and complexity of CPS cannot be easily handled by all the currently existing methods and techniques. Currently, it is not possible for a single M&S language or framework to adequately address all the challenges related to CPS [20].

In M&S, we are not only limited to the computational implementations of models. We distinguish between live simulations in which the model involves humans interacting with one another (role-playing, playacting, etc.); virtual simulations where the model is simulated by a fusion of humans and computer-generated experiences; and constructive simulations where the model is entirely implemented in a digital computer and may have increased levels of abstraction. Increasingly, we are mixing the three forms of simulation in what is commonly known as live-virtual-constructive (LVC) simulation [21, 22].

From a systems theoretic perspective, a CPS model is a hybrid system made up of both continuous and discrete systems. A continuous system (CS) is one that operates in continuous time and in which input, state, and output variables are all real values. A discrete (dynamic) system (DDS) is one that changes its state in a piece-wise constant event-based manner (which also included discrete-time systems as they are a special case of discrete event systems) [23]. A typical example of a hybrid system is a CPS in which the computation subsystem is discrete and a physical system is CS. The LVC environment also qualifies as a CPS, with live systems as CS, constructive systems as DDS, and virtual systems as a hybrid (containing both CS and DDS). At the fundamental level, there are various ways to model both timed and untimed discrete event systems, all of which can be transformed into, and studied within, the formal DEVS theory [8, 24–27].

Figure 1.4 associates each of the CPS constituent elements with the corresponding M&S paradigm and how it can be incorporated in the LVC environment.

Using M&S for CPS engineering is not straight forward due to the inherent complexities residing in both the modeling and the simulation activities. A recent panel explored the state-of- the-art of CPS modeling and the complexity associated in engineering intelligence, adaptation, and autonomy through M&S. The literature survey conducted in [28], enumerated the following active research areas and the associ-

CPS Contributor	M&S paradigm	LVC element
Sensors	Continuous, physics-based	L, V, C
Actuators	Continuous, physics-based	L, V, C
Hardware platform	Both continuous and discrete	L, V
Software controller	Discrete	V, C
Network	Discrete	L, V
End User	Discrete, agent-based	L, V, C
Critical Infrastructure	Both continuous and discrete (hybrid)	V, C
Ensemble, emergent behaviors	Discrete, agent-based	V, C

Fig. 1.4 CPS contributor and the associated M&S paradigm (reproduced from [15])

ated technologies for CPS modeling, and concluded that the need for a common formalism that can be applied by practitioners in the field is not yet fulfilled

- DEVS formalism: Strong mathematical foundation that supports multi-paradigm modeling, multi-perspective modeling, and complex adaptive systems modeling to handle emergent behaviors.
- Process algebra: Provides hybrid processes using multi-paradigm modeling. Models combine behavior on a continuous time scale with discrete state transition behavior at given points in time.
- Hybrid automata: Combines finite state machines with Ordinary Differential Equations (ODE) to account for non-deterministic finite states. Bond graphs are used to govern changes.
- Simulation languages: Combines discrete event and continuous system simulation languages. Involves modular design of hybrid languages, multiple abstraction levels combining different formalisms.
- Business processes: Use of standardized notation languages like Business Process Modeling Notation (BPMN) provide value in securing buy-in from the stakeholders in an efficient manner.
- Interface design for co-modeling: Functional Mock-up Interface (FMI) as a means of integration of various CPS components. DEVS can also be used as a common denominator in a vendor neutral manner.
- Model-driven approaches: Model transformation chains to arrive at a single formal model. Governance is required to develop such automation.
- Agent-based modeling: Paradigm to employ component models at scale with individual behaviors, to study ensemble effects.

The above-mentioned approaches and technologies allow the development of CPS models, albeit in a piece-wise manner. These model pieces and their definitions and specifications are dictated by the cross-domain CPS operational use-case. Assuming we now have a validated model (i.e., a model that has been deemed valid by the stakeholders), next comes the task of executing it on a computational platform, i.e., simulation. The piece-wise model composition sometimes does not directly translate into a monolithic simulation environment due to the confluence of both continuous and discrete elements in the hybrid system.

To design contemporary CPS through cloud-based M&S, two major challenges must be addressed. The first one is the design of a modeling language able to deal with the complexity of current CPS. This chapter has already stated in previous sections that none of the actual modeling languages alone can address all the challenges related to CPS modeling. The second one is to provide modern frameworks to perform CPS engineering, taking into account the distributed nature of CPS that usually integrate cloud computing aspects. In the following, we visit these two main challenges.

1.3.1   The Need for a Unified M&S Process

As stated above, high-level languages such as UML, SysML or MARTE have been used for modeling complex CPS. UML has been traditionally used for modeling software systems, and although it defines a syntax of model diagrams, it does not offer semantics for CPS modeling. The OMG SysML standard offers functionalities like requirements management, which is interesting for CPS, but does not provide mechanisms to define aspects of real-time embedded systems such as performance, energy consumption or non-functional constraints. MARTE, another OMG standard for real-time embedded systems, still suffers from not having detailed guidelines and semantics, a problem for its correct utilization. Taking into account Cloud computing modeling aspects, other modeling languages have been used like SoaML [29] or CloudML [30]. However, these approaches are still software-oriented and not able to define the complexity of current CPS.

In the search of a *Universal Modeling Language*, or better said, a *Unified Modeling and Simulation Process*, the initial foundations of the theory of modeling and simulation should be carefully revisited and taken into account. One of the most important theories in the last 50 years was pioneered by Zeigler in [31]. It defines a set of elements and systems specifications for model-driven systems engineering. In this regard, the DEVS theory was formulated on two orthogonal concepts. It distinguishes between system structure, i.e., how the system is internally constituted, and system behavior, i.e., how the system is externally manifested. This is the foundation of modular systems that have defined input and output interfaces through which all interaction with the environment occurs [8]. This hierarchy of systems specification has 5 levels as shown in Table 1.1, which is also related to some of the concepts as defined in the theory of modeling and simulation provided in [25].

Following a standard specification certainly facilitates the definition of a M&S Unified Process. Such a process must be based on an *open system* concept. An *open system* is a system that can exchange energy, material, and information with the outside world through its reconfigurable interfaces. Current CPS provide a dynamic environment analogous to a variable structure system. This M&S unified process must be able to perform the following aspects:

Table 1.1 Hierarchy of system specifications (adapted from [8])

Level	Name	System specification	Elements from the M&S framework
5	Coupled systems	Systems built from component systems with a coupling recipe	Model, simulator, experimental frame
4	I/O system structure	System with state and transitions to generate the behavior	Model, simulator, experimental frame
3	I/O function	Collection of input/output pairs partitioned according to initial state	Model, source system
2	I/O behavior	Collection of input/output pairs from external black box view	Model, source system
1	I/O frame	Input and output variables and ports together with values over a time base	Source system

1. Requirements specification using many Domain-Specific Languages (DSLs), like UML, SysML, Business Process Modeling Notation (BPMN), DoDAF, MoDAF, UAF, etc.
2. Platform Independent modeling (PIM) at lower levels of systems specification, using a specific modeling language based on a well-known M&S formalism.
3. Model Structures at a higher level of System resolution using the same modeling language.
4. Platform Specific Modeling and execution environment.
5. Automated Test Model generation using PIMs.
6. Cloud-based execution support that allows the deployment of complex M&S scenarios.
7. Interfacing of models with real-time systems where the model becomes the actual systems component.
8. Verification and Validation at every level of system specification and lifecycle development.

The application of M&S in CPS engineering using a unified process must be fully supported by a suite of modeling languages, i.e., domain-specific languages that preserve the semantics of the specific domain. This allows engineers to decouple the domain from the model and the model from the simulation framework. This provides many benefits, since models can be constructed independently of the M&S platform, and with DSLs, models can still preserve the domain semantics. A modeling language also facilitates the definition of transformation from DSLs to PIMs.

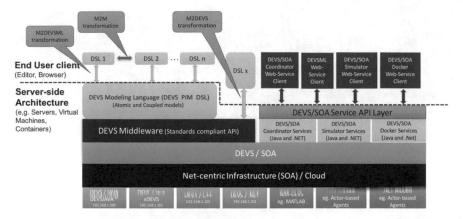

Fig. 1.5 DEVSML stack (reproduced from [9])

The DEVS Unified Process (DUNIP) [8], with the underlying DEVS Modeling Language (DEVSML) stack [9] (see Fig. 1.5), were designed to fulfill all these requirements, using as a foundation the DEVS application of discrete event dynamical systems theory.

Modeling languages that handle only the software aspects are not positioned to accurately define current CPS. Other languages more oriented to systems engineering have not reached a proper level of M&S maturity to take into account aspects related to the physical processes or distributed nature of contemporary CPS. For CPS M&S, effective semantic alignment is needed to support holistic M&S of complex and heterogeneous CPS. It requires both an M&S Unified Process and an expressive M&S Modeling Language that would yield an unambiguous model. Although there are some approaches at the research level that address these challenges, this is still an open problem that needs to be addressed with the help of an industry standard consortium.

1.3.2 Cloud Implications for CPS Engineering

Cloud-based CPS aims to integrate the paradigms of Cloud computing applied to CPS engineering. There are several examples of Cloud-based CPS such as the design of autonomous vehicles, smart homes, automated factories, HW auto-diagnosis and maintenance, smart grid, etc. In general, any system that follows the IoT paradigm can be considered a Cloud-based CPS.

As illustrated in Fig. 1.6, in a Cloud-based CPS architecture every physical *thing* may have a digital twin representative that can be hosted in the Edge, Cloud or Fog layer. In contemporary Cloud-based CPS, physical and corresponding cyber things do not have to be geographically concentrated. Everything (physical or cyber) must be identified by a unique ID. Additionally, although not represented in Fig. 1.6, physical

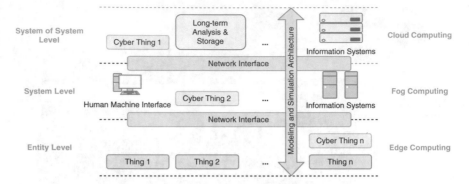

Fig. 1.6 Cloud-based CPS architecture

things are capable of maintaining discrete connections with the cyber things. The edge layer can implement small applications to provide real-time responses. Data are generated by things and consumed by humans through Human Machine Interfaces (HMI), which can also control and provide access to each thing, monitoring and diagnosing the state of the corresponding system. HMI is responsible for maintaining a real-time and collaborative control between the different elements, to keep the coordination of the physical world and the cyber world. Data can be processed in fog or cloud layers using information systems. The allocation of resources (cyber things, HMI, information systems, etc.) will depend on the different requirements of the whole infrastructure in terms of energy consumption, latency, performance, and accuracy. A cloud-based M&S architecture can facilitate the design and deployment of cloud-based CPS, acting as a substrate that is applicable across the entire Cloud-based CPS engineering process. However, this M&S architecture defines several technical challenges, enumerated below.

First, it must be a distributed architecture in nature. This aspect brings other key challenges [32]

1. Application-driven, scalable simulations of large, complex networks.
2. Exploitation of heterogeneous machine architectures.
3. Making parallel and distributed simulation broadly accessible through simpler model development and cloud computing platforms.
4. Online decision making using real-time distributed simulation.
5. Energy- and power-efficient parallel and distributed simulation.
6. Rapid composition of distributed simulations.

Second, the M&S architecture must incorporate co-simulation aspects related to CPS [33]. As in the Unified Modeling Process, here the proposed architecture must be able to combine the strengths of different CPS simulation environments, combining them into a co-simulation framework. These allow system engineers a true multidisciplinary M&S. This demands high level of interoperability, hardware- and software-in-the-loop, automatic code generation, and synthesis, etc.

Third, it must include verification and validation processes, with explicit methods for testing, validation and verification of legacy systems. As a result, these methods will have to be extended to address the scalability needs of Cloud-based CPS. It includes risk, trust, security analysis, etc.

Fourth, the M&S framework must provide support for virtualization, integrating virtualized CPS resources of the whole Cloud-based CPS architecture (both things or management resources). This can provide virtualization as software services so other real or virtual components can use them in co-simulation for data analysis, monitoring and controlling tasks, etc.

Finally, the problem of reliability, scalability, and energy efficiency must be addressed by the M&S architecture at several levels [20]: (1) at application-algorithm level, (2) at technology level; (3) at circuit-level, avoiding worst-case design; and (4) at the system level, using energy-efficient accelerators with build-in trade-off Quality-of-Service versus energy and minimum required sub-systems.

1.3.2.1   Current Efforts

Under the framework of industry or research projects or software tools, there have been some attempts to develop M&S methodologies to manage Cloud-based CPS engineering. In the MODAClouds project [34] the engineering team can model, develop, deploy, operate, monitor, and control cloud applications exploiting the benefits of working with multiple clouds, and fulfilling that the cloud infrastructure and services will always meet some user-defined business requirements. The INTO-CPS project [35] has created an integrated *tool chain* for comprehensive Model-Based Design of CPS. The tool chain supports multidisciplinary, collaborative modeling of CPS from requirements, through design, down to realization in hardware and software. This enables traceability at all stages of the development. Mittal and Risco-Martín [9] integrated Docker with the granular Service-Oriented Architecture Microservices paradigm and advanced the state-of-the-art in model and simulation interoperability in Cloud-based CPS (Fig. 1.5). They described the architecture incorporating DevOps methodologies using containerization technologies to develop cloud-based distributed simulation farm for Cloud-based CPS specified using the DEVS formalism. Another framework called Simulation, Experimentation, Analytics, and Test (SEAT) framework by The MITRE Corporation employs the docker technology and Continuous Integration/Continuous Delivery (CI/CD) pipelines to address integration and interoperability challenges inherent in the CPS M&S engineering [36]. While the DEVSML stack is focused more on transforming various models into a DEVS specification, SEAT focuses on the black box approach of bringing user-apps packaged in docker containers and providing them with a data model to interoperate with other docker containers. The simulation environment (only the constructive in LVC) is deployed as another container app. The DEVSML stack can be subsumed in the SEAT framework as it is docker compliant as well. Figure 1.7 shows the SEAT layered architecture framework.

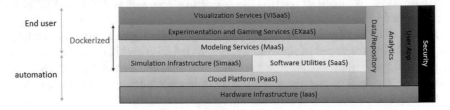

Fig. 1.7 SEAT layered architecture framework (reproduced from [36])

These recent developments bring together cloud technologies, co-simulation methodologies, verification, validation, virtualization, and hybrid modeling approaches to deliver an M&S substrate that is applicable across the entire CPS landscape. Performing Cloud-based CPS engineering through the support of an integrative M&S framework, along with a solid M&S process, will facilitate both the conception, design, and 5 C production of such complex systems.

1.4   Book Overview

This book brings together cloud simulation technologies and their application to CPS engineering. It is divided into four parts: Foundations, Methodology, Applications, and Reliability Issues.

The first part of the book, Part I—Foundations, puts together the foundational concepts in composability, distributed simulation, service-oriented systems engineering, and cloud computing. The second chapter by Andreas Tolk connects various fields of research to support solving composability challenges for effective CPS applications in the domain of cloud, edge, and fog computing. The third chapter by Saurabh Mittal and Doug Flournoy details an overview of various mechanisms for maintaining a consistent truth representation in distributed M&S systems. It also presents an architecture to implement mobile propertied agents in a cloud environment for a more robust CPS test and evaluation framework. The fourth chapter by Robert Siegfried provides an overview of the Allied Framework for M&S as a Service (MSaaS) for NATO and allied nations to demonstrate that MSaaS is capable of realizing the vision that M&S products, data, and processes are conveniently accessible to a large number of users in a cloud environment. The next chapter by Bo Hu Li et al. introduces a Cyber-Physical System Engineering Oriented Intelligent High-Performance Simulation Cloud (CPSEO-IHPSC). CPSEO-IHPSC provides support to access services related to an intelligent high-performance simulation resources, capabilities, and CPS products on demand.

The second part of the book, Part II—Methodology, brings together different methodologies to perform service composition, scheduling and the integration of nature-inspired modeling in Cloud-based CPS engineering. Chapter 6 by Lin Zhang et al. gives a literature review on Cloud-based simulation and proposes a service network-based method to implement service composition and scheduling in simulation. Chapter 7 by Tuncer Oren revises the definition of CPS from the point of view of the evolution of physical tools. It also elaborates on nature-inspired modeling and computing for simulation-based CPS engineering. Chapter 8 by Col. Rob Kewley provides a roadmap for simulation engineers and systems engineering integrators who would like to employ the DEVS Distributed Modeling Framework in their work in order to ease integration and to improve performance in the cloud. Chapter 9 by Daniel Dubois presents methods and an algorithm for simulation of discrete space-time partial differential equations in classical physics and relativistic quantum mechanics. The development of simulation-based CPS indeed evolves to quantum computing and the chapter presents computing tools that are well adapted to these future requirements of quantum computing.

The third part of the book, Part III—Applications, introduces a few real-world applications of cloud-based CPS engineering. The tenth chapter by Thomas Bitterman shows the design and implementation of a system that implements the simulation as a service model, based on the software as a service model. This system extends the software as a service principle to include high-performance computing hosted applications. Chapter 11 by Kevin Henares et al. presents an automated Cloud-based CPS engineering process for a robust migraine prediction system that allows the generation of alarms before the appearance of new pain episodes. Chapter 12 by Mayank Singh and Hemant Gupta reviews the history of the battery, how a virtual battery works, its application and security considerations in a CPS. The last chapter in this section by Matthew T. McMahon et al. details the development of the MITRE Elastic Goal-Directed simulation framework (MEG), designed to provide modelers and analysts with access to (1) cloud-enabled high-performance computing support, (2) a wide range of design of experiments methods, and (3) robust data processing and visualization.

The last part of the book, Part IV—Reliability issues, analyzes different aspects of reliability, truth, resilience, and ethical requirements of cloud-based CPS engineering. Chapter 14 by Md Ariful Haque et al. describes a cloud-based simulation platform for deriving cyber resilience metrics for a CPS. Chapter 15 by Sanja Lazarova-Molnar and Nader Mohamed gives a holistic overview of the reliability analysis of CPS. This chapter also identifies the impact that data and new data infrastructures may have on a CPS. Chapter 16 by Margaret Loper defines trust issues, including reliability, with CPS from a multi-dimensional perspective. It describes a set of research projects conducted that span the multiple dimensions of trust. The last chapter by Tuncer Oren describes several dimensions of reliability for CPS: (1) categories or reliability issues, (2) reliability and security aspects of computation, (3) reliability and failure avoidance in simulation, and (4) aspects of sources of errors.

1.5 Summary

CPS are complex systems with hardware, software, and networking components. The physical hardware components can be accessed and controlled by software over a local network or a geographical network as big as the Internet. Employing M&S for doing hardware-software co-engineering is an already solved problem with established methodologies for engineering closed-loop systems. The Internet era and the distributed networked aspect in CPS have introduced vulnerabilities both at IT and OT levels, which have made CPS engineering a challenging task. To further add to the complexity, CPS are multi-modal and multi-domain systems. There do not exist standard practices to perform multi-domain systems engineering. At best, software and systems engineering communities have developed common *lingua franca* and notations such as in UML, SysML or SoAML, but they do not span the entire landscape when it comes to application of these notations to hard engineering disciplines such as Electrical engineering.

M&S for complex systems engineering have developed approaches to build closed-loop systems using formal systems modeling concepts founded on mathematical Set theory. This affords validation and verification rigor in M&S solutions. However, the multi-domain nature of CPS requires the model transformation chains to be built from domain-specific models to a common reference model aligned with the mathematical foundation. While hybrid modeling addresses the model interoperability and semantic alignment, the simulation integration is addressed through the emerging co-simulation approaches.

Cloud-based M&S brings together the latest in Cloud and HPC computing for their use for M&S purposes. Incorporating cloud technologies for M&S solutions requires explicit cloud-systems engineering and the know-how to leverage new technologies for remote management, execution, and access. The emerging container technology, DevOps processes, CI/CD pipelines, remote deployment, and management offer automation and ease of use when done correctly. Community involvement at SISO CBMS SG and NATO MSG give evidence of the importance of MSaaS for next generation M&S solutions.

A new generation of architectures are in the making that brings together formal M&S, cloud technologies, hybrid modeling techniques, and co-simulation approaches for distributed CPS LVC solutions. This book compiles the state-of-the-art in the next generation cloud-based CPS M&S architectures that aid CPS engineering. We encourage you to continue your journey in the chapters ahead as you explore new vistas in M&S for CPS engineering in the cloud context.

Acknowledgements The author's affiliation with The MITRE Corporation is provided for identification purposes only, and is not intended to convey or imply MITRE's concurrence with, or support for, the positions, opinions or viewpoints expressed by the author(s). Approved for Public Release. Distribution Unlimited. Case Number 19-1916-7.
This work has been partially supported by the Education and Research Council of the Community of Madrid (Spain), under research grant P2018/TCS-4423.

References

1. Wolf W (2007) News briefs. IEEE Comput 40(11):104–105
2. Mittal S, Tolk A (2019) The complexity in application of modeling and simulation for cyber physical systems engineering. In: Complexity challenges in cyber physical systems: using modeling and simulation (M&S) to support intelligence, adaptation and autonomy. Wiley
3. Mittal S, Risco-Martín JL (2017) Simulation-based complex adaptive systems. In: Guide to simulation-based disciplines: advancing our computational future, pp 127–151
4. Fitzgerald J, Pierce K (2014) Collaborative design for embedded systems: co-modelling and co-simulation. In: Co-modelling and co-simulation in embedded systems design. Springer, pp 15–25
5. Booch G (2005) The unified modeling language user guide. Pearson Education
6. Friedenthal S, Moore A, Steiner R (2014) A practical guide to SysML: the systems modeling language. Morgan Kaufmann
7. Mittal S, Risco-Martín JL (2013) Netcentric system of systems engineering with DEVS unified process. CRC Press
8. Mittal S, Risco-Martín JL (2013) Model-driven systems engineering for netcentric system of systems with DEVS unified process. In: Proceedings of the 2013 Winter Simulation Conference (WSC 2013), pp 1140–1151
9. Mittal S, Risco-Martín JL (2017) DEVSML 3.0 stack: rapid deployment of DEVS farm in distributed cloud environment using microservices and containers. In: Proceedings of the 2017 Spring Simulation Multi-Conference (SpringSim 2017), pp 19:1–19:12
10. Alam KM, El Saddik A (2017) C2PS: a digital twin architecture reference model for the cloud-based cyber-physical systems. IEEE Access
11. Henares K, Pagán J, Ayala JL, Zapater M, Risco-Martín JL (2019) Cyber-physical systems design methodology for the prediction of symptomatic events in chronic diseases. In: Complexity challenges in cyber physical systems: using modeling and simulation M&S to support intelligence, adaptation and autonomy
12. Lee J, Bagheri B, Kao H-A (2015) A cyber-physical systems architecture for industry 4.0-based manufacturing systems. Manufact Lett 3:18–23
13. Sanders C (2019) Research into cloud-based simulation: a literature review (2019-siw-031). In: Simulation Innovation Workshop, SISO
14. Zeigler BP, Zhang L (2015) Service-oriented model engineering and simulation of system of systems engineering. In: Concepts and methodologies for modeling and simulation. Springer
15. Mittal S, Tolk A (eds) (2019) Complexity challenges in cyber physical system: using modeling and simulation M&S to support intelligence, adaptation and autonomy. Wiley
16. SISO (2018) Cloud Based Modeling and Simulation (CBMS) Study Group (SG). https://www.sisostds.org/StandardsActivities/StudyGroups/CBMSSG.aspx
17. Truong J, Wallace J, Mittal S, Kewley R (2019) Final report for the Cloud Based Modeling and Simulation Study Group (CBMS SG) SISO-REF-nnn-DRAFT. Simulation Interoperability Standards Organization, in review
18. Hannay JE, Berg T (2017) The NATO MSG-136 reference architecture for M&S as a service. In: Proceedings of NATO modelling and simulation group symposium on M&S technologies and standards for enabling alliance interoperability and pervasive M&S applications (STO-MP-MSG-149). NATO Science and Technology Organization
19. Jensen JC, Chang DH, Lee EA (2011) A model-based design methodology for cyber-physical systems. In: 2011 7th International Wireless Communications and Mobile Computing Conference, pp 1666–1671
20. Quadri I, Bagnato A, Brosse E, Sadovykh A (2015) Modeling methodologies for cyber-physical systems: research field study on inherent and future challenges. ADA USER 36(4):246
21. Hodson D, Hill R (2014) The art and science of live, virtual, and constructive simulation for test and analysis. J Defen Model Simul 11(2):77–89
22. Mittal S, Diallo S, Tolk A (2018) Emergent behavior in complex systems engineering: a modeling and simulation approach, vol 4. Wiley

23. Lee K, Hong J, Kim T (2015) System of systems approach to formal modeling of CPS for simulation-based analysis, vol 37. ETRI
24. Vangheluwe H (2000) DEVS as a common denominator for multi-formalism hybrid systems modelling. In: IEEE International Symposium on Computer-Aided Control System Design, pp 129–134
25. Zeigler BP, Praehofer H, Kim TG (2000) Theory of modeling and simulation. Integrating discrete event and continuous complex dynamic systems, 2 edn. Academic Press
26. Mittal S, Martin JLR (2016) DEVSML studio: a framework for integrating domain-specific languages for discrete and continuous hybrid systems into DEVS-based M&S environment. In: Proceedings of Summer Computer Simulation Conference
27. Traoré MK, Zacharewicz G, Duboz R, Zeigler B (2018) Modeling and simulation framework for value-based healthcare systems. Transactions of the SCS
28. Tolk A, Page E, Mittal S (2018) Hybrid simulation for cyber physical systems: state of the art and literature review. In: Proceedings of the Annual Simulation Symposium in Spring Simulation Conference
29. O. M. Group (2011) Service oriented architecture modeling language (SoaML). http://www.omg.org/spec/SoaML/
30. SINTEF (2014) CloudML. http://cloudml.org
31. Zeigler BP (1976) Theory of modeling and simulation. Wiley
32. Fujimoto RM (2016) Research challenges in parallel and distributed simulation. ACM Trans Model Comput Simul (TOMACS) 26(4):1–29
33. Mittal S, Zeigler BP (2017) The practice of modeling and simulation in cyber environments. In: The profession of modeling and simulation. Wiley
34. MODA clouds project (2020) http://multiclouddevops.com
35. Integrated tool chain for model-based design of cyber-physical systems (2020) http://into-cps.au.dk
36. Mittal S, Kasdaglis N, Harrell L, Wittman R, Gibson J, Rocca D (2020) Autonomous and composable m&s system of systems with the simulation, experimentation, analytics and testing (SEAT) framework. Winter Simulation Conference (Virtual)

José L. Risco Martín is an Associate Professor of the Computer Science Faculty at Universidad Complutense de Madrid, Spain, and head of the Department of Computer Architecture and Automation at the same University. He received his M.Sc. and Ph.D. degrees in Physics from Universidad Complutense de Madrid, Spain in 1998 and 2004, respectively. His research interests focus on Computer Simulation and Optimization, with emphasis on Discrete Event Modeling and Simulation, Parallel and Distributed Simulation, Artificial Intelligence in Modeling and Optimization and Feature Engineering. In these fields, he has co-authored more than 150 publications in prestigious journals and conferences, several book chapters, and three Spanish patents. He has received the SCS Outstanding Service Award in 2017, and the HiPEAC Technology Transfer Award in 2018. He is associate editor of SIMULATION: Trans. of Soc. Mod. and Sim. Int., and has organized several Modeling and Simulation conferences like DS-RT, ANSS, SCSC, Summer-Sim or SpringSim. He is ACM Member and SCS Senior Member. Prof. José L. Risco Martín has participated in more than 15 research projects and more than 10 contracts with industry. He has elaborated simulation and optimization models for companies like Airbus, Repsol or ENAGAS. He has been CTO of DUNIP Technologies LLC, and co-founder of BrainGuard SL.

Saurabh Mittal is the Chief Scientist for Simulation, Experimentation and Gaming Department at The MITRE Corporation and the President for the Society for the Society for Modeling and Simulation (SCS) International. He served as the Director-at-Large for SCS from 2017-2020. He holds a Ph.D. and MS in Electrical and Computer Engineering with dual minors in Systems and Industrial Engineering, and Management and Information Systems from the University of Arizona, Tucson. He has co-authored over 100 publications as book chapters, journal articles, and conference proceedings including 4 books, covering topics in the areas of complex systems, system of systems, complex adaptive systems, emergent behavior, modeling and simulation (M&S),

and M&S-based systems engineering across many disciplines. He serves on many international conference program/technical committees, as a referee for prestigious scholastic journals and on the editorial boards of Transactions of SCS and Journal of Defense M&S. He is a recipient of Herculean Effort Leadership award from the University of Arizona, US DoD's highest civilian contractor recognition: Golden Eagle award, and Outstanding Service and Professional Contribution awards from SCS.

Chapter 2
Composability Challenges for Effective Cyber Physical Systems Applications in the Domain of Cloud, Edge, and Fog Computing

Andreas Tolk

Abstract The cloud computing paradigm allows ubiquitous access to resources and computational services. These shared assets can be accessed with limited administrative constraints, can rapidly be configured, and provide a huge shared pool of easily accessible information. Often, a significant portion of the computational services is provided by assets at the edge of these clouds, such as by the computational components of cyber physical systems, resulting in the closely related paradigm of edge computing. If smart devices are used to provide similar functionality at the edge of the cloud, these compositions are often referred to as fog computing. In all these cases, bringing data intensive systems with multi-modality together requires more than technical communications. A common information sphere must allow the homogenous access to heterogeneous information structures, often not simply manifested in different facets and viewpoints, but in conceptually different worldviews. This chapter provides an evaluation of the challenges and a survey of available concepts applicable to cope with these challenges. The central idea is the rigorous separation of propertied concepts and processes that are working on these concepts, allowing to unambiguously identify complementary and competitive views, both needed for the successful application of cyber physical systems in complex environments.

2.1 Introduction

Within just a couple of decades, we have witnessed dramatic changes not only in computing paradigms and how computers are applied. Information technology continues not only to influence our lives but is also now tightly interwoven with our lives in a way that was hardly imaginable only a few years ago. We are wearing smartwatches that monitor our health parameters. We use smart devices in our homes that allow us to remotely control all settings from our phones. We download videos

A. Tolk (✉)
The MITRE Corporation, 903 Enterprise Parkway #200, Hampton, VA 23666, USA
e-mail: atolk@mitre.org

© Springer Nature Switzerland AG 2020   25
J. L. Risco Martín et al. (eds.), *Simulation for Cyber-Physical Systems Engineering*,
Simulation Foundations, Methods and Applications,
https://doi.org/10.1007/978-3-030-51909-4_2

on demand and play video games with friends in other houses, cities, or even countries. We use the same information technology to perpetually monitor production
processes and use the derived data to continuously optimize them.

This development started with the mainframe computers of the 1950–1970 era.
The introduction of personal computers and their continuing improvement became
a disrupting technology that brought information technologies to every household.
While personal computers were stand-alone solutions at first, the introduction of
networks in general, and bringing the Internet to every household, prepared the way
to a fully connected world. Cloud computing focused first on the increasing number
of data suddenly available by providing data centers and supporting functionality,
but quickly included many other on demand services. The need to access them
quickly and provide access to the services with low latency resulted in a shift of
services from the center of the cloud to the access points, resulting in the concept
of edge computing. The continuous rise of smart devices and wireless and mobile
technology to connect with them resulted in the requirement to bring the cloud
"closer to the ground," supporting wireless access of millions of devices [3]. This
latest paradigm is called fog computing. If components of the fog computing by
themselves provide computational power, such as smart sensor, some publications
talk about mist computing. Figure 2.1 shows this development.

The second major development of concern in this chapter is the rise of cyber physical systems (CPS), which are generally defined as a new generation of systems with
integrated computational and physical capabilities that can interact with humans—as
well as with other CPS—through many new modalities [2]. They perform monitoring
and control tasks on different levels of autonomy, reaching from systems that support
a human user to autonomous systems like drones or autonomous cars, or industrial
devices. As a rule, they are sensor- and communication-enabled, and they are often
using wireless technology. When they are cloud-enabled, they can not only utilize the
data and services provided by the cloud, but they can also provide their own services
to others, allowing interesting solutions, such as smart cars and smart traffic lights to
support each other by making the traffic flow better. Therefore, CPS are promising

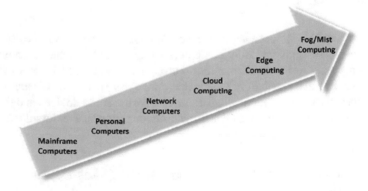

Fig. 2.1 Development of computing paradigms

candidates for the utilization of fog and edge computing users as well as providers of services.

However, in order to do so, a common understanding of the nature of the services is needed. While interoperability addresses the exchange of information and the utilization in the receiving system, composability addresses the alignments of underlying concepts as well, resulting in the consistent representation of truth in the participating systems [37].

This chapter will connect these various fields of research to support solving such composability challenges for effective CPS applications in the domain of cloud, edge, and fog computing. After these computing paradigms are defined in more detail, the epistemology of computational functions will map the recent composability research results to the CPS domain to prepare the recommendations on assuring the conceptual alignment and consistency required.

2.2   Cloud, Edge, and Fog Computing

As discussed in the introduction, the three computing paradigms described here in more detail are not mutually exclusive, but they support each other. The edge computing paradigm enhanced the cloud, and fog computing enhances edge computing. The literature points out that the sheer number of comprised systems, services, and access points is continuously growing. While the cloud contains thousands of data centers and services, the edge has millions of access points, and this number increases to billions to serve the many smart devices reached out to by the fog, which can be enhanced by mist computing on these edge devices. Figure 2.2 shows the interplay of the data center and other cloud services in the center, edge computing allowing access to critical components, and fog computing providing access and interconnecting the many different smart systems and devices, including CPS.

The community did not yet agree on unambiguous definitions, and several industry papers refer to the smart devices as the edge layer or edge devices, which may lead to confusion. Furthermore, there is no clear hierarchy between edge and fog computing, but they overlap and provide alternative access points, as discussed from the user perspective in [22], as they provide complementary capabilities to extend the cloud. An alternative interpretation discussed in the community is that edge computing is more a concept while fog computing using smart Internet of Things (IoT) devices to implement this concept. Within this chapter, edge computing is used to refer to cloud services provided at the edge of the cloud, with fog computing being the next extensions (and mist computing being an extension of the fog computing paradigm to smart components of IoT devices).

The purpose of these following sections is not an exhaustive presentation of the computational paradigms. The interested reader is referred to more detailed discussions in readily available publications on these topics, among many others [5], for cloud computing, [36] for edge computing, and [4] for fog computing. Instead, these

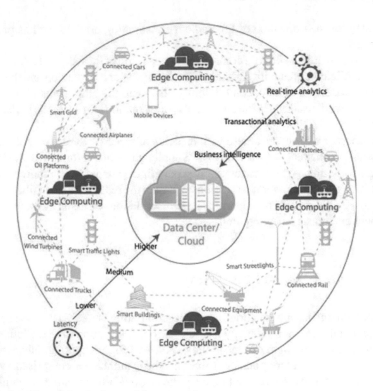

Fig. 2.2 Interplay of cloud, edge, and fog computing

sections will describe the main characteristics that are of interest for the integration of CPS as smart systems utilizing and contributing to cloud, edge, and fog computing, respectively.

2.2.1 Cloud Computing

It is hard to set a start date on the idea to address a set of network functionality as cloud computing. Some sources are pointing to the work of Compaq Computers in the late nineties [33], others are crediting AT&T [16]. The use of clouds to represent not further specified functionality provided by networks goes back to the early days of computers. It quickly became common practice to use the term cloud computing for the concept to use services provided by mostly anonymous servers reachable via a network instead of providing the critical functionality on the local machine. When exactly this happened may never be satisfactorily answered.

The business success, however, can be traced back to Amazon creating subsidiary *Amazon Web Services* and the *Elastic Compute Cloud* (EC2) in 2006, as well as Google, following with the Google App Engine in 2008. Both business solutions offered to outsource services, in the beginning predominantly data storage and evaluation services, so that companies did not buy expensive local solutions instead.

The service provided by clouds usually falls into one of these three categories. All of them are usually pay-as-you-go offers [25]:

- Software as a Service (SaaS) is the most common category. The application is owned by the providers and is offered via the network to be utilized by the user without the need to own a copy of the software.
- Platform as a Service (PaaS) provides whatever is required for building and delivering a web-based application. The user utilizes services needed for software development, provisioning, and hosting
- Infrastructure as a Service (IaaS) provides all types of computing resources, including servers, networks, storage, etc.

One of the main concerns of using cloud-based solutions is security, i.e., the security of data in the cloud, but also security when using the services not to become a victim of cybersecurity attacks. Many cloud services providers have built-in security solutions for prevention, detection, and restoration processes. In addition, initiatives like the Open Web Application Security Project (OWASP), as well as an increasing number of security standards, by the government and industry help to address such concerns. An overview of challenges and solutions has been compiled by [10].

2.2.2 Edge Computing

With the increasing popularity of cloud-based solutions, the number of users increased as well. In addition, while the big solutions in the cloud supported business decisions quite well, the need for real-time analytics increased with the shift of user types for the cloud: the more smart devices were introduced to the Internet of things (IoT), the more need for low latency quick solutions arose as well.

This changed the paradigm and the architecture, as more computing power and resources had to move from the center of the cloud to its edge. Real-time—or near real-time—requirements created an event horizon regarding how far away a server can be located in order to be still able to provide the required information timely. An example of how to compute such event horizons is given in [26]. Edge computing is addressing these needs by bringing computational power and needed data closer to the user, i.e., to the edge of the cloud, its access point. It does not need to wait for a remote cloud server or other centralized systems for processing requests. This also reduces data traffic and the creation of bottlenecks within the cloud but may result in the need for redundant implementations at various access points.

A technology of particular interest to users of mobile technology is described in [15]. The author describes how close by resource-rich nodes can be found and utilized by mobile devices, such as used by first responders, when being confronted by otherwise limited connectivity and computing resources. The available resources build a temporary mini-cloud—or cloudlet—that bundles resources for a geographically and temporarily bound and limited task.

Another challenge was created by the great diversity of data formats and protocols supported by the growing number of smart IoT devices, as well as the vast amount of data they provide due to their sensors and other components. Having special data processors available right at the edge where they are needed can take the protocols used there into account and avoids sending data via long connections through the cloud to remote servers.

2.2.3   Fog Computing

The term fog computing was coined by Cisco [4]. Like edge computing, it provides services for computation, storage, and networking between smart IoT devices and traditional cloud services typically located at the edge of the network. To do this, the participating devices build so-called fog nodes to bring computing and storage services to the smart devices. In contrast to edge computing, where the services are provided at the edge of the cloud but still are part of the cloud, a fog nodes use local area networks or networks based on wireless networks to provide these often special services to the local users, mainly smart IoT devices. The fog nodes can make use of cloud services to complement their own capability and feed their results back into the cloud for further analysis and evaluation, but first and foremost they provide real-time analysis for quick actions. They also may encapsulate sensitive data.

The constraints on fog nodes are few. Principally, any device providing computing, storage, or networking services with access to the cloud can become a fog node. Smart CPS interconnected via wireless communication can form an ad hoc sensor network, the systems hosting the sensors collecting data of current energy use within the PowerGrid can also provide real-time analysis of the overall usage of energy and reconfigure the PowerGrid to better serve the observed constellation [17].

With the growing number of sensors and computational devices on CPS, cloud- and fog-based architectures are also applied for the various systems providing the computational functionality and their input, such as sensors, actuators, and others. In other words, to provide a multitude of computational functionality paired with a significant quantity of sensors, cloud and fog like concepts are applied to govern the CPS itself. This application is usually referred to as mist computing.

2.2.4   Cyber Physical Systems and Cloud, Edge, and Fog Solutions

There are many examples of CPS utilizing cloud computing support in various domains, such as vehicular technology [41], health [48], or industrial applications [9]. A connection to the topic of fog computing from the general IoT perspective with a focus on CPS has been given in [8], with many ongoing industry activities described in [11].

In most of the related work, CPS are mainly smart IoT devices that utilize the services provided in the cloud, edge, or fog computation environments. As CPS often provide a myriad of computational capabilities, and wireless communication devices belong often to the many modalities to use for the exchange of information with humans and others CPS, the question arises how to make this potential usable to other systems as well. The main argument against such efforts is the multitude of formats, protocols, and technologies utilized in the various CPS implementations, but also the differences in underlying ideology [34]. There are simply too many options and standards to allow for a common approach. In this section, the focus is cloud, edge, and fog applications. There are not many use cases where the information provided by individual sensors embedded in a mist computing environment is utilized directly, which means without the knowledge and control of the embedding system. For the sharing of information, mist computing principles are currently not yet relevant, and if they become relevant, the same principals discussed for fog computing are applicable and can be extended accordingly by interpreting the component as a system itself.

Within the next section, the epistemology of computational functions, as they are used in CPS, will help to understand the underlying challenge, as well as provides a way to overcome these current hurdles.

2.3   Providing Computational Capability

Computers are ubiquitously not only changing and improving our daily lives, but also the way we conduct scientific experiments to gain new knowledge. As discussed in [14], computable functions implement effective procedures as algorithms that map a finite range of natural numbers as input and map them to a natural number of a finite domain. While the increase in computational capability and capacity pushes the borders of what is practically achievable by information processing all the time, there are theoretically well established limits that we will never be able to cross. It is likely human nature that young computer scientists often show an unbridled optimism permeating their research and prediction. Similar to researchers in the era of artificial intelligence, who predicted that computers will soon outsmart humans, discover new mathematical theories, and will influence most scientific work, as discussed in [32], developers of cloud, edge, and fog architectures foresee new breakthroughs based on this technology as well, such as in [35].

It would be disappointing if scientists and researchers in their field were not enthusiastic about their research and its application potential. However, as discussed in [40], many researchers in the domain of interest for this chapter—CPS—are not aware that their methods have significant overlaps with the concepts of modeling and simulation. The epistemological constraints of the simulation are limits for the computational capabilities of CPS as well. If CPS are used to provide capabilities in support of edge and fog computing, the same limits that constrain the composition of simulation systems do exist. If multi-modality requires different facets of the situated environment to be represented in various formats, they still need to be conceptually consistent with each other.

2.3.1 Models as the Reality of Computational Functions

In a recent project on CPS, the importance of computational capabilities regarding intelligence, adaptation, and autonomy aspects of these new CPS was elaborated [28]. There are multiple domains requiring computational capability as well, as many activities require the calculation and computation. Within the context of this chapter, pure data manipulation or control computations are not in the focus. Those computational capabilities are important for CPS to function properly, but they have well-specified domains and regions, leaving little room for interpretation.

The computational capabilities supporting the interpretation of sensor information while creating a perception of the current situation, the identification or generation of alternative courses of actions, the interpolation of the current perceived situation for the prediction of possible future states when evaluating the various action, and collaborative planning with other CPS are different. They all are based on models, which are task-driven simplifications and abstractions of a perception of reality, in other words, formal representations of conceptualizations. Computer simulation engineers developed various methods to support the conceptual alignment of different models, when the implementing simulation systems shall be composed, many of them were standardized in contributions to the Distributed Simulation Engineering and Execution Process [18], several of them can be applied to address CPS interoperability as well. The following examples motivate why simulation solutions are relevant to CPS computational capability interoperability challenges as well.

The first example copes with the challenge of how machines can perceive the world and has been published in [46]. For every machine that needs to understand the situated environment it is operating in, three main steps necessarily must be conducted, namely (1) sensing, (2) perceiving, and (3) understanding.

- Sensing describes the observation of the situated environment by the systems using sensors. There are many types of sensors, such as acoustic, chemical, electromagnetic, thermal, or optical. For sensing to be successful, the sensor must be able to recognize certain attributes of the observed object, the observed object must expose these attributes, and the situated environment must not expose the same attributes (also known as the target-background-noise-ratio). If any of these three constraints is violated, the object cannot be sensed by the sensor and remains hidden to the observer.
- Once these sets of attributes are sensed, they are collected into a perception of the situated environment. The perception is based on a model of what can be observed. Data types, accuracy, and resolution of this model reflect the technical specifications of the sensors used to populate it. Usually, the bundling of

attribute observations into objects happens here as well, but sometimes this task is conducted in the third step.

- Understanding the observed situation completes the processes of perceiving the situation by mapping the perceived observations with the observations conducted so far. Very important is the mapping of the perception of the observed object to the known object types used by the system to capture is understanding. A CPS participating in traffic may have detailed models of all expected car types to match the observations to. It may be able to recognize make and year based on characteristic properties that are observed. However, if not all required characteristics are observed, a valid mapping may not be accepted, also known as a type I error: valid results are not accepted.
- However, if an object is observed that is not captured in this set of recognizable ones, problems occur. There is no reference in the knowledge base describing an elephant. It can classify the observation as "unknown," or the similarity to a known object is so close that it leads to a wrong match, also known as type II error: nonvalid results are accepted.

Besides its sensors, a CPS generally can receive additional information through messages from other systems, or also via one of its multimodal interfaces from human team members. The same steps are performed as well, as the message must be received, the content understood, and the resulting information used to add to the perceived situation. The process of generating a perceived situation based on existing information and additional information from a high variety of sources, such as sensors and messages, is supported by the mathematics underlying multisensory fusion methods, such as captured in [24]. Furthermore, a priori knowledge can help to support the matching and mapping activities, e.g., using Bayes theorem to deduct like systems not only based on an observed characteristic attribute, but also on the likelihood to observe a certain system based on the overall capacities. Many numerical methods are given in [23], that are applicable here as well.

To wrap up this motivational section, a short summary of [38], will show the more general applicability of insights of the epistemology of simulation services to the computational capabilities enabling computational intelligence for autonomous systems, such as CPS. The following figure shows the topology of intelligent software agents as used in agent-based models in direct comparison with the topology of autonomous systems.

The left side of Fig. 2.3, shows the principle topology of many intelligent software agents. These agents perceive their environment, are socially capable, make sense of their situation, make decisions, and act. They can learn by adapting their knowledge base from observations of the effects of their actions. They can plan alone or in collaboration with others. The right side shows an autonomous system. The signal processing to the control unit plays a major role, and so does power supply and power management. The physical capabilities in the form of actuators, manipulators, and locomotion are also often explicitly handled to better allow to compute how to best act in any given, perceived situation. This implied similarity of both topologies motivates that the control unit of autonomous systems and CPS can benefit from research results

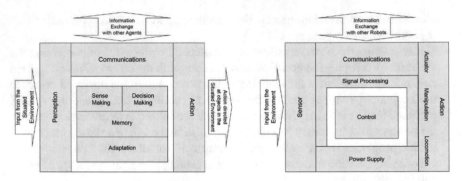

Fig. 2.3 Topology of intelligent software agents versus autonomous systems

of intelligent software agents. All this requires, however, that the various models used can operate together, which is not a trivial task, as will be motivated in the following sections.

2.3.2 Interoperability Versus Composability

Besides solving many challenges of operation systems in the real environment, including power supply optimization and physical constraints in locomotion, manipulation, and actuators, the CPS community focused so far successfully on the alignment of the various MAC protocols, as compiled in [43] and physical control networks as described in [6], some of them extending the web of things paradigms [13]. The work of [30], focuses on the embedded software system challenge. All these publications contribute significantly to the interconnection of CPS, using the cloud or the fog discussed before. All these solutions are pivotal for the successful exchange of data between the CPS.

However, we are not only facing a huge variety of systems, data formats, and protocols, but the underlying assumptions and constraints need to be understood as well. Currently, we mainly exchange data, not yet information or knowledge, as captured in the well-known data-information-knowledge-wisdom pyramid [1]. The rigor with which mathematical formalism already can manipulate the data should not cover the fact that each of these data is based on models which by themselves are task-driven simplifications and abstractions of a perception of reality. This requires conceptual alignment, and the topic of conceptual consistency will be covered shortly in its own section.

To better cope with these issues, the M&S community differentiates between interoperability of simulation and composability of models [37]. Interoperability is the ability to exchange information and to use the data exchanged in the receiving system. Interoperability can be engineered into a system or a service after definition and implementation. Alternative data representations can be mediated into each other if the constraints are understood. Only when data must be disaggregated, which requires that information that got lost in the aggregation process be reinserted, the engineer has the problem from where to extract this needed information, but often heuristics can be applied that lead to satisfactory results. Composability is different from interoperability. Composability is the consistent representation of truth among the computational representation in all participating systems. It extends the ideas of interoperability by adding the pragmatic level to cover what happens within the receiving system based on the received information. In contrast to interoperability, composability cannot be engineered into a system after the fact. Composability requires often significant changes to the computational capabilities of the participating systems. Recent studies within the IoT domain introduced the idea of pragmatic interoperability, which is a necessary step towards composability [31]. Inconsistent versions of truth are not allowed. However, the next section will show that this should not exclude different facets or interpretations, like providing alternative views or methods.

2.3.3 Complementary and Competing Models

When it comes to computational representation, diversity can actually be a good thing, if it is not used to create inconsistent compositions. The M&S community differentiates between complementary models and competing models.

In complementary models, the scope of each of the models is different to allow to focus on specific facets of a challenge that needs to be solved by several systems. One possibility is to use geographical areas of responsibility, so that all systems are taking care of their area first, so that no conflicts arise. Another possibility is to focus on different types of components of the problem. First, responding CPS systems may comprise of systems specialized to extinguish fire, others can remove debris, and another type can provide first aid to victims of an accident. In this case, the systems provide different capabilities that are all needed to accomplish all tasks, but not every system needs to have all functions needed to provide them. It is also possible that CPS systems must compete for a limited set of resources, but that is not the kind of competition that is discussed here. If the competing CPS share common concepts and have a common understanding of the situation, they are not competing models.

Competing models provide an alternative view to solve identical challenges. They often result from different views from the programming experts, or from different underlying assumptions and constraints that rule how to solve a problem. In a complex environment, it is often no longer possible to decide which solution will provide the optimal success, and in different parts of the solution space different solutions may turn out to be favored. It is also possible that the definition of success in the form of measures of merit change over time, leading to the necessity of a reevaluation of former solutions and solution types. If two competing models are composed of a computational solution, they introduce contradictions. Depending on which model is used for the solution, the results will differ. This is not the expected behavior of the computational functionality of CPS.

Nonetheless, such competing or alternative solutions can have advantages. The example of using several prediction models for future paths of hurricanes is a well-known example: As the various models are based on different assumptions and constraint, and as they also may introduce different new insights or data points needed for the better calibration of the model, they all differ regarding path, strength, and uncertainty. Competing models usually result from vague and incomplete knowledge about the problem. Therefore, taking all recommendations into consideration is a good practice for human decision makers, so CPS should be enabled to do the same.

In every case, it is essential to ensure the conceptual consistency evaluation of provided computational compositions, which allows to identify complementary views, as well as alternative solutions. This requires a mathematically rigorous approach that allows to express different scopes, resolutions, and structures represented in the different solutions and provided for the mapping of them to a common reference model. The next section will provide an overview of relevant research that can allow for successful CPS cooperation in a complex environment, enabled by cloud or fog support.

2.4   Conceptual Consistency

The research presented in this section extends the findings documented in [39]. The research presented in this reference focused on the use of cloud concepts to allow composable M&S services. Using the results of the last section, they can be extended towards computational functionality provided or utilized by CPS in cloud, edge, and fog domains.

The last section showed that most computational capabilities of interest in the scope of this chapter is based on assumptions and constraints. They are model-based, using different abstraction levels and make different simplifications, depending on the tasks the developers originally intended to support. As such, they are close to simulation functions, which are also based on models. The M&S community introduced the concepts of interoperability of simulation systems and composability of models. As discussed, composability has been defined as the consistent representation of truth among the computational representation in all participating systems [37].

In the mathematical and computational sense, this means that two simulation systems produce an equivalent solution space of possible states within the projection of common solutions [12]. A practical interpretation of these findings is that if two simulation systems are simulating the same entity, they must use equivalent representations of the entity and its attributes, or not represent them at all. The models underlying the computational functions of the CPS must be equivalent where they overlap, which follows from Robinson's Consistency Theorem and Łoś Theorem of model theory. A simplified explanation for these two theorems is that Robinson's Consistency Theorem states that two models are conceptually aligned if and only if their intersection is consistent: there is only one interpretation of truth valid in both models, and as it is possible that two models are using heterogeneous modeling approaches or languages, the Łoś Theorem expands such different representation through the Cartesian product and defines filters that allow the comparison in a common equivalent representation, generalizing Robinson's Consistency Theorem to heterogeneous representations of common concepts.

If the computational functionality of CPS are based on models using different abstraction levels, data mediation must be able to map them to each other without loss of information. For example, if in a first responder scenario, the first CPS represents the number of possible casualties as the total number of victims, while a second CPS differentiates between males and females, and adults and children, conceptual consistency requires that at every moment the sum of male adults, female adults, male children, and female children equals the total number of victims. If one system captures the blood type of individuals, the other does not, then there should be no representation of the blood type implied for the second system. In [44], a smart emergency response system prototype is presented that was created by a team of nine organizations from industry and academia.

These research insights on the need for conceptual alignment of model-based applications, including heterogeneously developed computational functionality of a group of collaborating CPS, have two major implications. First, we need to clearly represent all entities of the shared space between all composed systems in a way that enforces consistency in its representation. The first subsection will provide related research. Second, a common reference model will be needed to capture the various views, as well as how to map between them. This will be done in the last subsection.

2.4.1 Data and Processes

One of the main accomplishments of software engineering of the recent years was the introduction of object-oriented languages. The use of objects made the use of code more intuitive and allowed to hide detailed implementations within the objects, making the code more stable and the reuse more secure. The clear structure provided modules that encapsulated data and associated processes.

However, in the context of simulation, encapsulation can easily lead to the lack of transparency needed to ensure composability. The reason is that conceptual alignment requires understanding in all three semiotic categories: syntax, semantics, and pragmatics. Treating the computational capability of a CPS as an object that can provide its functionality via the well-specified interfaces can support the perfect alignment of syntax and even semantic, but as the implementation behind identical APIs can differ significantly, the use of information—the pragmatic category—is not aligned, resulting in inconsistencies and challenges during the execution. For example, if a group of CPS supports a first response operation of firefighters, they may all be able to support saving victims from a burning house. However, the exact steps, the tactics, techniques, and procedures that are followed, can differ. If the commanding and orchestrating entity is not aware of these different perspectives and methods, the CPSs can negatively interfere with each other, even endangering the success of the whole operation.

One of the resulting recommendations captured in [39], was the explicit capturing of data in mobile agents to ensure that the entity represented by this data is always consistent. This still allows for competing services, but one entity cannot be manipulated at the same time by two competing services, resulting in an inconsistent representation. The OMG Data-Distribution Service for Real-Time Systems (DDS), an open international middleware standard directly addressing publish-subscribe communications for real-time and embedded systems, is based on the same idea to ensure interoperability by enforcing consistency [20]. This is necessary, but not sufficient to address the challenges derived from the examples above. We also need an ontological representation of the computational capability that can be used to ensure full transparency of the pragmatic aspects as well.

2.4.2   Ontological Representations

The basics of the ontological representations are based on the ideas published in summary in [45]. As Zeigler points out, mathematical systems theory provides a framework for representing and studying dynamical systems, such as pioneered by Wymore [42] and others. As it is unlikely that all CPS that are interconnected via cloud or fog communications use the same data structures, languages, or even operating systems, a metamodel approach is needed to capture all semiotic aspects of their computational capabilities. The system entity structure (SES) method provides the means to do this [47]. The application of SES for developing ontologies for architectural specifications for the system of systems is elaborated in [27]. A recent example of an SES application in the context of the risk assessment framework for CPS and IoT can be found in [29].

SES was developed to be a formal ontology framework to capture system aspects and their properties. When fully specified, the resulting SES model formally describes a solution set of all possible permutations and combinations resulting from the use cases for such a system. The use case models are also referred to as pruned entity structure (PES), as only the elements and attributes needed for the use case are captured. As the resulting description is formal, it is machine-readable and implementation independent. As it describes all aspects of the system, it is a "universal language" to describe systems in a heterogeneous, complex environment, such as CPS being interconnected using cloud or fog methods. If all CPS from the example of the first response operation of firefighters describe their capabilities accordingly, the commanding element can take advantage of their diversity instead of being surprised by the developments in the scenario. The SES method also provides a graphical interpretation that is helpful for human–machine communications. An example of how to use the same ideas to describe complex scenarios in implementation agnostic form is given in [19]. Another aspect supporting the SES method is that it allows to derive implementation specific artifacts from the specifications, which allows to adapt and learn, as discussed earlier in the chapter.

2.5 Summary and Discussion

Cloud, edge, and fog computing provides readily available communication to support the exchange of data in an unprecedented richness. However, in order to make use of data, it needs to be put into context to become information, and the information needs to be embedded in the procedure to become knowledge [1]. CPSs provide a plethora of computational capabilities via multimodal interfaces. In order for them to cooperate, they need to be able to communicate their capabilities in an implementation agnostic form. A common, formal representation provides the necessary transparency for composability, the consistent interpretation of truth, without exposing all implementation details. Such a formal representation also allows to identify competing interpretations, which still can be useful by providing alternative viewpoints and developments.

Using the topological similarities between intelligent software agents, as they are used in agent-based models, and autonomous systems, such as CPSs are, motivates to look at recent M&S research results to support the full utilization of cloud, edge, and fog computing. The importance of composability as a concept beyond interoperability is one of the directly applicable insight.

Using the mission to be supported as the common denominator for all participating systems, in form of mobile data agents as recommended in [39], or by applying ideas as captured in DDS, can help to gain conceptual alignment. SES adds the pragmatic component as well.

Obviously, these recommendations are neither complete nor exclusive. However, they support visions as captured for domain-specific solutions, such as captured in [21], for industrial CPS. Ideas of context-aware CPS in a situated environment, such as published in [41], can be supported as well. SES can become a unifier, as a formal, machine-readable, implementation agnostic specification.

As already called for in [40], more M&S engineers have to engage with CPS engineers to share research results and ideas, and vice versa. Perceptions and decisions of CPS are based on meaningful abstractions and simplifications of reality, in other words: they are based on models. When a CPS is doing interpolations and predictions to decide on the best path of action, its computational capabilities becomes de facto simulations. The computational constraints of cloud, edge, fog, and mist computing must, therefore, be understood in the context of distributed simulation, as well as distributed decision systems.

This chapter hopefully showed some examples of common topics of interest and made the case that M&S methods will play a pivotal role in reaching the next levels of CPS capabilities in the era of cloud and fog computing. We just have to share our results better, avoiding the mistakes described in Chen and Crilly [7], where scientists working on synthetic biology conceptually had a lot to share with colleagues in the domain of swarm robotics, but as both team were so deeply rooted in the terminology and epistemology of their home sciences, they did not perceive the reusability of their research results. CPS and M&S engineers need to work closer together, including sensor and computational experts to ensure appropriate decisions and solutions in these highly complex situations.

Acknowledgements and Disclaimer Part of the underlying research leading to this chapter was supported by the MITRE Innovation Program. The content was discussed with many colleagues, and I am thankful for the contributions and recommendations for improvement. Among these colleagues, I want to particularly thank foremost Dr. Saurabh Mittal, Dr. Ernest Page, Dr. Justin Brunelle, and Mr. Douglas Flournoy.

The author's affiliation with The MITRE Corporation is provided for identification purposes only and is not intended to convey or imply MITRE's concurrence with, or support for, the positions, opinions, or viewpoints expressed by the author. This paper has been approved for Public Release; Distribution Unlimited; Case Number 19-01906-9.

References

1. Ackoff R (1989) From data to wisdom. J Appl Syst Anal 16:3–9
2. Baheti R, Gill H (2011) Cyber-physical systems. Impact Control Technol 12:161–166
3. Bonomi F, Milito R, Zhu J, Addepalli S (2012) Fog computing and its role in the internet of things. In: Proceedings of the first edition of the workshop on mobile cloud computing. ACM, Helsinki, pp 13–16
4. Bonomi F, Milito R, Natarajan P, Zhu J (2014) Fog computing: a platform for Internet of things and analytics. In: Bessis N, Dobre C (eds) Big data and internet of things: a roadmap for smart environments, pp 169–186. Springer International Publishing, Cham
5. Buyya R, Broberg J, Goscinski A (2011) Cloud computing: principles and paradigms. Wiley, Hoboken, NJ
6. Cai Y, Deyu Q (2015) Control protocols design for cyber-physical systems. In: Advanced information technology, electronic and automation control conference (IAEAC). IEEE, Chongqing, China, pp 668–671
7. Chen C-C, Crilly N (2016) Describing complex design practices with a cross-domain framework: learning from synthetic biology and swarm robotics. Res Eng Design 27(3):291–305
8. Chiang M, Zhang T (2016) Fog and IoT: an overview of research opportunities. IEEE Internet Things J 3(6):854–864
9. Colombo AW, Bangemann T, Karnouskos S, Delsing J, Stluka P, Harrison R, Jammes F, Lastra JL (2014) Industrial cloud-based cyber-physical systems. The IMC-AESOP approach, p 22
10. Coppolino L, D'Antonio S, Mazzeo G, Romano L (2017) Cloud security: emerging threats and current solutions. Comput Electr Eng 59:126–140
11. de Brito MS, Hoque S, Steinke R, Willner A, Magedanz T (2018) Application of the fog computing paradigm to smart factories and cyber-physical systems. Trans Emerg Telecommun Technol 29(4):e3184
12. Diallo SY, Padilla JJ, Gore R, Herencia-Zapana H, Tolk A (2014) Toward a formalism of modeling and simulation using model theory. Complexity 19(3):56–63
13. Dillon T, Zhuge H, Wu C, Singh J, Chang E (2011) Web-of-things (WoT) framework for cyber-physical systems. Concurr Comput Pract Exp 23:905–923
14. Dowek G (2011) Proofs and algorithms: an introduction to logic and computability. Springer, London

15. Echeverría S, Root J, Bradshaw B, Lewis G (2014) On-demand VM provisioning for cloudlet-based cyber-foraging in resource-constrained environments. In: 6th international conference on mobile computing, applications and services. IEEE, Austin, TX, pp 116–124
16. Hernandez D (2014) Tech time warp of the week: watch AT&T invent cloud computing in 1994. Accessed Aug 2019. https://www.wired.com/2014/05/tech-time-warp-cloud-is-born/
17. IEEE (2013) IEEE smart grid vision for computing: 2030 and beyond. IEEE Smart Grid Research, New York
18. IEEE (2010) Recommended practice for distributed simulation engineering and execution process (DSEEP). IEEE
19. Jafer S, Durak U (2017) Tackling the complexity of simulation scenario development in aviation. In: Proceedings of the symposium on modeling and simulation of complexity in intelligent, adaptive and autonomous systems. Society for Computer Simulation International, San Diego, CA. Paper 4
20. Joshi R, Castellote GP (2000) A comparison and mapping of data distribution service and high-level architecture. Realtime Innovations Inc, Whitepaper, Sunnyvale, CA
21. Karnouskos S, Colombo AW, Bangemann T, Manninen K, Camp R, Tilly M, Sikora M et al (2014) The IMC-AESOP architecture for cloud-based industrial cyber-physical systems. In: Colombo AW, Bangemann T, Karnouskos S, Delsing J, Stluka P, Harrison R, Jammes F, Lastra JL (eds) Industrial cloud-based cyber-physical systems: the IMC-AESOP approach. Springer International Publishing, Cham, Switzerland, pp 49–88
22. Klonoff DC (2017) Fog computing and edge computing architectures for processing data from diabetes devices connected to the medical internet of things. Diabetis Sci Technol 11(4):647–652
23. Kruse R, Schwecke E, Heinsohn J (1991) Uncertainty and vagueness in knowledge based systems: numerical methods. Springer, Berlin, Germany
24. Liggins II, Martin DH, Llinas J (2017) Handbook of multisensor data fusion: theory and practice. CRC Press, New York, NY
25. Mell P, Grance T (2011) The NIST definition of cloud computing. Special Publication 800-145. National Institute of Standards and Technology, Gaithersburg, MD
26. Millar JR, Hodson DD, Seymour R (2016) Deriving LVC state synchronization parameters from interaction requirements. In: Proceedings of the IEEE/ACM 20th international symposium on distributed simulation and real time applications. IEEE, London, pp 85–91
27. Mittal S, Martin JLR (2013) Netcentric system of systems with DEVS unified process. CRC Press, Boca Raton, FL
28. Mittal S, Tolk A (2019) The complexity in application of modeling and simulation for cyber physical systems engineering. In: Mittal S, Tolk A (eds) Complexity challenges in cyber physical systems, Wiley, Hoboken, NJ, pp 3–25
29. Mittal S, Cane SA, Schmidt C, Harris RB, Tufarolo J (2019) Taming complexity and risk in internet of things (IoT) ecosystem using system entity structure (SES) modeling. In: Mittal S, Tolk A (eds) Complexity challenges in cyber physical systems: using modeling and simulation (M&S) to support intelligence, adaptation and autonomy. Wiley, Hoboken, NJ, pp 163–190
30. Mosterman PJ, Zander J (2016) Cyber-physical systems challenges: a needs analysis for collaborating embedded software systems. Softw Syst Model 15(1):5–16
31. Muniz MH, David JMN, Braga R, Campos F, Stroele V (2019) Pragmatic interoperability in IoT: a systematic mapping study. In: 25th Brazilian symposium on multimedia and web systems. ACM, Rio de Janeiro, Brazil, pp 73–80
32. Newell A, Simon HA (1961) GPS: a program that simulates human thought. Report No. P-2257. RAND, Santa Barbara, CA
33. Regalado A (2011) Who coined "cloud computing"?. https://www.technologyreview.com/s/425970/who-coinedcloud-computing/. Accessed Aug 2019
34. Sehgal VK, Patrick A, Rajpoot L (2014) A comparative study of cyber physical cloud, cloud of sensors and internet of things: their ideology, similarities and differences. In: IEEE international advance computing conference. IEEE, Gurgaon, India, pp 708–716
35. Shi W, Dustdar S (2016) The promise of edge computing. Computer 49(5):78–81

36. Shi W, Cao J, Zhang Q, Li Y, Xu L (2016) Edge computing: vision and challenges. IEEE Internet Things 3(5):637–646
37. Taylor SJE, Khan A, Morse KL, Tolk A, Yilmaz L, Zander J, Mosterman PJ (2015) Grand challenges for modeling and simulation: simulation everywhere-from cyberinfrastructure to clouds to citizens. Simulation 91(7):648–665
38. Tolk A (2015) Merging two worlds: agent-based simulation methods for autonomous systems. In: Williams AP, Scharre PD (eds) Systems with autonomous capabilities: issues for defence policy makers. NATO Innovations in Capability Development Publication Series, Norfolk, VA, pp 291–317
39. Tolk A, Mittal S (2014) A necessary paradigm change to enable composable cloud-based M&S services. In: Proceedings of the winter simulation conference. IEEE, Piscataway, NJ, pp 356–366
40. Tolk A, Page EH, Mittal S (2018) Hybrid simulation for cyber physical systems: state of the art and a literature review. In: Proceedings of the annual simulation symposium. Society for Modeling and Simulation, Baltimore, MD. Article 10
41. Wan J, Zhang D, Zhao S, Yang LT, Lloret J (2014) Context-aware vehicular cyber-physical systems with cloud support: architecture, challenges, and solutions. IEEE Commun Mag 52(8):106–113
42. Wymore AW (1967) Mathematical theory of systems engineering. Wiley, Hoboken, NJ
43. Xia F, Rahim A (2015) MAC protocols for cyber-physical systems. Springer, Cham, Switzerland
44. Zander J, Mosterman PJ, Padir T, Wan Y, Fu S (2015) Cyber-physical systems can make emergency response smart. Proc Eng 107:312–318
45. Zeigler BP (1990) Object-oriented simulation with hierarchical, modular models: intelligent agents and endomorphic systems. Academic Press, San Diego, CA
46. Zeigler BP (1986) Toward a simulation methodology for variable structure modeling. In: Elzas MS, Oren TI, Zeigler BP (eds) Modeling and simulation methodology in the artificial intelligence era. North Holland, Amsterdam, Netherlands, pp 195–210
47. Zeigler BP, Sarjoughian HS (2013) System entity structure basics. In: Zeigler BP, Sarjoughian HS, Duboz R, Soulié JC (eds) Guide to modeling and simulation of systems of systems. Springer, London, UK, pp 27–37
48. Zhang Y, Qiu M, Tsai CW, Hassan MM, Alamri A (2015) Health-CPS: healthcare cyber-physical system assisted by cloud and big data. IEEE Syst J 11(1):88–95

Andreas Tolk is a Senior Principal Computer Scientist at The MITRE Corporation in Hampton, VA, and adjunct Full Professor at Old Dominion University in Norfolk, VA. He holds a Ph.D. and M.Sc. in Computer Science from the University of the Federal Armed Forces of Germany. His research interests include computational and epistemological foundations and constraints of model-based solutions in computational sciences and their application in support of model-based systems engineering, including the integration of simulation methods and tools into the systems engineering education and best practices. He published more than 250 peer reviewed journal articles, book chapters, and conference papers, and edited 12 textbooks and compendia on Systems Engineering and Modeling and Simulation topics. He is a Fellow of the Society for Modeling and Simulation (SCS) and Senior Member of (IEEE) and the Association for Computing Machinery (ACM) and received multiple awards, including professional distinguished contribution awards from SCS and ACM.

Chapter 3
Truth Management with Concept-Driven Agent Architecture in Distributed Modeling and Simulation for Cyber Physical Systems Engineering

Saurabh Mittal and Douglas Flournoy

Abstract Cyber Physical Systems (CPS) are inherently distributed in nature; that is, the constituent systems are separated by geographical distances. This implies that there is always latency in communication exchanges between the system components. Various mechanisms such as frequency of updates, dead reckoning, and communication reliability are used to address the latency issue. In addition, there are computational, physics, communication protocol, and projection accuracy limitations, which constrain the solution space. The semantic interoperability issue in a cyber physical system requires that concepts and their associated data objects be aligned. When CPS are put in a distributed simulation environment without semantic alignment, the problems are compounded. Earlier work in the area of Mobile Propertied Agents (MPAs) provides a conceptual framework to engineer a central core of objects that are shared across various participants of such a System of Systems (SoS) for maintaining universal truth. This chapter will provide an overview of various mechanisms for maintaining a consistent truth representation in distributed Modeling and Simulation (M&S) systems and will present an architecture to implement MPAs in a cloud environment for a more robust CPS Test and Evaluation framework.

3.1 Introduction

Cyber Physical Systems (CPS), according to a definition provided by the National Science Foundation (NSF), are hybrid networked cyber and engineered physical elements co-designed to create adaptive and predictive systems for enhanced performance. These systems are built from, and depend upon, the seamless integration of computation and physical components. Advances in CPS are expected to enable

S. Mittal (✉)
The MITRE Corporation, 1051 Channingway Drive, Fairborn, OH 45324, USA
e-mail: smittal@mitre.org

D. Flournoy
The MITRE Corporation, 202 Burlington Road, Bedford, MA 01730, USA
e-mail: rflourno@mitre.org

© Springer Nature Switzerland AG 2020
J. L. Risco Martín et al. (eds.), *Simulation for Cyber-Physical Systems Engineering*,
Simulation Foundations, Methods and Applications,
https://doi.org/10.1007/978-3-030-51909-4_3

43

capability, adaptability, scalability, resiliency, safety, security, and usability that will expand the horizons of these critical systems.

CPS engineering is an activity that brings these elements together in an operational scenario. Sometimes, an operational scenario may span multiple domains, for example, Smart Grid incorporating Power critical infrastructure and Water infrastructure. CPS engineering requires a consistent model of operations that need to be supported by the compositions of various CPS contributors. CPS engineering lacks tools to design and experiment within a lab setting. How does one develop a repeatable engineering methodology to evaluate ensemble behaviors and emergent behaviors when larger systems involving critical infrastructure cannot be brought into a lab setting [1, 2].

CPS are Systems of Systems (SoS) wherein geographically separated constituent systems may have independent managerial and operational control and evolutionary trajectories resulting in emergent behaviors. The employment of M&S practices for SoS is well established [3–5]. Decomposing the SoS in Live, Virtual, and Constructive (LVC) elements provides a tractable mechanism to perform M&S for SoS engineering. This requires distributed simulation technology to support the LVC systems engineering as an LVC simulation system is an SoS itself. It must be emphasized that an SoS is a system classification, and a deployed SoS, such as CPS, and an LVC simulation SoS are two different SoS. While the first is user-oriented, the LVC is M&S-focused and has its own challenges and strategies for conducting distributed simulation and providing a testbed for SoS Test and Evaluation (T&E).

One of the major challenges of an LVC testbed is the management of truth data. Truth data is fundamentally the Time, State, and Position (TSP) of various constituent models, hardware/software systems, and simulation infrastructure. Due to the geographical separation of subsystems in an LVC SoS, the TSP is maintained by each of the constituent components and needs to be synchronized as SoS subsystems are integrated and a simulation execution progresses in time. The Distributed Simulation community has worked on truth management for over 30 years and has employed various strategies and standardized mechanisms to address the problem. While the community solves the time and position synchronization problem using data engineering methods, the concept of state in the TSP triad does not have a standardized methodology. The community does solve the SoS integration problem at the syntactic data level but does not provide solutions for semantic interoperability wherein a concept, such as *state* of a system, is a complex variable, and may have multiple layers of attributes and the subsystems may only care for only a subset of attributes. This is acceptable if the SoS domain is singular, such as Army, Air Force, and Navy, where the concepts between the subsystems are well established. In a multi-domain system, wherein, an SoS tries to bring in concepts from ground, maritime, air, cyber, etc., one needs semantic interoperability at the concept-definition level. CPS, as stated earlier, is another example of such a multi-domain SoS.

This chapter builds on the concept of mobile propertied agents (MPA) in a concept-driven agent architecture (CDAA) [6] and describes a reference architecture implementation for truth management in a distributed cloud-based CDAA for CPS engineering.

The chapter is organized as follows: Sect. 3.2 presents the SoS considerations for CPS and describes how CPS engineering needs to incorporate Live, Virtual, and Constructive (LVC) elements. Section 3.3 describes the CPS M&S state of the art. Section 3.4 provides an overview of distributed simulation concepts, challenges, and truth management strategies. Section 3.5 describes the concept-driven agent architecture, mobile propertied agents, and a reference architecture implementation. Section 3.6 discusses the cloud implications for truth management for the reference architecture. Section 3.7 applies the architecture to CPS engineering. Section 3.8 concludes the chapter.

3.2 SoS Nature of CPS

While the focus of CPS is both on computation and physical devices, it belongs to the class of super complex systems in a man-made world, where labels such as System of Systems (SoS), Complex Adaptive Systems (CAS), and Cyber CAS (CyCAS) are used interchangeably [7]. All of them are multi-agent systems. The constituting agents are goal-oriented with incomplete information at any given moment and interact among themselves and with the environment.

A typical CPS comprises the following components:

- Sensors,
- Actuators,
- Hardware platforms (that host sensors and actuators),
- Software interfaces (that access hardware directly or remotely through a cyber environment),
- Computational software environments (that may act both as a controller or service provider),
- Networked environments (that allows communication across geographical distances),
- End-user autonomy (that allows a CPS to be used as a passive system or an active interactive system),
- Critical infrastructures (water, power, etc., that provide the domain of operation and operational use case),
- Ensemble behaviors, and
- Emergent behaviors.

Figure 3.1 shows various aspects of CPS divided into Left-Hand Side (LHS) and Right-Hand Side (RHS). LHS consists of a collection of users, systems (both hardware and software), and devices (physical platforms). Traditional systems engineering practices and end-user use cases can be developed in the LHS. The RHS shows aspects related to infrastructures. Fundamentally, they can be characterized into Information Technology (IT) and Operational Technology (OT). Between the LHS and RHS is the network/cyber environment that allows information exchange

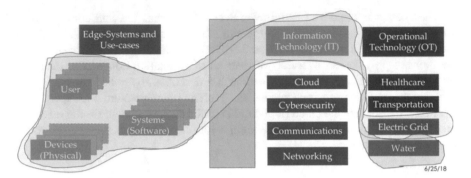

Fig. 3.1 CPS Landscape [1, 2]

between the two. With the network spanning large geographical distances, the presence of many entities/agents and their concurrent interactions in the CPS result in the ensemble and emergent behaviors. The "infrastructure-in-a-box" is largely unavailable but can be brought to bear with various existing domain simulators in an integrated simulation environment.

SoS is characterized by the constituent systems under independent operational and managerial control, geographical separation between the constituent systems, independent evolutionary roadmap, and the holistic emergent behavior arising out of spatiotemporal interactions of the constituent systems [8]. A CPS qualifies this definition very clearly as various manifestations of CPS such as consumer Internet of Things (IoT) and Industrial IoT have the essential elements. Any deployed CPS is a conglomeration of numerous vendors providing hardware, software, and services to enable a certain CPS capability. Much of the operational considerations hide behind open architecture implementation and protocol standards that provide vendor/developer-agnostic functionalities. Establishing control mechanisms in an SoS [9] is an essential aspect of CPS engineering. Despite having a standards-based approach, the CPS engineering is plagued with a lack of methodologies to establish efficient control mechanisms when various CPS components try to limit emergent behaviors. Reproducing these emergent behaviors in a simulation testbed is a nontrivial endeavor [10, 26].

3.3 CPS Modeling and Simulation

From a systems theoretic perspective, a CPS model is a hybrid system made up of both continuous and discrete systems. A continuous system (CS) is one that operates in continuous time and in which input, state, and output variables are all real values. A discrete (dynamic) system (DDS) is one that changes its state in a piecewise constant event-based manner (which also includes discrete-time systems as they are a special case of discrete event systems) [11]. A typical example of a hybrid system is a CPS in which the computation subsystem is discrete and a physical system is a CS.

The increase in overlapping CPS capabilities in a multitude of domains also introduces a level of complexity unprecedented in other engineered systems. The cross-sector deployment and usage introduces risk that may have cascaded impacts in a highly networked environment. The M&S discipline has supported the development of complex systems since its inception. In M&S, we are not only limited to the computational implementations of models. We distinguish between live simulations in which the model involves humans interacting with one another (role-playing, play-acting, etc.), virtual simulations where the model is simulated by a fusion of humans and computer-generated experiences, and constructive simulations where the model is entirely implemented in a digital computer and may have increased levels of abstraction and varying levels of autonomy. Increasingly, we are mixing the three forms of simulation in what is commonly known as live-virtual-constructive (LVC) simulation [12]. LVC simulations are used mainly for training but they can be adapted for the type of experimentation/exploration needed to investigate emergent behavior [7], as shown by the cyclical process in Fig. 3.2, elaborated in [13]. Figure 3.3 associates each of the CPS constituent elements with the corresponding M&S paradigm and how it can be incorporated in the LVC environment.

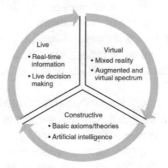

Fig. 3.2 Experimental LVC approach for generating emergence [24]

CPS Contributor	M&S paradigm	LVC element
Sensors	Continuous, physics-based	L, V, C
Actuators	Continuous, physics-based	L, V, C
Hardware platform	Both continuous and discrete	L, V
Software controller	Discrete	V, C
Network	Discrete	L, V
End User	Discrete, agent-based	L, V, C
Critical Infrastructure	Both continuous and discrete (hybrid)	V, C
Ensemble, emergent behaviors	Discrete, agent-based	V, C

Fig. 3.3 CPS contributor and the associated M&S paradigm

Using M&S for CPS engineering is not straightforward due to the inherent complexities residing in both the modeling and the simulation activities. Performing a CPS simulation requires that a CPS model be first built. The literature survey conducted in [14, 15] enumerates the following active research areas and the associated technologies for CPS modeling and has concluded that the need for a common formalism that can be applied by practitioners in the field is not yet fulfilled:

- Discrete Event Systems (DEVS) formalism,
- Process algebra,
- Hybrid automata,
- Simulation languages,
- Business processes,
- Interface design for co-modeling,
- Model-driven approaches, and
- Agent-based modeling.

The abovementioned approaches and technologies allow the development of CPS models, albeit in a piecewise manner. These model pieces and their definitions and specifications are dictated by the cross-domain CPS operational use case. The piecewise model composition sometimes does not directly translate into a monolithic simulation environment due to the confluence of both the continuous and discrete systems in the hybrid system. In the literature survey [14] as well as in many discussions with the experts [15], the use of co-simulation was identified as the preferred course of action in support of CPS for development, testing, and eventually training. Co-simulation is the co-existence of independent simulators to support a common model [16]. However, co-simulation still has to address the distributed nature of the CPS M&S solution.

At the fundamental level, there are various ways to model both timed and untimed discrete event systems, all of which can be transformed to, and studied within, the formal Discrete Event Systems (DEVS) theory [17, 4, 18–20, 27]. The recent developments bring together cloud technologies, co-simulation methodologies in a distributed environment, and hybrid modeling approaches to deliver an M&S substrate that is applicable across the entire CPS landscape (Fig. 3.4) [2]. The LHS in Fig. 3.4 employs traditional Systems Engineering practices, and it provides the context use case for CPS applications. The RHS provides various domain simulators and employs IT and OT to provide "infrastructure-in-a-box" through LVC architectures. To bridge LHS and RHS, emerging disciplines like Machine Learning and Data Science will need to be employed to understand data-driven approaches that tackle emergent behaviors when LHS and RHS interact in a parallel distributed discrete event co-simulation environment.

While CPS modeling does have options, CPS simulation must leverage hybrid simulation techniques coupled with distributed simulation technologies to engineer a robust distributed co-simulation environment. The co-simulation methodology when applied within the SoS context becomes a problem that the Distributed Simulation community has been dealing with for the past 30 years. The lessons learned from the Distributed Simulation community must be taken into account when disparate and

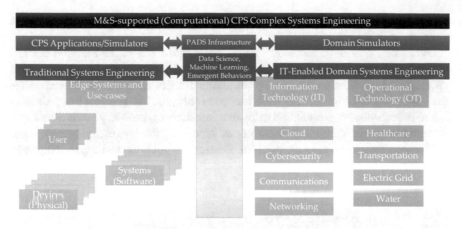

Fig. 3.4 M&S Supported (Computational) CPS Engineering testbed perspective

orthogonal domains in CPS are brought together in a distributed co-simulation environment. Because the problem of CPS hybrid simulation has now been classified as a distributed simulation problem, let us review the challenges that must be overcome for CPS M&S engineering.

3.4 Distributed Simulation Considerations for SoS

Applications in the area of Distributed Simulation engineering present challenges similar to those faced by CPS Engineers. Hence, insights gained by the Distributed Simulation community in solving these challenges can be leveraged to enhance CPS efforts. This section draws these parallels by first providing an overview of Distributed Simulation and its SoS nature. Then, a brief history of key Distributed Simulation challenges and solution approaches is provided. Finally, emphasis is placed on truth management challenges in Distributed Simulation, as the solutions to these challenges are particularly apropos to the latency issues in CPS.

3.4.1 Distributed Simulation Overview

A distributed simulation is comprised of 2 or more simulation applications exchanging data over a network as they execute a common scenario. The use of distributed simulations has increased steadily in recent years as, for cost and other reasons, simulation engineers choose to meet project requirements by connecting multiple existing simulations together rather than building all the necessary functionality into a new simulation for managing TSP data. Sometimes, that is the only possible solution as the simulators may involve physical hardware that cannot co-locate with the other simulators.

Over time, a large de facto community of practice has formed around the use of distributed simulations. Standards for exchanging data between simulations have emerged, and solutions to challenging simulation integration issues have been developed and shared. Much of this work has been advanced to meet military simulation requirements, but simulationists from other domain areas have discovered and benefited from the community's efforts as well. Regardless of the domain area, distributed simulations are now used for a variety of analysis, testing, training, and experimentation purposes.

3.4.2 The SoS Nature of Distributed Simulation

A distributed simulation is itself a system of "systems" (in this case, software applications). Each component application is included in the larger distributed system because it provides certain capabilities in a manner the other components cannot. On the other hand, each simulation component is dependent on the others for information in order to fully represent the simulated scenario, requiring simulations to exchange data and respond appropriately to data from other simulations. In this sharing of information between the components, distributed simulations present all the interoperability and timing challenges that face systems of systems.

3.4.3 Distributed Simulation Challenges and Solution Approaches

In distributed simulations, some of the ways these SoS-like interoperability and timing challenges manifest themselves include

1. **Runtime performance**. All simulation components must be able to "keep up with" a common simulation clock during execution. For distributed simulations that run in "real time" (that is, a second's worth of simulation time is processed within one second of actual "wall clock" time), the simulations must be able to send and receive all necessary messages, along with taking care of processing these messages and any other internal model calculations, without "falling behind". For distributed simulations that support large military training exercises, this can mean timely messaging and processing for tens of thousands of simulated platforms.
2. **Unambiguous messaging**. The data messages between components must be constructed such that the contents of all fields within the messages are syntactically and semantically aligned.

3. ***Repeatability and reproducibility***. For analytical constructive simulations, it may be desirable to achieve identical results from multiple runs with the same initial conditions. However, this may not be true in two cases: (1) A stochastic simulation may be desired to see a range of results. For repeatability, the random number seed is included as one of the starting conditions. (2) Any human-in-the-loop solution will not be repeatable due to the possibility of a human participant responding differently or at different times from one run to the next. For distributed simulations, this requires (1) the components to be triggered in a consistent manner, (2) all messages between components arrive timely at their destination, and (3) the ordering of all messages between components is preserved.

4. ***Clock synchronization***. In many analysis use cases, distributed simulations must run faster than real time (i.e., many, many runs are required for statistical significance) or it is known that they will run slower than real time (i.e., extremely compute-intensive subject matter). In these cases, a clock control messaging and processing algorithm must be implemented to keep all components' simulated clocks in synchronization with one another.

Middleware solutions developed by and for the distributed simulation community over many years have helped to address these challenges. The Distributed Interactive Simulation (DIS) protocol was an early (and still existing) standard that addressed ambiguity and some performance issues for military simulations by rigidly defining a set of message types and broadcasting all messages to all simulation components. Later, the High-Level Architecture (HLA) and Test and Training Enabling Architecture (TENA) standards provided additional capabilities. HLA provided Time Management capabilities to facilitate results' repeatability and runtime clock synchronization. Both TENA and HLA offered publish and subscribe services to conserve message traffic over the network.

These solutions have been reused with success for many years for distributed simulations where components exist on a Local Area Network (LAN) or custom Wide Area Network (WAN) configuration. The emerging trend toward Cloud Computing may be a cause for reassessing some aspects of these and other middleware solutions used in the distributed simulation community.

3.4.4 Truth Management Approaches in Distributed Simulation

In addition to the challenges mentioned above, a fundamental Distributed Simulation issue is the proper sharing of TSP data among simulation components for the objects being simulated, or *truth management*. Sharing of truth data across components is what facilitates key cross-component interactions, such as

- a missile flying in one simulation intercepts (or misses) a helicopter in another simulation,

- a radar operated by one simulation detects an aircraft flown by another simulation, causing a tracker software in a 3rd simulation component to create and/or update a track on that aircraft.

Through much trial and error, the Distributed Simulation community has learned that effective truth management involves a balance, or "tuning," of the following three primary factors including

1. *Update Frequency*: Frequency of state data updates,
2. *Reliable Message Propagation*: Reliability of state data communications between simulation components, and
3. *State Projection*: Use, including configuration, of a predictive consistency algorithm known as dead reckoning.

The relationship between these primary factors needs to be established in light of the following limitations:

I. *Computational limitation*: If all objects' truth data could be updated continuously (that is, infinitely high update frequency), and simulation components had the computing capability to handle the continuous processing necessary to support this, a consistent scenario picture would be achieved at all times across all distributed components. In the missile example cited above, the importance of accurate and up-to-date proximity information across multiple simulations is paramount.

II. *Physics limitation*: Due to a dynamically changing combination of network throughput limitations and latencies and component processing limitations, there are limits on how often state data updates can be sent and received/processed. For example, if the simulation components are separated by a few thousand miles, the limits of physics are hit as the electromagnetic signal takes nonzero propagation time.

III. *Communication Protocol limitation*: Network throughput and latency limitations can be partially relieved by using less reliable (i.e., UDP) messaging protocols instead of more reliable protocols (i.e., TCP). However, there is a performance versus message loss trade-off to be considered here.

IV. *Projection accuracy limitation*: Dead-reckoning can be used to predict the movement of simulated objects between state data updates so as to lessen the impact of infrequent or lost updates. Thresholds can be used to tune the dead-reckoning algorithm to achieve the appropriate balance of location accuracy and network load [21]. DIS supports a dead-reckoning approach that, if used, assumes local predictive model implementations at each simulation component.

To summarize, distributed simulation technologies address the technical challenges of update frequency, message propagation delays, and state projections within the limitations of computational implementation, physics, standards applicability, and accuracy of projecting states to ensure consistent TSPs between the constituent system components.

3.5   Conceptual Alignment and Reference Architecture for Truth Management

The Distributed Simulation community performs simulation systems engineering to address the three technical issues of update frequency, reliable message propagation, and state projection. The end-user operational and test community in an SoS LVC setting is more concerned with four conceptual semantic issues [6] such as

1. *Simulation ownership*: Distributed simulation in an LVC setting relies on standards like DIS, HLA or TENA. Employing simulation standards in a distributed cloud-based M&S solution warrants the use of simulation services that hide simulation specific solutions that the Distributed Simulation community has solved. The Modeling and Simulation as a Service (MSaasS) paradigm [22, 23] hides model and simulation infrastructure behind a Service Layer. The additional layer introduces more complexity in a distributed SoS model, and managing ownership at both the modeling and simulation levels for the entire SoS becomes more complicated as the Service Layer may consolidate service implementations in a service-oriented architecture and it remains hidden how a service API gets realized.

2. *Shared state within SoS*: Each federation in a distributed simulation maintains its own state. In a cloud-based netcentric environment, it remains unclear if the Service Layer abstraction continues to perform in a similar manner within an event-driven Service-Oriented Architecture (SOA). As Mittal and Martin [18] point out, there is no notion of an agency in Event-Driven Architecture (EDA) when implemented over SOA. The state is encapsulated in the message exchange. This conflicts with the federated LVC solution that the Distributed Simulation community has addressed.

3. *Partial observability problem within a multi-agent system*: Most of the simulated LVC systems are Multi-Agent Systems (MAS). With two complexity levels introduced by black-box federations that hide the internal state and the Service Layer abstractions that hide the implementation, these MASs that relied on sharing individual states have to overcome the partial observability with these two additional constraints when they interact through a simulated environment which is also behind a service interface. The issue of conceptual alignment becomes center stage as misalignment at the conceptual level will remain hidden behind the service interface, resulting in an erroneous system.

4. *Knowledge-base of constituent systems*: In any cloud-based solution, the knowledge-base of each of the federates must be conceptually aligned at the ontology level so it facilitates model composability. The service interface must use harmonized ontologies that intend to engineer a multi-domain solution, for example, bringing Army and Air Force nomenclatures into a single SoS model.

3.5.1 Mobile Propertied Agents (MPAs) and Concept-Driven Agent Architecture (CDAA)

An MPA is defined as an agent that encapsulates a semantic concept, its associated properties (by way of syntactic data elements) and provides interfaces to manipulate the properties by external services [6]. In addition, an MPA contains a state-machine to record the current state of the encapsulated properties as it gets dynamically invoked by the external federates who want to use the semantic concept. Instead of allowing multiple representations of the same concept at various places, each concept is represented by exactly one MPA. If a service needs the concept, it floats to the Service Layer and becomes an MPA. The behavior of an MPA may be either discrete event or discrete time depending upon the larger infrastructure the MPA is part of. Regardless, the objective behind MPA is to have a state that remains consistent and/or gets updated as it is accessed by different simulation federates. Discrete event foundation is the most likely case as managing the global clock may be prohibitive in Cloud-based M&S. Within EDA, an MPA can maintain the standard Greenwich Mean Time (GMT) representation as the underlying enterprise cloud-based infrastructure must be based on at least one Time Zone. Any invocation of an MPA results in an event and consequently, any property update within the MPA also results in an event. An MPA may also contain platform-independent code (e.g., in XML/JSON) that gets executed at the external federate. In an advanced case of MPA design, the MPA will be under a supervisory control to implement, and governance and security policies. An MPA facilitates semantic interoperability and implements ontological concepts that can maintain the state of their properties as they are accessed and utilized by external simulation federates. Consequently, a semantic concept is now, theoretically a system and being modular, can be integrated with other systems (external federates) in an enterprise context such as in the Cloud.

An MPA is hosted within a Complex Event Processing (CEP) Cloud infrastructure (e.g., Esper and TIBCO) in close conjunction with a controlling middleware that provides various transformers/adapters to enable invocation of an MPA by external federates using the SOA. MPAs ensure the consistent representation of truth regarding the challenges derived from the need to align the data.

A Concept-Driven Agent Architecture (CDAA) is an architecture for a multi-agent system (MAS) that is guided by the concepts that the agents utilize to perform their function. It is an SoS of modular concept-as-a-system wherein the concepts are modeled as MPAs interoperating with other systems in a netcentric environment. A parallel distributed Cloud-based M&S involving MPAs shall have the following components, as shown in Fig. 3.5:

1. *Event cloud that hosts MPAs*: An event cloud is the central concept in an EDA. Architecturally speaking, it is a database accessible through a specific Event Query Language (EQL) with near real-time response times for mission-critical systems. This acts as a shared memory or blackboard for MPAs. Here, various patterns and spatiotemporal relations can be defined that utilize knowledge within the MPAs and keep MPAs consistent. It can manage the lifecycle of an MPA.

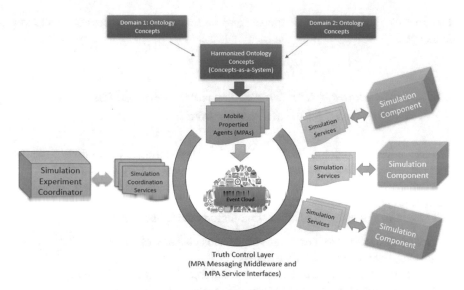

Fig. 3.5 Concept-driven Agent Architecture (CDAA) with MPA Cloud

2. *Truth Control Layer* (TCL): This middleware ensures that MPA invocations remain consistent. It acts as a knowledge-broker in the semantic domain and interfaces between the Event cloud and simulation federates (elaborated in Sect. 3.5.3).
3. *Simulation event coordinator*: This is a component that manages a simulation exercise/event. It deploys requirements, employs assets, executes the simulation, orchestrates the service composition, and assembles the simulation results in various formats. It implements a simulation protocol (an algorithm) that manages time and preserves causality between various simulation federates.
4. *Simulation Services*: These are the simulation applications that are made available as services. They can very well be domain agnostic simulation engines or formal Systems Theory-based M&S kernels (e.g., DEVSVM) that are engaged with the Simulation Coordinator through the simulation protocol. They manipulate the states of MPAs.
5. *Enterprise Service Bus(es)/message queues/RTIs*: These buses interface between TCL and the participating federates. They implement various message transformers to align the ontology gaps and facilitate conceptual alignment.

As can be clearly seen, CDAA focuses more on the semantic interoperability and composability of MPAs toward a simulation solution in a Cloud-based M&S infrastructure. All composable M&S cloud services are memoryless, as the required state information needed is provided by MPAs representing the participating concepts. While event cloud, TCL, and runtime messaging services cooperate to bring the right MPAs to the simulation services, the simulation coordinator orchestrates that services are provided in the right order and at the right time, and as such ensures the consistent representation of truth regarding temporal and process orchestration challenges.

3.5.2 Mathematical Considerations for a Parallel Distributed Cloud-Based M&S Infrastructure

Because of the unavoidable distributed M&S data synchronization limitations listed above in Sect. 3.4, the community has worked out solutions for parametrically tuning simulation systems to keep these errors within use-case-driven bounds. One such solution [21] is summarized here.

The realities of distributed simulation include the following:

1. state data shared across multiple simulations via network messaging will diverge to some extent between data updates, and
2. scenario events generated by one simulation will be received and processed at some slightly later time by the receiving simulations.

Given this, the notion of "plausibility limits" is introduced to define just how much state divergence (spatial error) and/or event error tolerance is acceptable. For example, distributed simulations involving collisions or high-speed weapon engagements may have extremely small plausibility (error) limits.

Given a defined plausibility limit l, the maximum inter-update period p between position data updates from one entity to another entity can be calculated. The mathematics is described in detail in [21]. Suppose l is this maximum error tolerance (plausibility limit). Then given the updating entity's maximum acceleration a_{max} and the network propagation delay d, the update rate p is provided by

$$p = \sqrt{\frac{2*l}{\|a_{max}\|} - d}$$

Building on the above equation and realizing there is variability associated with both d and a_{max} in a distributed simulation deployed over a Wide Area Network (WAN), the probability of the plausibility limit l being exceeded can also be calculated. If the probability of exceeding the plausibility limits is unacceptable, measures can be taken to attempt to bring these probabilities into compliance with use-case requirements. Data update rates can be increased as long as the associated increase in network traffic does not impact simulation performance. If simulation components are distributed by long distances over a WAN, the components can be located closer or possibly brought together in a LAN. Plausibility exceedance calculations can be done for these and other proposed solutions to estimate the benefits before implementing the solutions.

The work by Millar et al. [21] clearly defines the relationship among the spatial error tolerance, the rate of change of velocity of an entity, and network propagation delay to arrive at the update frequency for entity position in an LVC context. In addition, they developed mathematics on the probability of exceeding the error tolerance. We shall now discuss how these results can help us architect an MPA-based solution for truth management in a cloud-based computing infrastructure.

3.5.3 Reference Architecture Implementation

As can be seen from the previous section, plausibility limits bear a relationship to the simulation infrastructure as well as to the second-order rate of the change of the entity's position. We extend the notion of position to the broader concept of MPA. We apply the plausibility limit concept to the "state" of an MPA. Here, the state is more than a spatial position and may include a collection of attributes/variables. Consequently, the inter-update rate required by an MPA entity becomes a configurable parameter in the deployment infrastructure for an MPA-based solution.

Figure 3.6 shows the MPA/CDAA layered architecture. The lowest MPA Communication Channel layer is the networking infrastructure incorporating network hardware channels. The MPA Hosting Infrastructure layer incorporates cloud-based infrastructure, including the event clouds for complex event processing. The MPA (Data Object) layer defines the syntactic layer of the system that contains various MPA data structures and event taxonomy to be hosted on a cloud CEP infrastructure. The Ontology layer specifies the interrelationships between various MPA concepts. The Services layer makes available the MPAs as a service. It accesses the ontology and utilizes the inherent MPAs and their relationships/interactions through web services. Both the Ontology and Service layers constitute the semantic layer of the CDAA. The Orchestration layer specifies the invocation of various services to generate the behavior of the system based on MPA interactions. This constitutes the pragmatic layer of the architecture.

The implementation of the MPA/CDAA in a cloud computing infrastructure is done in the following manner. The Ontology layer incorporating the MPA data objects is hosted in an event cloud and a run-time infrastructure for data exchange. The

Fig. 3.6 MPA/CDAA
Layered architecture
framework

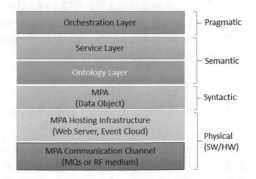

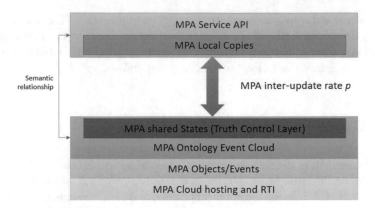

Fig. 3.7 MPA implementation architecture

ontology is no longer a static entity but a dynamic entity that maintains all of the MPAs' true states. Both the event cloud and RTI may be separated by a geographical distance within an LVC context. The MPA Services layer will communicate with the MPA event cloud that houses the MPA's true copy with the MPA Service API. The event cloud, as described in Sect. 3.5, is the TCL. The MPA Service maintains a local copy of the MPA which gets synchronized in a periodic or event-based manner with the TCL. This synchronization rate can aptly be derived from the plausibility limit considerations, as in the previous section, for an MPA. This architecture proposes MPA-driven update rates. Figure 3.7 shows the interaction between the MPA Service layer and the MPA TCL hosting the MPA ontology layer.

As the MPA/CDAA will be applied to real-time and virtual-time (faster than real-time) simulation systems, the MPA is required to have its own lifecycle. In virtual-time systems, MPAs can also advance the simulation clock as all the message propagation happens in logical time. It can be synchronized in zero logical time taking nonzero real time between the local and true copies. In real-time systems, and especially, in real-time distributed systems, the MPA inter-update rate p determines the MPA sync time. Figure 3.8 shows the MPA lifecycle for virtual-time systems, and Fig. 3.9 shows the lifecycle for real-time systems. An MPA is initialized in *passive* state. When an owner is assigned for an MPA, it transitions to *active* state. The solid transitions are triggered when an external event occurs that warrants MPA attention. The dotted transitions are internal transitions that MPA undergoes to keep the local and true copies in sync.

3.6   Cloud Implications for Plausible Solution

This section introduces the truth management considerations for implementing an MPA-based architecture solution. TSP MPA implications and trade-offs are discussed as they relate to the four limiting factors: computational, physics, communication

Fig. 3.8 MPA lifecycle for
Virtual-time systems

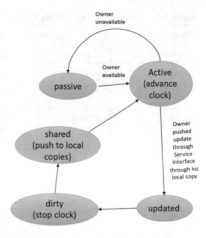

Fig. 3.9 MPA lifecycle for
real-time systems

protocol, and projection accuracy. These considerations are summarized in the MPA
columns of Table 3.1.

Time: Time considerations in an MPA architecture include management of timing
inconsistencies and rate adjustments. Time's role in the four truth management
limitations manifests as follows:

- *Computational.* As update frequencies are adjusted higher, eventually the simulation CPUs will reach their processing capacity and will not be able to keep up
  with the number of incoming updates.
- *Physics.* The time it takes for data transmissions to travel geographic distances is
  limited by the laws of physics. If the distance between simulations is long enough
  that latencies are in the hundreds of milliseconds, high update rates can approach
  and exceed these limits.
- *Communication Protocol.* The choice of network protocol used to transmit data
  impacts transmission latencies. Data packets occasionally collide when traversing

Table 3.1 MPA Truth Management aspects with distributed simulation limitations

Limitations	MPA Truth Management aspects		
	Time	State	Position
Computational	Inter-update frequency (Tc = p)	Number of variables and complexity of state representation	Kinematics
Physics	Latency (Tl)	Fidelity/resolution trade-off (ideally, high fidelity and low resolution)	Kinematics
Communication Protocol	Reliability (Tr)	Performance/reliability trade-off for simulation/scenario-critical MPA	Simulation critical MPA needs to be reliable. Scenario-based MPA may/may not need to be reliable
Projection accuracy	Source of error accumulation (a combination of Tc, Tl, and Tr)	High fidelity, reliable propagation, and low error accumulation in state using dead-reckoning algorithms	Covered within state considerations

a network, causing transmissions to fail. Reliable TCP-based communications handle this by requiring receipt confirmation, an extra step that generates additional latency. "Fire-and-forget" UDP-based communications provide much lower latency at the risk of dropping messages.

- *Projection Accuracy.* Projection accuracy is impacted by timing in a combination of all the above latency and update rate considerations.

State: State considerations in an MPA architecture are driven by (a) the complexity of the state representations needed, (b) fidelity/resolution requirements, and (c) message reliability requirements for the simulation use case. Fidelity requirements refer to the requirements that align MPA as close as possible to the real world. Resolution requirements refer to the requirements that a computational representation of an MPA would need. Intuitively, high fidelity and high resolution are resource exhaustive, so a balance is required. These requirements map to the limiting factors as follows:

- *Computational.* The structure of the state data—that is, the number and complexity of the state variables to be updated—factors into the computational load required to package/send and receive/process the data.
- *Physics.* Scenario specifics drive state fidelity and resolution requirements. The physics of simulated platforms that travel and maneuver faster than others (e.g., aircraft and missiles) drive up both fidelity and resolution requirements. Conversely, slower moving platforms can be simulated at an appropriately high fidelity with lower update resolution.

- *Communication Protocol.* There is a trade-off between performance and reliability for scenario-critical MPAs. Highly critical MPA state and execution management values must be sent reliably so as not to risk data loss during execution.
- *Projection Accuracy.* Tunable Dead-Reckoning algorithms have been developed to extrapolate state data as accurately as possible, considering not only the simulated motion of entities but also expected network latencies and update rates.

Position: Position is an especially demanding subset of simulation state data with respect to the distributed simulation limiting factors. Considerations for Position in an MPA architecture are as follows:

- *Computational and Physics.* In a distributed simulation, the MPA architecture needs to manage the computational load of position updates, and the network delays incurred by the update messages, so that the kinematics of scenario events (e.g., missile strikes) aren't degraded by inconsistencies.
- *Communication Protocol.* In many distributed simulation cases, position updates by entity-specific MPAs will comprise the vast majority of messages sent. Depending on the use case, the risk of dropping occasional position updates for some or all of the platforms may be worth the performance benefits of having those platforms' MPAs send the updates via UDP instead of TCP.
- *Projection Accuracy.* As discussed above in the state data considerations, Dead Reckoning is used within the simulation community to extrapolate the positions of entities consistently across distributed simulation components. These algorithms can be leveraged in MPA architectures.

From a cloud deployment of MPA/CDAA perspective, MPAs need to be partitioned accordingly, so that the simulation execution can take into account the LVC considerations. If the simulation is all constructive, then CDAA simulation execution can be deployed in the cloud without any impact on the Time factor. The state and the position factors are still dependent on the computational limitations. For constructive simulations, we are engineering a virtual-time system (Fig. 3.8). When the simulation is both virtual and constructive (VC), the virtual components, e.g., cockpit simulators, may be geographically separated. All the TSP considerations identified in Table 3.1 will need to be considered. The partitioning of MPA local copy (geographical location of the virtual component) with the shared MPA true copy (in an event cloud) must be architected carefully, such that the TSPs are optimized. The deployed simulation architecture may still be virtual-time system, if not constrained by any real-time usage requirements for any virtual components. When the simulation is LVC, the simulation is real time and all the considerations defined in Table 3.1 come into play. The deployment architecture must account for novel MPA partitioning strategies to mitigate the limitations.

3.7   MPA/CDAA Applied to CPS M&S

Figure 3.3 associates CPS elements with LVC nomenclatures. It also assigns the elements to different modeling paradigms such as discrete, continuous, and hybrid. Figure 3.4 puts together CPS application domain in reference to a multi-domain execution environment, incorporating critical infrastructure domains like Power, Water, Transportation, etc. It warrants a co-simulation environment wherein various domain simulators need to be brought in to support a CPS simulation. Bringing together knowledge from multiple domains into a single application ontology, such as CPS, is a nontrivial endeavor (Chap. 23, [4]) and to the best of our knowledge, the present work wherein MPA/CDAA architecture can align semantic concepts, leads the way.

A CPS-in-a-box is a testbed that brings in the needed domain simulators in a co-simulation environment and aligns various MPAs in a CDAA. One such effort in the works is documented in [24], extended from Pratt et al. [25]. In this work, a constructive digital twin of a smart thermostat was co-simulated with a virtual Power domain simulator (e.g., GridLAB-D), and a constructive home simulator housing the digital twin model was aligned with the GridLAB-D's house agent concept. While the architecture did not apply MPA concepts, it did align the house concept such that the house agent defined in GridLAB-D is further extended in the developed IoT System model to apply to a cyber inclusion use case for an Internet-of-Things application.

CPS-in-a-box must be extensible to go through the constructive, virtual, and live systems integration progression path through the harmonized ontology concepts defined in a holistic manner for a multi-domain solution. The simulation system through this progression must account for TSP factors, and how they need to be addressed for distributed cloud-based simulation systems. The MPA/CDAA architecture approach brings the conceptual alignment to the forefront of CPS testbed engineering due to the inclusion of a multi-domain knowledge base, a necessity in any CPS solution.

3.8   Conclusions and Future Work

CPS are multi-domain SoS that need to bring together concepts from different domains in an integrated system for successful operation. This requires that data between the different domains is aligned before many orthogonal system domains are integrated as an SoS in a CPS. The M&S community is aware of the issues arising from such an integration and such issues fall into the semantic interoperability category. While there are data and simulation standards available for SoS engineering in a single domain, the semantic challenge of concept alignment in a multi-domain solution is an open research problem. The use of ontologies to perform semantic alignment is one possible course of action to bring conceptual alignment between two knowledge bases.

Any CPS engineering testbed will likely employ M&S-based methods. Being classified as an LVC SoS, problems inherent in an LVC SoS simulation system also are applicable to such a testbed. Adding the conceptual alignment in a multi-domain SoS adds further complexity and requires new approaches to semantic interoperability. MPA and CDAA reference architecture implementation brings conceptual alignment at the center of such a multi-domain integration. Aligning concepts from different domains in a pragmatic ontology for a CPS use case and deploying in an event cloud for shared access by the constituent systems is a plausible approach to perform multi-domain systems engineering.

Managing truth (TSP) in a cloud based CDAA implementation incorporates 30 years of techniques, technologies, and procedures from the Distributed Simulation community and the recent Cloud-based Systems engineering community. This chapter discussed the challenges in truth management and the MPA/CDAA reference architecture implementation that addresses truth management in a cloud-based deployment. We described the MPA/CDAA concept in detail and the mathematical underpinnings behind the synchronization of local copies of TSP MPAs with the true copy of TSP MPAs in the event cloud. We also addressed four types of limitations: computational, physics, communication protocol, and projection accuracy, inherent in distributed simulation SoS and their impact on a cloud-based CDAA.

Indeed, the MPA/CDAA reference architecture implementation must be realized in a CPS Solution Architecture before the approach can be deemed valid. This chapter contributed the foundational theory and principles behind an MPA/CDAA-based architecture applied to M&S-based CPS engineering.

References

1. Mittal S, Tolk A (eds) (2019) Complexity challenges in cyber physical systems: Using modeling and simulation (M&S) to support intelligence, adaptation and autonomy. Wiley, Hoboken, NJ
2. Mittal S, Tolk A (2019) The complexity in application of modeling and simulation for cyber physical systems engineering. In: Mittal S, Tolk A (eds) Complexity challenges in cyber physical systems: Using modeling and simulation (M&S) to support intelligence, adaptation and autonomy. Wiley, Hoboken, NJ
3. Mittal S, Zeigler BP, Martin JLR, Sahin F, Jamshidi M (2008) Modeling and simulation for system of systems engineering. In: Jamshidi M (ed) System of systems engineering for 21st century. Wiley, NJ
4. Mittal S, Martin J (2013) Netcentric system of systems engineering with DEVS Unified Process. CRC Press, Boca Raton, FL USA
5. Rainey LB, Tolk A (2015) Modeling and simulation support to system of systems engineering applications. Wiley, Hoboken, NJ
6. Tolk A, Mittal S (2014) A necessary paradigm change to enable composable cloud-based M&S services. In: Proceedings of the Winter Simulation Conference
7. Mittal S (2019) New frontiers in complex systems engineering: The case of synthetic emergence. In: Sokolowski J, Durak U, Mustafee N, Tolk A (eds) Summer of Simulation: 50 years of seminal computer simulation research. Springer, AG

8. Maier M (2015) The role of modeling and simulation in system of systems development. In: Rainey LB, Tolk A (eds) Modeling and simulation support for system of systems engineering applications. Wiley, Hoboken, NJ

9. Mittal S, Rainey LB (2015) Harnessing emergent behavior: The design and control of emergent behavior in system of systems engineering. In: Proceedings of Summer Computer Simulation Conference, Chicago, IL

10. Mittal S, Martin JLR (2017) Simulation-based complex adaptive systems. In: Mittal S, Durak U, Oren T (eds) Guide to simulation-based disciplines: Advancing our computational future. Springer

11. Lee KH, Hong JH, Kim TG (2015) System of systems approach to formal modeling of CPS for simulation-based analysis. ETRI J 37(1)

12. Hodson DD, Hill RR (2014) The art and science of Live, Virtual, and Constructive simulation for test and analysis. J Defense Model Simul 11(2):77–89

13. Diallo S, Mittal S, Tolk A (2017) Research agenda for next generation complex systems engineering. In: Mittal S, Diallo S, Tolk A (eds) Emergent behavior in complex systems engineering: A modeling and simulation approach. Wiley, Hoboken, NJ

14. Tolk A, Page EH, Mittal S (2018a) Hybrid simulation for cyber physical systems: State of the art and a literature review. In: Frydenlund E, Jafer S, Kavak (eds) Proceedings of the Annual Simulation Symposium. Society for Modeling and Simulation International, San Diego, CA, pp 10:1–10:12

15. Tolk A, Barros F, D'Ambrogio A, Rajhans A, Mosterman PJ, Shetty SS, Traor'e MK, Vangheluwe H, Yilmaz L (2018b) Hybrid simulation for cyber physical systems: A panel on where are we going regarding complexity, intelligence, and adaptability of CPS using simulation. In: Mittal S, Martin JLR (eds) Proceedings of the Symposium on M&S of Complexity in intelligent, adaptive and autonomous systems. Society for Modeling and Simulation International

16. Mittal S, Zeigler BP (2017) Theory and practice of M&S in cyber environments. In: Tolk A, Oren T (eds) The profession of modeling and simulation: Discipline, ethics, education, vocation, societies and economics. Wiley, Hoboken, NJ, USA

17. Vangheluwe HL (2000) DEVS as a common denominator for multi-formalism hybrid systems modelling. In: IEEE International Symposium on Computer-aided Control System Design. IEEE, pp 129–134

18. Mittal S, Martin JLR (2013) Model-driven systems engineering for netcentric system of systems with DEVS unified process. In: Proceedings of the 2013 Winter Simulation Conference, pp 1140–1151

19. Mittal S, Martin JLR (2016) DEVSML Studio. A framework for integrating domain-specific languages for discrete and continuous hybrid systems into DEVS-based M&S environment. Summer Computer Simulation Conference, Montreal, Canada

20. Traoré MK, Zacharewicz G, Duboz R, Zeigler BP (2018) Modeling and simulation framework for value-based healthcare systems. Simulation 95(6):481–497

21. Millar JR, Hodson DD, Seymour R (2016) Deriving LVC state synchronization parameters from interaction requirements. In: IEEE/ACM 20th International Symposium on Distributed Simulation and Real-time Applications

22. Hannay JE, Berg T (2017) The NATO MSG-136 Reference architecture for M&S as a Service. M&S Technologies and Standards

23. Siegfried R, Berg TV, Altinalev T, Arabaci G (2018) Modelling and simulation as a service (MSaaS) rapid deployment of interoperable and credible simulation environments

24. Mittal S, Tolk A, Pyles A, Balen N, Bergollo K (2019) Digital twin modeling, co-simulation and cyber use-case inclusion methodology for IoT systems. In: Proceedings of the 2019 Winter Simulation Conference

25. Pratt A, Ruth M, Krishnamurthy D, Sparn B, Lunacek M, Jones W, Mittal S, Wu H, Marks J (2017) Hardware-in-the-Loop simulation of a distribution system with air conditioners under model predictive control. In: IEEE Power and Energy Society General Meeting. Institute of Electrical and Electronics Engineers, Inc

26. Mittal S, Diallo S, Tolk A (2017) Emergent behavior in complex systems engineering: A modeling and simulation approach. Wiley, NJ
27. Zeigler BP, Praehofer H, Kim T (2000) Theory of modeling and simulation: Integrating discrete event and continuous complex dynamic systems, Academic Press

Saurabh Mittal is the Chief Scientist for the Simulation, Experimentation, and Gaming Department at the MITRE Corporation. He is on the editorial boards of Transactions of the Society of Modeling and Simulation (SCS) International and the Journal of Defense Modeling & Simulation. He received both Ph.D. and M.S. in Electrical and Computer Engineering from the University of Arizona, Tucson. He can be reached at smittal@mitre.org.

Douglas Flournoy is a Principal M&S Staff Member and Deputy Chief Scientist for the Simulation, Experimentation, and Gaming Department at the MITRE Corporation. He received an M.S. in Operations Research from George Washington University, Washington. His research interests and work experience include data-driven simulation, distributed simulation, and simulation-based experimentation and analysis. He can be reached at rflourno@mitre.org.

Chapter 4
Implementing the Modelling and Simulation as a Service (MSaaS) Paradigm

Robert Siegfried

Abstract Simulation is used widely within the defence sector to support training, capability development, mission rehearsal and decision support in acquisition processes. Consequently, Modelling and Simulation (M&S) has become a critical capability and M&S products are highly valuable resources that need to be conveniently accessible by a large number of users as often as possible. M&S as a Service (MSaaS) combines service orientation and the provision of M&S applications via the as-a-service model of cloud computing to enable more flexible simulation environments that can be deployed and executed on demand. The NATO Modelling and Simulation Group (NMSG) investigates MSaaS with the aim of providing the technical and organizational foundations to establish the Allied Framework for M&S as a Service. The Allied Framework for M&S as a Service is the common approach of NATO and nations towards implementing MSaaS and is defined by an Operational Concept Document, a Technical Reference Architecture, and supporting Governance Policies. This chapter provides an overview of the Allied Framework for MSaaS and initial experimentation results to demonstrate that MSaaS is capable of realizing the vision that M&S products, data and processes are conveniently accessible to a large number of users whenever and wherever needed.

4.1 Introduction

NATO and the Nations use distributed simulation environments for various purposes, such as training, mission rehearsal and decision support in acquisition processes. Consequently, modelling and simulation (M&S) has become a critical technology for the coalition and its nations. Achieving interoperability between participating simulation systems and ensuring the credibility of results currently requires often enormous efforts with regards to time, personnel and budget.

R. Siegfried (✉)
Aditerna GmbH, Riemerling/Munich, Germany
e-mail: robert.siegfried@aditerna.de

© Springer Nature Switzerland AG 2020
J. L. Risco Martín et al. (eds.), *Simulation for Cyber-Physical Systems Engineering*,
Simulation Foundations, Methods and Applications,
https://doi.org/10.1007/978-3-030-51909-4_4

The NATO Modelling and Simulation Group (NMSG) is a part of the NATO Science and Technology Organization (STO). The mission of the NMSG is to promote cooperation among Alliance bodies, NATO, and partner nations to maximize the effective utilization of M&S. Primary mission areas include: M&S standardization, education and associated science and technology. The NMSG mission is guided by the NATO Modelling and Simulation Masterplan (NMSMP) [1]. The NMSMP vision is to "Exploit M&S to its full potential across NATO and the Nations to enhance both operational and cost-effectiveness". This vision will be achieved through a cooperative effort guided by the following principles:

- Synergy: leverage and share the existing NATO and national M&S capabilities.
- Interoperability: direct the development of common M&S standards and services for simulation interoperability and foster interoperability between Command and Control (C2) and simulation.
- Reuse: Increase the visibility, accessibility and awareness of M&S assets to foster sharing across all NATO M&S application areas.

The NMSG is the Delegated Tasking Authority for NATO M&S interoperability standards. This is the rationale for the close relationship between NMSG and the Simulation Interoperability Standards Organization (SISO), which was formalized in a Technical Cooperation Agreement signed in July 2007 and renewed in 2019.

Recent technical developments in the area of cloud computing technology and service-oriented architecture (SOA) may offer opportunities to better utilize M&S capabilities in order to satisfy NATO critical needs. *M&S as a Service (MSaaS)* is a concept that includes service orientation and the provision of M&S applications via the as-a-service model of cloud computing to enable composable simulation environments that can be deployed rapidly and on demand.

One of the technical working groups under the NMSG is MSG-136 ("Modelling and Simulation as a Service—Rapid deployment of interoperable and credible simulation environments") [2]. This group investigated the concept of MSaaS with the aim of providing the technical and organizational foundations for a future permanent service-based *Allied Framework for MSaaS* within NATO and partner nations. NATO MSG-136 started its three-year term of work in November 2014 and finished in November 2017. MSaaS is looking to provide a strategic approach to deliver simulation coherently against the NMSMP vision and guiding principles.

This chapter provides an overview of the activities performed by MSG-136 and presents the results achieved, from the following perspectives:

- *Operational concept* of MSaaS: how it works from the user point of view;
- *Technical concept* of MSaaS: reference architecture, services metadata, and engineering process;
- *Governance concept* and roadmap for MSaaS within NATO.

MSG-136 proposed an incremental development and implementation strategy for the Allied Framework for M&S as a Service. The incremental approach facilitates a smooth transition in the adoption of an Allied Framework for M&S as a Service and describes a route that will incrementally build an Allied Framework for M&S as a Service. The current NMSG efforts are executed by the technical working group MSG-164 and are described at the end of this chapter.

4.1.1 Terminology

M&S products are highly valuable to NATO and military organizations and it is essential that M&S products, data and processes are conveniently accessible to a large number of users as often as possible. Therefore, a new *M&S ecosystem* is required where M&S products can be accessed simultaneously and spontaneously by a large number of users for their individual purposes. This "as a Service" paradigm has to support stand-alone use, as well as the integration of multiple simulated and real systems into a unified simulation environment whenever the need arises.

This chapter uses the term *service* always in the sense of *M&S service*, unless stated otherwise, using the following definition:

An **M&S service** is a specific M&S-related capability delivered by a provider to one or more consumers according to well defined contracts including service level agreements (SLA) and interfaces (cp. [19]).

The provided capability is implemented in a (distributed) system and/or organization.

M&S as a Service (MSaaS) is an enterprise-level approach for discovery, composition, execution and management of M&S services.

4.1.2 Allied Framework for MSaaS

The *Allied Framework for MSaaS* is the common approach of NATO and Nations towards implementing MSaaS and is defined by the following documents:

- *Operational Concept Document*: The Operational Concept Document (OCD) describes the intended use, key capabilities and desired effects of the Allied Framework for MSaaS from a user's perspective.
- *Technical Reference Architecture*: The Technical Reference Architecture describes the architectural building blocks and patterns for realizing MSaaS capabilities.
- *Governance Policies*: The Governance Policies identify MSaaS stakeholders, relationships and provide guidance for implementing and maintaining the Allied Framework for MSaaS.

The above-mentioned documents define the blueprint for individual organizations to implement MSaaS. However, specific implementations—i.e. solutions—may be different for each organization.

4.1.3   Chapter Overview

This chapter is structured as follows. Section 4.2 discusses the Operational Concept for the Allied Framework for MSaaS. The purpose of the operational concept is to inform relevant stakeholders on how the framework will function in practice. The capabilities and key characteristics of the proposed framework are discussed, as well as the interactions of the users. Section 4.3 presents the technical concept of the Allied Framework for MSaaS. The technical concept is described in three volumes: Reference Architecture, Services Discovery, and Engineering Process. Section 4.4 discusses the governance concept. This covers roles, policies, processes and standards for the management of the Allied Framework for MSaaS within NATO. Section 4.5 provides an overview of the experimentation performed. This includes experimentation to explore and test enabling technology for architecture building blocks from the reference architecture, and experimentation to test solutions for certain types of simulation services. Section 4.6 provides an overview of the evaluation activities performed. Section 4.7 discusses the next steps and the incremental development and implementation strategy for the Allied Framework for MSaaS. And finally, Sect. 4.8 provides a summary and conclusions.

4.2   Operational Concept

4.2.1   MSaaS from the User Perspective

MSaaS enables users to discover new opportunities for training and working together and enables users to enhance their operational effectiveness, saving costs and efforts in the process. By pooling individual user's requirements and bundling individual requests in larger procurement efforts, the position of buying authorities against industrial providers is strengthened.

MSaaS aims to provide the user with discoverable M&S services that are readily available on demand and deliver a choice of applications in a flexible and adaptive manner. It offers advantages over the existing stove-piped M&S paradigm in which the users are highly dependent on a limited amount of industry partners and subject matter experts.

The MSaaS concept is illustrated in Fig. 4.1. MSaaS is an enterprise-level approach for discovery, composition, execution and management of M&S services. MSaaS provides the linking element between M&S services that are provided by a community of stakeholders to be shared and the users that are actually utilizing these capabilities for their individual and organizational needs.

The Allied Framework for MSaaS defines user-facing capabilities (front-end) and underlying technical infrastructure (back-end). The front-end is called the MSaaS Portal. The front-end provides access to a large variety of M&S capabilities from which the users are able to select the services that best suit their requirements,

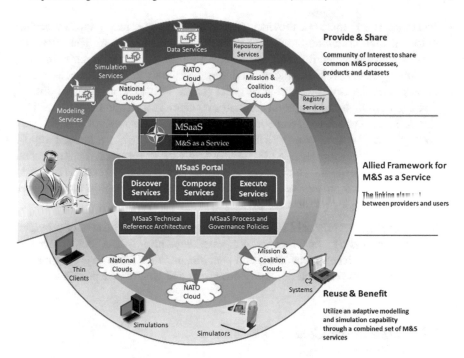

Fig. 4.1 MSaaS Concept

and track the experiences and lessons learned of other users. The users are able to discover, compose and execute M&S services through the front-end, which is the central access point that guides them through the process:

- **Discover**: The Allied Framework for MSaaS provides a mechanism for users to search and discover M&S services and assets (e.g. Data, Services, Models, Federations and Scenarios). A registry is used to catalogue available content from NATO, National, Industry and Academic organizations. This registry provides useful information on available services and assets in a manner that the user is able to assess their suitability to meet a particular requirement (i.e. user rating, requirements, simulation specific information and verification and validation information). The registry also points to a repository (or owner) where that simulation service or asset is stored and can be obtained, including business model information (i.e. license fees, pay per use costs).
- **Compose**: The Framework provides the ability to compose discovered services to perform a given simulation use case. Initially, it is envisaged that simulation services will be composed through existing simulation architectures and protocols (e.g. using DIS, HLA, DDS) and can be readily executed on demand (i.e. with no set-up time). In the longer term, distributed simulation technology will evolve, enabling further automation of discovery, composition and execution than is possible today.

- **Execute**: The Framework provides the ability to deploy the composed services automatically on a cloud-based or local computing infrastructure. The automated deployment and execution allows to exploit the benefits of cloud computing (e.g. scalability, resilience). Once deployed and executed the M&S services can be accessed on demand by a range of users (Live, Virtual, Constructive) directly through a simulator (e.g. a flight simulator consuming a weapon effects service), through a C2 system (e.g. embedded route planning functionality that utilizes a route planning service) or may be provided by a thin client or by a dedicated application (e.g. a decision support system utilizing various services like terrain data service, intelligence information service, etc.). The execution services support a range of business models and are able to provide data relevant to those models (i.e. capture usage data for a pay per use business model).

The Allied Framework for MSaaS is the linking element between service providers and users by providing a coherent and integrated capability with a Technical Reference Architecture, recommendations and specifications for discovery, composition and execution of services and necessary processes and governance policies.

4.2.2 Operational Concept Document

The purpose of the Operational Concept Document (OCD) for the Allied Framework for MSaaS is to inform relevant stakeholders on how the framework will function in practice. The capabilities and key characteristics of the proposed framework are included in the OCD, as well as how stakeholders will interact with the system.

Specifically, the main goals of the OCD are to inform the operational stakeholders on how to evolve from their current operational stove-piped systems to the Allied Framework for MSaaS. It also serves as a platform for stakeholders to collaboratively adapt their understanding of the system's operation as new developments, requirements or challenges arise. Therefore, the OCD is written in the common language of all interested parties.

4.2.3 Vision Statement and Goals

The *MSaaS Vision Statement* is defined as:

"M&S products, data and processes are conveniently accessible and available on demand to all users in order to enhance operational effectiveness." [3]

To achieve the MSaaS Vision Statement the following MSaaS goals are defined:

1. To provide a framework that enables credible and effective M&S services by providing a common, consistent, seamless and fit for purpose M&S capability that is reusable and scalable in a distributed environment.

2. To make M&S services available on demand to a large number of users through scheduling and computing management. Users can dynamically provision computing resources, such as server time and network storage, as needed, without requiring human interaction. Quick deployment of the customer solution is possible since the desired services are already installed, configured and online.
3. To make M&S services available in an efficient and cost-effective way, convenient short set-up time and low maintenance costs for the community of users will be available and to increase efficiency by automating efforts.
4. To provide the required level of agility to enable convenient and rapid integration of capabilities, MSaaS offers the ability to evolve systems by rapid provisioning of resources, configuration management, deployment and migration of legacy systems. It is also tied to business dynamics of M&S that allow for the discovery and use of new services beyond the users' current configuration.

4.3 Technical Concept

The technical concept comprises several volumes

– Volume 1: MSaaS Technical Reference Architecture: discusses layers, architecture building blocks and architectural patterns [3].
– Volume 2: MSaaS Discovery Service and Metadata: discusses services metadata and metadata for services discovery [4].
– Volume 3: MSaaS Engineering Process: discusses a services-oriented overlay for the DSEEP [5].

This section will focus primarily on the MSaaS Reference Architecture (RA) and briefly explain the other volumes.

4.3.1 MSaaS Reference Architecture

Principles
The MSaaS Reference Architecture (RA) is defined with a number of principles in mind. These principles are similar to the Open Group SOA Reference Architecture (SOA RA) [2] key principles and are the starting point for the architecture work by MSG-136. The principles are
The MSaaS RA:

1. Should be a generic solution that is vendor-neutral.
2. Should be modular, consisting of building blocks which may be separated and recombined.
3. Should be extendable, allowing the addition of more specific capabilities, building blocks, and other attributes.

4. Must be compliant with NATO policies and standards (such as AMSP-01 [6] and STANAG 4603 [7]).
5. Must facilitate integration with existing M&S systems.
6. Should be capable of being instantiated to produce

   (a) Intermediary architectures
   (b) Solution architectures

7. Should address multiple stakeholder perspectives.

Architecture Concepts

An architecture can generally be described at different levels of abstraction and the term *reference architecture* is typically used for a more abstract form of architecture. The purpose of the MSaaS RA is to provide a template for the development of an MSaaS *intermediate architecture* or of one or more specific MSaaS *solution architectures*. The MSaaS RA provides guidelines, options and constraints for making design decisions with regards to MSaaS solution architecture and solution implementation.

The MSaaS RA uses several concepts for describing the architecture. These concepts and their relationships are illustrated in Fig. 4.2.

The MSaaS RA defines a number of *capabilities* in the form of *architecture building blocks* and organizes these capabilities in so-called *layers*. An architecture building block captures, amongst others, requirements, applicable standards, relationships with other building blocks, related architectural patterns and references to (examples of) enabling technology. The particular connection between architecture building blocks that recur consistently in order to solve certain classes of problems is called a *pattern*. A pattern describes how architecture building blocks can be put together for creating proven solution architectures. The *enabling technology* provides means for the technical realization of an architecture building block.

The MSaaS RA layers are modelled after the SOA RA layers [2], while the content of each layer in terms of architecture building blocks is supplied by the NATO C3 Taxonomy [8].

Fig. 4.2 Reference architecture concepts

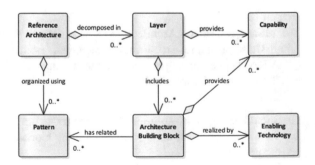

Fig. 4.3 Reference
architecture layers

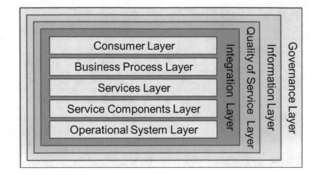

Layers and Architecture Building Blocks

The MSaaS RA is decomposed in layers, similar to the SOA RA layering structure, and each layer includes a set of architecture building blocks that provide some capability. The 9 layers are illustrated in Fig. 4.3. Some of the layers are cross-cutting layers. For example, the architecture building blocks in the Quality of Service Layer affect the building blocks in the Operational System Layer up to the Integration Layer.

Note that the SOA RA layers are presented from technical infrastructure layers to consumer-facing layers in that order. Also, some naming may cause confusion between C3 Taxonomy users and the SOA RA users. For example, the Operational Systems Layer does not refer to the defence operations that the C3 Taxonomy's Operational Capabilities layer does, but rather to the operational run-time capabilities in a SOA.

The architecture building blocks per layer are shown in Table 4.1.

The architecture building blocks are aligned with the NATO C3 Taxonomy and necessary changes will be recommended.

As an example, the *Business Process Layer* provides the capabilities to compose and execute a simulation, and contains the following architecture building blocks:

- *M&S Composition Services*: compose a simulation environment from individual services that together meet the objectives of the simulation environment.
- *M&S Simulation Control Services*: provide input to, control, and collect output from a simulation execution.
- *M&S Scenario Services*: manage the simulation of scenarios.

Each of these architecture building blocks has associated requirements and other attributes. As an example, some requirements for the M&S Composition Services are listed in Table 4.2.

The architecture building blocks of the MSaaS RA are organized in a taxonomy, in line with the NATO C3 Taxonomy (see Fig. 4.4). Most of the architecture building blocks in Table 4.1, fall under the M&S Enabling Services, providing capabilities to create a simulation environment in which M&S Specific Services are brought together to fulfil the purpose of that simulation environment. M&S Specific Services are mostly Simulation Services and Composed Simulation Services, such as Synthetic Environment Services, Route Planning Services, or Report Generation Services.

Table 4.1 Layers and architecture building blocks

Layer	Architecture building blocks
Operational Systems Layer	• Infrastructure Services • Communication Services
Service Components Layer	• SOA Platform Services
Services Layer	• M&S Specific Services
Business Process Layer	• M&S Composition Services • M&S Simulation Control Services • M&S Scenario Services
Consumer Layer	• M&S User Applications • NATO User Applications
Integration Layer	• M&S Message-Oriented Middleware Services • M&S Mediation Services
Quality of Service Layer	• SOA Platform SMC Services • M&S Security Services • M&S Certification Services
Information Layer	• M&S Information Registry Services
Governance Layer	• M&S Repository Services • Metadata Repository Services

Table 4.2 M&S Composition Services requirements

Function	Requirements
Manage Lifecycle	1. The M&S Composition Services shall provide the means to define a parameterized simulation composition 2. The M&S Composition Services shall provide the means to update, delete and retrieve a defined simulation composition
Execute Composition	3. The M&S Composition Services shall provide the means to start the execution of a simulation composition, and to provide composition parameter values 4. The M&S Composition Services shall provide the means to orchestrate, restart and stop the execution of a simulation composition
Programmatic Interfaces	5. The M&S Composition Services shall provide APIs to the Manage Lifecycle and Execute Composition functionality

Architectural Patterns

The architectural patterns show how architecture building blocks in the MSaaS RA are related, can be combined, how they interact, and what information is generally exchanged. The architectural patterns serve as a reference for solution architectures and design patterns for solution architectures. An initial set of architectural patterns is documented, but the idea is that the architecture building blocks, as well as the architectural patterns, are governed as a "living document" and will evolve further as knowledge is gained and as technology evolves.

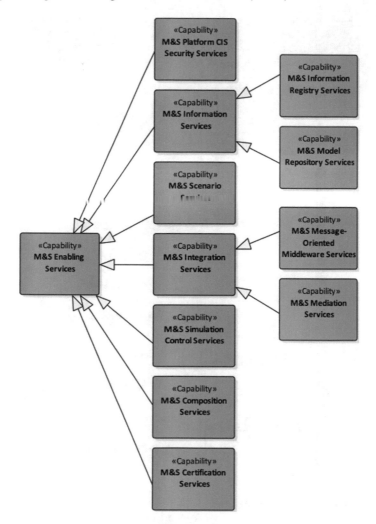

Fig. 4.4 Taxonomy of architecture building blocks

Figure 4.5 illustrates one example of an architectural pattern, in relation to the M&S Composition Services mentioned earlier.

In this example, a user composes a simulation environment using an M&S Composer Application. This application, in turn, employs the capabilities of M&S Composition Services and the M&S Model Repository Services. This pattern provides support for the definition, update, retrieval and deletion of compositions. The M&S Composer Application is user-facing, while the other architecture building blocks operate "behind the scene". The interactions in the figure also imply requirements on each architecture building block.

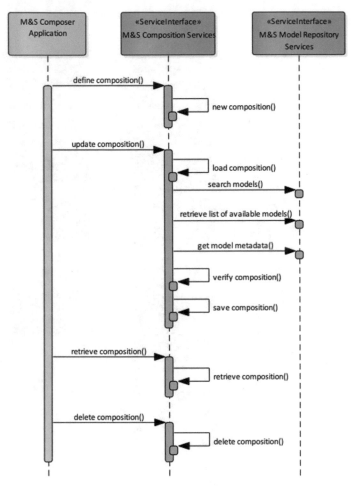

Fig. 4.5 Example of an architectural pattern

4.3.2 MSaaS Discovery Service and Metadata

Technical volume 2 [4], discusses information and standards related to the description of services and exchange of metadata. More specifically, it

- provides an overview of standards related to services discovery and services interface description, and
- presents national initiatives related to the exchange of services metadata, and to information models that support the (automated) composition, deployment and execution of simulation environments.

This volume relates to several architecture building blocks in the MSaaS RA, such as the M&S Composition Services for automated composition, deployment and execution; and the M&S Model Repository Services for metadata standards.

4.3.3 MSaaS Engineering Process

Technical volume 3 [5] discusses a service-oriented overlay for the Distributed Simulation Engineering and Execution Process (DSEEP) [9], by adding an overlay for a service-oriented implementation strategy (besides HLA, DIS and TENA). If the MSaaS capabilities are to be realized, simulation engineers must have a well developed and documented process for bringing them into being. The OCD identifies this requirement in Fig. 4.6.

The MSaaS Engineering Process (MSaaS-EP) defines that process [5]. However, it is important for engineers to understand the larger set of engineering environments and processes in which it is executed.

The MSaaS-EP is executed within an existing MSaaS Implementation, the specific realization of M&S as a Service by a certain organization as defined in the Operational Concept Document. An MSaaS Implementation includes both technical and organizational aspects. The Allied Framework for Modelling and Simulation as a Service (MSaaS) Governance Policies establish policies that guide the development

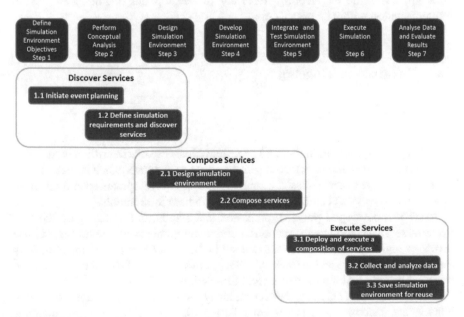

Fig. 4.6 This figure, taken from the MSaaS Operational Concept Document, shows the alignment of engineering activities with the DSEEP. [18]

of an MSaaS implementation. [10] This implementation will include implementations of M&S Enabling Services which provide capabilities to create a simulation in which M&S Services and M&S User Applications are brought together to fulfil the purpose of that simulation.

The MSaaS-EP is executed to build Composed Simulation Services compliant with the MSaaS Reference Architecture (RA). The MSaaS RA defines a set of architectural building blocks and architectural patterns to support the MSaaS-EP.

The services used during the MSaaS-EP to construct a composed simulation service are catalogued using the M&S Registry Services. Volume 2 to of the MSaaS technical documentation, Discovery Service and Metadata, defines the data standards that allow service discovery in an MSaaS implementation.

The MSaaS-EP mirrors the IEEE Recommended Practice for Distributed Simulation Engineering and Execution Process (DSEEP). The documentation of the MSaaS-EP assumes engineering knowledge of the DSEEP and it will only address the MSaaS-specific engineering considerations during DSEEP execution.

If the MSaaS-EP is executed in a multi-architecture environment, it will also mirror the DSEEP Multi-Architecture Overlay (DMAO). The documentation of the MSaaS-EP also assumes knowledge of the DMAO. As multi-architecture compositions are discussed, it will only address MSaaS-specific engineering considerations in the context of the DMAO.

In short, the MSaaS engineering process covers the engineering necessary to develop composed simulation services within an MSaaS implementation, but it does not cover the engineering necessary to develop and maintain an MSaaS implementation within an organization.

4.4 Governance Concept

4.4.1 Governance and Roles

A challenging aspect of establishing a persistent capability like the Allied Framework for MSaaS is to develop an effective governance model. Governance ensures that all of the independent service-based efforts (i.e. design, development, deployment, or operation of a service) combined will meet customer requirements.

MSG-136 developed policies, processes and standards for managing the lifecycle of services, service acquisitions, service components and registries, service providers and consumers. These are defined in the *Allied Framework for Modelling and Simulation as a Service (MSaaS) Governance Policies* [10], and are intended to be published as Allied Modelling and Simulation Publication AMSP-02.

The NMSG is the delegated NATO authority for M&S standards and procedures. Nations are encouraged to use the standards nationally or in other multinational collaborations. After completion of the MSG-136 task group, the NMSG M&S Military Operational Requirements Subgroup (MORS) will become the custodian of the governance policies. MORS is the custodian of best practices with regards to the use of M&S in the training domain and in other domains. The governance policies will be submitted to MORS for future maintenance, updates and dissemination with respect to the operational needs of NATO agencies and national stakeholders.

The NMSG M&S Standards Subgroup (MS3) will become custodian of the MSaaS Technical Reference Architecture [3], and is responsible for the maintenance of the MSaaS technical aspects and standards documents.

4.4.2 General Policies

The general policies for instituting governance mechanisms of MSaaS-based solutions are:

- An MSaaS implementation shall conform to the governance policies as identified and established by the governance document.
- An MSaaS solution architecture shall comply with the MSaaS Technical Reference Architecture (see Sect. 4.3, Technical Concept).
- Any M&S service shall conform to the practices and recommendations for Integration, Verification and Compliance Testing as defined by NATO MSG-134 [11].

The ability to effectively manage all stages of the service lifecycle is fundamental to the success of governing M&S services. The Service Lifecycle Management Process as defined in [12], contains a set of controlled and well defined activities performed at each stage for all versions of a given service. Table 4.3 lists the sequential service provider lifecycle stages.

Table 4.3 Service provider lifecycle stages

Lifecycle stage	Description
Proposed	The proposed service's needs are identified and assessed as to whether needs can be met through the use of services
Definition	The service's requirements are gathered and the design is produced based on these requirements
Development	The service specifications are developed and the service is built
Verification	The service is inspected and/or tested to confirm it is of sufficient quality, complies with the prescribed set of standards and regulations, and is approved for use
Production	The service is available for use by its intended consumers
Deprecated	The service can no longer be used by new consumers
Retired	The service is removed from the Allied Framework and is no longer used

All service providers shall define levels for each service (e.g. regarding availability, etc.). Service Providers and users shall agree on a Service Level Agreement (SLA) prior to usage. Obviously, service providers are required to indicate the
forecasted retirement date of a specific version of a service.

4.4.3 Security Policies

The approach to ensuring security is intrinsically related to the cloud computing
service model (SaaS, PaaS, or IaaS) and to the deployment model (Public, Private,
Hybrid, or Community) that best fits the Consumer's missions and security requirements. The Consumer has to evaluate the particular security requirements in the
specific architectural context, and map them to proper security controls and practices
in technical, operational and management classes. Even though the Cloud Security
Reference Architecture [13], inherits a rich body of knowledge of general network
security and information security, both in theory and in practice, it also addresses the
cloud-specific security requirements triggered by characteristics unique to the cloud,
such as decreased visibility and control by consumers. Cloud security frameworks
including information management within an infrastructure shall support the cloud
implementers, providers and consumers [14]. However, MSG-136 recognizes that a
more tailored approach may be needed to exploit MSaaS specific capabilities and
proposes to develop additional guidelines as part of follow-on work.

4.4.4 Compliance Policies

Compliance testing of individual components of a NATO or multinational simulation environment is the ultimate responsibility of the participating organizations. Currently, NMSG and its support office (MSCO) do not provide compliancy
testing services or facilities. Some existing HLA certification tools and services
cover only basic testing (i.e. HLA Rules, Interface Specification and Object Model
Template (OMT) compliance) and do not provide in-depth functional testing that is
needed to support federation integration and validation. The available tools are also
outdated. The current NMSG activity MSG-134 is addressing the next generation of
compliancy testing and certification needs for HLA [11].

4.5 Experimentation

MSG-136 performed several experiments to test enabling technology for MSaaS.
Two strands of experimentation were performed: (1) experimentation to explore
and test enabling technology for architecture building blocks from the reference

architecture and (2) experimentation to test solutions for certain types of Simulation Services. Test cases were defined, tests performed and test results recorded in an experimentation report [15]. A brief overview of the experimentation and test cases follows below.

4.5.1   Explore and Test Enabling Technology

Most test cases in this strand of experimentation evolve around container technology as the enabling technology for a number of architecture building blocks. This technology enables M&S Enabling Services and M&S Specific Services to run on a local host, as well as in a cloud environment.

The experiment environment that was used for the test cases is illustrated in the following figure. The experiment environment is a collection of private clouds and a common cloud. The common cloud is Amazon Web Service (AWS), sponsored by NATO CSO (Fig. 4.7).

Common components are

- A private Docker Registry and a web-based front-end for the exchange of Docker container images (provided by NLD);
- A private GitHub repository for the description of container images in the Docker Registry, and for the exchange of software, configuration files and other developmental data (provided by the USA).

The Docker Registry contains several container images for containerized HLA federates, from which various compositions can be created for the different test cases. Many of these images have been created following the design patterns in [16].

Test cases include:

- Container networking: explore different container networking models for connecting containerized HLA federate applications.
- Containerization of HLA federates: evaluate approaches in containerizing HLA federate applications (see also [16]).

Fig. 4.7 Illustration of experiment environment

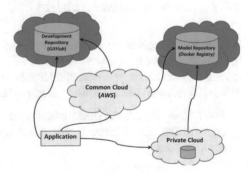

- Metadata Repositories and Discovery: Demonstrate the interoperation of repositories across nations.
- Simulation Composition: explore automated composition and execution of services.
- Container Orchestration Environments: evaluate two popular container orchestration environments for M&S (see also [17]).

4.5.2 Test Solutions for Simulation Services

Tests cases in this strand of experimentation concern the interoperation of applications with certain types of Simulation Services. Test cases include

- Computer Generated Forces (CGF)—Synthetic Environment Service: connect a CGF simulator to a Synthetic Environment Service to request environment data in various formats.
- C2 Application—Route Planning Service: connect a C2 Application to a Route Planning Service to request route planning information.

4.6 Evaluation

The evaluation activities focus on whether MSaaS will reduce costs and integration time for creating a new instance of a simulation environment, compared to what it costs today. What is the main advantage of having an MSaaS-based solution? The premise of the evaluation activities is to answer this objectively based on the measurements performed and data collected. The evaluation activities of MSG-136 are currently ongoing and will be included in the MSG-136 Final Report.

4.7 Implementation Strategy and Next Steps

4.7.1 Implementation Strategy

Service-based approaches rely on a high degree of standardization and automation in order to achieve their goals. Therefore, the development and implementation of a recommended set of supporting standards is a key output of the reference architecture. MSG-136 research has identified the importance of the following capabilities:

- *M&S Composition Services*: create and execute a simulation composition. A composition can be created from individual simulation services or from smaller compositions.

- *M&S Repository Services*: store, retrieve and manage simulation service components and associated metadata that implement and provide simulation services, in particular, metadata for automated composition.
- *M&S Security Services*: implement and enforce security policies for M&S services.

MSG-136 proposes an incremental development and implementation strategy for the Allied Framework for M&S as a Service. The incremental approach facilitates a smooth transition in the adoption of an Allied Framework for M&S as a Service and describes a route that will incrementally build an Allied Framework for M&S as a Service.

The proposed strategy also provides a method to control the rate of expansion of the new framework permitting the iterative development and training of processes and procedures. Finally, it permits those nations that have been early adopters of an Allied Framework for M&S as a Service and have national capabilities to accrue additional benefits from their investments and highlight the benefits as well as providing lessons learned and advice to those nations considering similar investments.

As illustrated in Fig. 4.8, the implementation strategy is broken down into three phases:

1. Phase 1 "Initial Concept Development". The Initial Concept Development (2015 until the end of 2017) was executed by NMSG-136 and focused on concept development and initial experimentation. For this period an MSaaS Portal and individual M&S services were provided by individual members of MSG-136 for trial use.
2. Phase 2 "Specification and Validation". From 2018–2021, MSG-164 will mature MSaaS in an operationally relevant environment and conduct necessary research and development efforts to evolve and extend the initial concepts as developed by MSG-136. This phase includes the development of suitable STANAGs or STANRECs, and moving from prototype implementation to operationally usable and mature systems.
3. Phase 3 "Implementation". By 2025, Full Operational Capability (FOC) is achieved which includes the adaptation of many existing simulation related

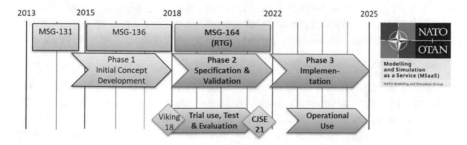

Fig. 4.8 MSaaS implementation strategy

services to the MSaaS Reference Architecture. This is achieved primarily by adding services to the Allied Framework for M&S as a Service.

4.7.2 Next Steps

The next steps in defining and evolving the Allied Framework for MSaaS are executed by MSG-164 (see the previous section). MSG-164 kicked off in February 2018 and will finish in 2021. Building upon the Allied Framework for M&S as a Service developed by MSG-136 this activity focusses on three main objectives

1. To advance and to promote the operational readiness of M&S as a Service.
2. To align national efforts and to share national experiences in establishing MSaaS capabilities.
3. To investigate critical research and development topics to further enhance MSaaS benefits.

MSG-164 will specify and test an MSaaS infrastructure that is suitable for use in an operationally relevant environment and will support continued MSaaS experimentation and evaluation efforts. This activity will also deliver a Technical Report and recommendations with regards to the organizational perspective of introducing MSaaS in NATO and in the Nations.

To address the objectives, MSG-164 will cover the following topics:

1. Demonstrate MSaaS application in an operationally relevant environment through operational experimentation as part of exercises and integration into simulation applications (like simulation-based capability development). Annual participation in CWIX to develop MSaaS to maturity through a phased approach.
2. Maintain and enlarge the MSaaS Community of Interest.
3. Establish interim governance structure and collect experiences with respect to. MSaaS governance.
4. Collect and share experiences in establishing MSaaS capabilities and providing M&S services.
5. Conduct research on M&S-specific service discovery and service composition.
6. Conduct research and development activities on M&S-specific federated cloud environments, federated identity management and cyber secure communications.
7. Conduct research on enabling services like scenario specification services, etc.

Additionally, MSG-164 will

1. Act as the governance body for the Allied Framework for M&S as a Service, maintaining and updating (if needed) the therein included documents, i.e. AMSP-02 (MSaaS Governance Policies), the MSaaS Operational Concept Description and the MSaaS Technical Reference Architecture) with associated technical documents.
2. Collaborate with international standards bodies (like SISO, IEEE, etc.).
3. Inform and engage stakeholders in NATO, Academia, and Industry about MSaaS.

4.8 Summary and Conclusions

The concept of *M&S as a Service* (MSaaS)—including discovery, composition, execution and management of M&S services—has been investigated and matured by the NATO Modelling and Simulation Group (NMSG) over the last years. The NMSG has approached MSaaS from different perspectives

- *Operational concept* of MSaaS: how it works from the user point of view;
- *Technical concept* of MSaaS: technical reference architecture, services discovery metadata and engineering process;
- *Governance concept* and roadmap for MSaaS within NATO.

Technical implementations of MSaaS have been developed and evaluated in several experiments and demonstrations. MSG-136 has also proposed an MSaaS governance approach to enable long-term maintainability of an MSaaS ecosystem that may be used by NATO and the Nations. The conclusion is that MSaaS is a promising innovation towards more accessible and more cost-effective M&S capabilities.

The participating nations and NATO organizations are currently implementing MSaaS using cloud technology, based on the MSG-136 and MSG-164 research and experimentation efforts and to inform the user community. MSG-64 will further investigate a number of areas including discovery and composability of M&S services; and will also address security aspects of cloud-based M&S solutions in more detail. Ultimately, MSG-164 seeks to mature MSaaS and to demonstrate the operational readiness of MSaaS through participation in actual military exercises.

The NMSG will continue to participate in the SISO Cloud-based M&S Study Group and share its approach and experiences. The goal is that our work will contribute to a set of open standards and recommendations for MSaaS.

Acknowledgements The author would like to acknowledge the contributions of the MSG-136 and MSG-164 team members—specifically Tom van den Berg, Chris McGroarty and Jon Lloyd—for the ideas and results that are included in this chapter.

References

1. NATO Modelling and Simulation Master Plan NMSMP v2.0 (AC/323/NMSG(2012)-015). https://www.sto.nato.int/NATODocs/NATO
2. https://www.sto.nato.int/Pages/activitieslisting.aspx?FilterField1=ACTIVITY_NUMBER& FilterValue1=MSG-136
3. NATO STO (2018) Operational Concept Document (OCD) for the allied framework for M&S as a service. AC/323(MSG-136)TP/830. STO-TR-MSG-136-Part-III, 16 January 2018
4. NATO STO (2018) Modelling and simulation as a service, Volume 1: MSaaS technical reference architecture. AC/323(MSG-136)TP/831. STO-TR-MSG-136-Part-IV, 16 January 2018
5. NATO STO (2016) Modelling and simulation as a service, Volume 2: MSaaS discovery service and metadata. AC/323(MSG-136)TP/832. STO-TR-MSG-136-Part-V, 16 January 2018

 6. SOA Reference Architecture, C119, Open Group Standard (2011)
 7. NATO AMSP-01 (NATO Modeling and Simulation Standards Profile), Edition C version 1, NATO Standardization Office, March 2015
 8. STANAG 4603 Edition 2, Modeling and Simulation Architecture Standards for Technical Interoperability: HLA, NATO Standardization Office, 17 February 2015
 9. C3 Classification Taxonomy, Baseline 1.0, NATO Allied Command Transformation (ACT) C4ISR Technology and Human Factors (THF) Branch, 15 June 2012
10. van den Berg TW, Cramp A (2017) Container orchestration environments for M&S; 2017-SIW-006; SISO SIW Fall 2017
11. IEEE 1730-2010: Recommended Practice for Distributed Simulation Engineering and Execution Process (DSEEP)
12. NATO Distributed Simulation Architecture & Design, Compliance Testing and Certification. STO Technical Report STO-TR-MSG-134. To be published
13. NATO STO (2018) MSaaS concept and reference architecture evaluation report. AC/323(MSG-136)TP/829. STO-TR-MSG-136-Part-II, 16 January 2018
14. Federal Aviation Administration (FAA): "System Wide Information Management (SWIM) Governance Policies", Version 2.0, 12 March 2014
15. NATO STO (2018) Modelling and simulation as a service, Volume 3: MSaaS engineering process. AC/323(MSG-136)TP/833. STO-TR-MSG-136-Part-VI, 16 January 2018
16. NATO Consultation, Command and Control Board (C3B): "NATO Cloud Computing Policy", AC/322-D(2016)0001, 7 January 2016
17. van den Berg TW, Cramp A, Siegel B (2016) Guidelines and best practices for using Docker in support of HLA federations. 2016-SIW-031; SISO SIW Fall 2016
18. NATO STO (2018) Allied Framework for Modelling and Simulation as a Service (MSaaS)—Governance Policies, 16 January 2018
19. National Institute of Standards and Technology (2013) NIST Cloud Computing Security Reference Architecture, NIST Special Publication 500-299, Draft, 15 May 2013

Robert Siegfried is Senior Consultant for IT/M&S projects and Managing Director of Aditerna GmbH, a company providing specialized services and consulting in this area. He earned his doctorate in modelling and simulation at the Universität der Bundeswehr München. For the German Armed Forces he has worked on documentation guidelines, model management systems, distributed simulation test beds and process models. He is an active member of NATO MSG-128 ("Mission Training through Distributed Simulation") and is co-chair of NATO MSG-136 ("Modelling and Simulation as a Service (MSaaS)—Rapid deployment of interoperable and credible simulation environments") and MSG-164 ("MSaaS Phase 2"). He is actively involved in multiple SISO working groups and is a member of the SISO Executive Committee.

Chapter 5
Cyber-Physical System Engineering Oriented Intelligent High Performance Simulation Cloud

Bo Hu Li, Xudong Chai, Ting Yu Lin, Chen Yang, Baocun Hou, Yang Liu, Xiao Song, and Hui Gong

Abstract The chapter introduces a Cyber-Physical System Engineering Oriented Intelligent High Performance Simulation Cloud (CPSEO-IHPSC) developed by the author's team. CPSEO-IHPSC is a kind of intelligent high performance simulation cloud system with the characteristics of digitization, networking, cloud, and intelligence, supporting user to access services of intelligent high performance simulation resource, capability, and CPS product on demand, anytime and anywhere, based on

B. H. Li (✉) · X. Song
State Key Laboratory of Intelligent Manufacturing System Technology, Beijing Institute of Electronic System Engineering, Beijing 100854, China
e-mail: libohu@tsinghua.edu.cn

X. Song
e-mail: songxiao@buaa.edu.cn

B. H. Li · T. Y. Lin
School of Automatic Science and Electrical Engineering, Beihang University, Beijing 100191, China
e-mail: lintingyu2003@foxmail.com

Beijing Complex Product Advanced Manufacturing Engineering Research Center, Beijing Simulation Center, Beijing 100854, China

X. Chai · B. Hou · Y. Liu
Aerospace Cloud Science and Technology Co., Ltd, Beijing 100041, China
e-mail: xdchai@263.net

B. Hou
e-mail: houbaocun_tz@sina.com

Y. Liu
e-mail: 289503799@qq.com

C. Yang
School of Computer Science and Technology, Beijing Institute of Technology, Beijing 100081, China
e-mail: yangchen666@bit.edu.cn

H. Gong
CASIC Intelligence Industry Development Co., Ltd, Beijing, China 100033
e-mail: gonghui@aiidc.com.cn

© Springer Nature Switzerland AG 2020
J. L. Risco Martín et al. (eds.), *Simulation for Cyber-Physical Systems Engineering*,
Simulation Foundations, Methods and Applications,
https://doi.org/10.1007/978-3-030-51909-4_5

the ubiquitous network. It could promote the CPS engineering activities to be more efficient, optimal, economic, flexible, and safer. Firstly, the CPSEO-IHPSC's connotation, architecture, and bodies of knowledge are proposed. Then, the author team's preliminary research results are introduced including the CPSEO-IHPSC's preliminary prototype developed, some key technologies solved, and application examples implemented in intelligent manufacturing system engineering and smart city system engineering. Finally, the proposal for developing CPSEO-IHPSC in the "new Internet + cloud computing + big data + new AI +" era is discussed.

Keywords Intelligent high performance simulation cloud · Cyber-Physical system engineering · Modeling and simulation

5.1 Introduction

5.1.1 Connotation of CPS

The term Cyber-Physical Systems (CPS) was proposed by the National Science Foundation of the United States in 2006 [1]. Afterward, it was interpreted by scholars in different countries or institutions with a focus on different aspects [2–5].

Based on the existing research results with respect to the connotation of CPS, this paper interprets CPS as follows:

CPS is a class of modeling and data driven systems that integrate human, physical space and information/cyberspace by means of advanced information, communication, and intelligence and control technologies, and that enable elements involved therein such as human, machine, object, environment and information to perceive, learn, analyze, decide and execute in an intelligent and autonomous manner so as to achieve optimized operation for a particular goal in a given time and space.

Wherein human refers to the participant of CPS; information/cyberspace indicates the space composed by the network/Internet, data, information, and knowledge; the physical space is the interrelated space composed by machine, materials, environment, etc. CPS is designed to enable systems to perform functions such as computing, communication, precise control, and remote cooperation and autonomy. With a variety of autonomous control systems and information service systems built via the network/Internet, it enables the organic coordination among human, physical space, and information/cyberspace, and, in particular, enables the feedback and control to the physical world by means of communication, computing, and intelligence after sensing the physical world.

Based on an existed schematic diagram of the CPS composition [6], an expanded schematic diagram of the CPS composition is given by the author's team. CPS may fall, in terms of its size, into two classes: system level and system of system (group of systems) level.

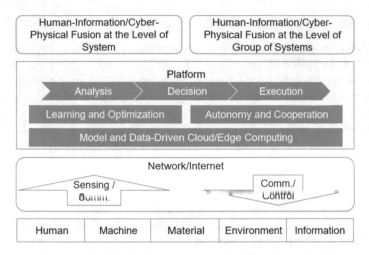

Fig. 5.1 Schematic diagram of the CPS composition

Wherein the CPS at the level of a group of systems is a class of complex system, such as intelligent manufacturing system, smart city system, IoT system, and complex military system (Fig. 5.1).

5.1.2 Connotation of CPS Engineering

CPS engineering refers to a class of systems engineering that conducts full-life cycle activities for CPS, such as scientific argument, research, analysis, design, production, management, testing, operation, training, evaluation, sale, service, destruction, and so on.

5.1.3 Challenges of CPS Engineering for Modern Modeling and Simulation

Modern modeling and simulation (M&S), which is a critical technology for CPS engineering, plays its role through the full-life cycle activities of CPS engineering. As the so-called "digital twin", it is a digitally created virtual model or replica of a (physical or logical) CPS entity, which realizes and supports the CPS and various optimizations in its processes by virtual-real interactive feedback, data fusion, and analysis, iterative decision optimization [7].

CPS engineering poses severe new challenges to modern M&S technology in three aspects: simulation modeling theory and method, simulation system theory and technology, and simulation application engineering theory and technology.

(1) New challenges with modeling theory and methods.

Challenges can be divided into two categories. The first category is concerned with object-oriented modeling (i.e., primary modeling). New systematic modeling methods are required for representation of such complex mechanism, composition, interaction, and behavior (that are continuous, discrete, qualitative/decision-making, and optimized) of human, machine, object, and environment in the CPS engineering. In particular, efforts should be made to develop the modeling technology for systematic perception and prediction based on nonmechanistic approaches such as big data, deep learning, etc. The second category is about modeling/algorithm (i.e., the secondary modeling) for the simulators. New simulation algorithms/methods are required for fully considering the architecture and software and hardware characteristics of the CPS and simulation systems, so as to build high-efficiency multi-level parallel simulation for various object-oriented models.

(2) New challenges with theory and technology of simulation systems.

Challenges are mainly posed to the following five supportive technologies. The first is intelligent simulation cloud. Since human, machine, object, and environment of the CPS are distributed and heterogeneous, a variety of resources and capacities should be virtualized, encapsulated, and managed as services using service-oriented architecture. The resulted services can be accessed on demand by users remotely and collaboratively through the Internet and other forms of networks, which means mathematical, human-in-the-loop, hardware-in-the-loop/embedded simulation can be conducted remotely in the intelligent simulation cloud. The second is intelligent virtual prototype engineering, supporting heterogeneous integration and parallel simulation optimization of multidisciplinary virtual prototypes of complex objects in CPS, as well as integrated optimization of human/organization, management and technology, information flow, knowledge flow, control flow and service flow in the whole system, and the full life systems engineering life cycle. The third is an intelligent problem-oriented simulation language. As for M&S of the CPS engineering problems, a description language which is used in the form very close to the original form of the CPS system studied and then compiled for the automatic call of the related algorithm, function, and model libraries to perform high performance simulation. The fourth is to build intelligent high performance modeling and simulation system for edge computing technology, because some simulation/computing needs to be conducted at the front-end equipment in some CPS engineering efforts to ensure the real-time response. In addition, it may also be necessary to support collaborative solution and analysis between the high performance computing center and mass front-end intelligent computing devices. The fifth is the research on intelligent cross-media visualization technology, which is aimed to provide intelligent, high performance, user-friendly visualization technology based on artificial intelligence for virtual scene computing and virtual-real integration in various CPS engineering systems.

(3) New challenges with simulation application engineering technology.

There are three types of technologies/methods involved. The first is the verification, validation, and accreditation (VV&A) method for intelligent simulation models, including the accreditation of the primary modeling, the accuracy of algorithm execution, and the simulation results. The second is the intelligent management, analysis, and evaluation of simulation results. A large number of simulation applications demanding fast simulation and prediction of the possible behavior patterns and performance of the overall system call for efficient parallel simulation, and efficient acquisition, management, and analysis of simulation results. The third is big data-based intelligent analysis and evaluation technology. For CPS engineering applications, factors such as the actual complexity of human, machine, object, and environment, as well as the restrictions of various operating equipment, should be taken into account. Consideration should also be given to research on application modes and technologies such as big data access and storage management, big data cloud, big data analysis and decision-making, big data visualization, and result evaluation.

In view of the connotation of CPS engineering and the challenges posed for modern modeling and simulation, the authors propose that new research contents on modeling and simulation technology for CPS engineering should include the following as shown in Fig. 5.2: (1) new modeling theories and methods for CPS engineering, (2) supportive technologies for CPS engineering-oriented simulation systems, and (3) new simulation system application engineering technology for CPS

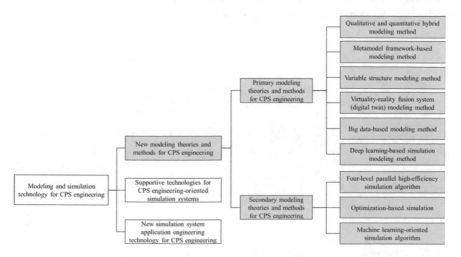

Fig. 5.2 New modeling theories and methods for CPS engineering

engineering. In the following paragraphs, we will elaborate on the research contents in detail.

New modeling theories and methods for CPS engineering as shown in Fig. 5.2 consist of: (1) The primary modeling theories and methods for CPS engineering, which are used to model the objects without considering the simulation execution factors and include qualitative and quantitative hybrid modeling method, metamodel framework-based modeling method, variable structure modeling method, virtual-reality fusion system (digital twin) modeling method, big data-based modeling method, and deep learning-based simulation modeling method; (2) Secondary modeling theories and methods for CPS engineering, including four-level parallel high-efficiency simulation algorithm, optimization-based simulation, and machine learning-oriented simulation algorithm.

The supportive technologies for CPS engineering-oriented simulation systems as shown in Fig. 5.3 include (1) Intelligent simulation cloud for CPS engineering, which can support the intelligent simulation in the cloud; (2) Multidisciplinary virtual proto-typing engineering for complex products based on virtuality-reality fusion; (3) High performance intelligent simulation language for complex systems; (4) Intelligent simulation system based on edge computing; (5) Intelligent cross-media visualiza-tion technology; (6) Universal interconnection interface and intelligent specialized parts technology. The universal interconnection interface technology is expected to promote the interaction between "everything" and interoperability, while the intelli-gent specialized parts technology should support the accelerated simulation for some big complex models.

New simulation system application engineering technology for CPS engineering as shown in Fig. 5.4 includes: (1) Intelligent simulation model checking, verification, validation and accreditation (VV&A) technology; (2) Intelligent system simulation experiment result management, analysis, and evaluation technology; (3) Big data intelligence analysis and evaluation techniques.

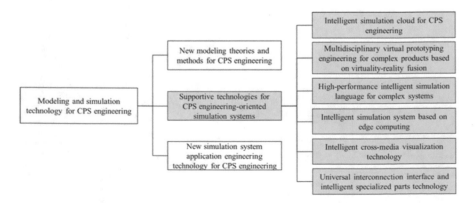

Fig. 5.3 Supporting technologies for CPS engineering-oriented simulation systems

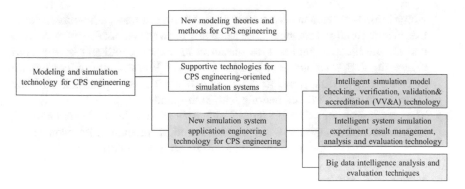

Fig. 5.4 New simulation system application engineering technology for CPS engineering

5.2 Cyber-Physical System Engineering Oriented Intelligent High Performance Simulation Cloud (CPSEO-IHPSC) for CPS Engineering

Cyber-Physical System Engineering Oriented Intelligent High Performance Simulation Cloud is an intelligent high performance simulation cloud system which integrates the new simulation modeling theory and method, the simulation system theory and technology, and the simulation application engineering theory and technology for CPS engineering. The connotation, architecture, and technical system of CPSEO-IHPSC are presented below.

5.2.1 Connotation of Cyber-Physical System Engineering Oriented Intelligent High Performance Simulation Cloud (CPSEO-IHPSC) for CPS Engineering:

Generally, CPSEO-IHPSC is a digital, networked, cloud-based and intelligent high performance simulation cloud system that is driven by the CPS engineering demands, based on the ubiquitous Internet, and built to allow users the on demand access from any place at any moment to high performance simulation resources, capabilities and CPS products and services, which enable CPS engineering to be completed in an efficient, high-quality, economical, green, flexible, and safe manner. Specifically,

(1) Technology for CPSEO-IHPSC: Based on ubiquitous network, five kinds of technology are deeply integrated, such as the new simulation technology, high performance simulation computing technology, information and communication technology, artificial intelligent technology, and CPS application technology, to form an intelligent high performance service cloud (interconnected service system) comprising simulation resources, capacities, and CPS products,

which is user-centered and uniformly operated in a digitized, networked, cloud-based, and intelligent manner. The users are allowed, via intelligent terminals and the intelligent cloud service platform, to access intelligent high performance simulation resources [8], capacities, and CPS product services (shown in Sect. 5.2.2) anytime, anywhere, and on demand, thus completing the full life cycle activities of CPS engineering with high-quality.

(2) The paradigm of CPSEO-IHPSC is a new mode of user-centered, human/machine/material/environment/information integrated, interconnected, service-based, collaborative, personalized, flexible, and intelligent;

(3) The ecosystem of CPSEO-IHPSC is characterized by ubiquitous interconnection, data-driven, shared services, cross-border integration, autonomous intelligence, and mass innovation;

(4) Technical characteristics of CPSEO-IHPSC are embodied in autonomous and intelligent perception, interconnection, collaboration, learning, analysis, cognition, decision-making, control and execution on the part of human, machine, material, environment, and information involved in the whole system and full life-cycle activities of CPS engineering.

(5) Implementation of CPSEO-IHPSC is to integrate and optimize the following six elements and six flows in the whole system and the full-life cycle of CPS engineering with the above-mentioned means. The six elements are human/organization, technology/equipment, operation and management, data, material, and capital, while the six flows are talent flow, technology flow, management flow, data flow, logistics, and capital flow.

(6) The aim of CPSEO-IHPSC is to enable the CPS engineering in an efficient, high-quality, economical, green, flexible, and safe manner so as to improve the competitiveness for CPS research and development (R&D).

5.2.2 CPSEO-IHPSC Architecture

The layers are explained as follows (Fig. 5.5):

(1) Intelligent high performance simulation resources/capacities/product layer includes

  (1) Soft resources for intelligent high performance simulation, such as various simulation models in CPS engineering, (big) data, software, information and knowledge, and so on;

  (2) Hard-resources/systems for intelligent high performance simulation, such as (large) intelligent high performance simulation hard-equipment/computing equipment/simulation test equipment, machines/equipment in the physical world, environment, and so on;

  (3) Intelligent high performance simulation capacities, such professional capacities in CPS engineering as scientific demonstration, design,

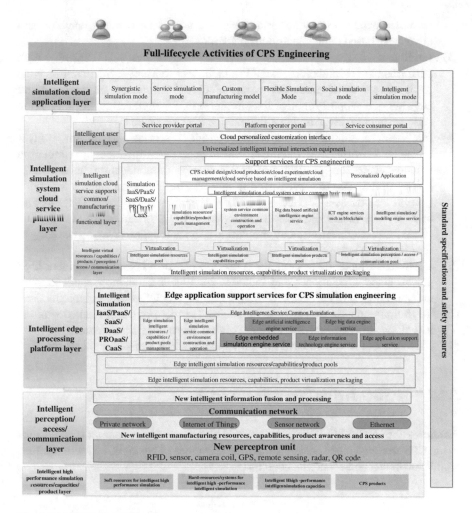

Fig. 5.5 CPSEO-IHPSC architecture

production, simulation, experiment, management, sale, (product) operation, (product) maintenance, integration, and other professional capaci-
ties(including human/knowledge, organization, capital, performance, reputation, resources, processes, products, and so on);

(4) CPS products, including digital, networked, cloud-based, and intelligent
CPS products, such as intelligent manufacturing system for accessing
intelligent high performance simulation cloud, smart city system, and so
on.

(2) Intelligent perception/access/communication layer includes perceptron, access
unit, communication network, information fusion/processing which provides
intelligent approaches for new intelligent high performance simulation

resources/capacities/product awareness and access on demand, anytime and anywhere, based on the ubiquitous network.

(3) Intelligent edge processing platform layer of intelligent high performance simulation cloud includes virtual intelligent edge resources/capacities/products/perception/access/communication pools, cloud edge service support for edge fog simulation/CPS engineering fog simulation service function layer, and intelligent edge user interface layer.

(4) Intelligent simulation system cloud service platform layer includes virtual intelligent high performance simulation resources/capacities/products/perception/access/communication pools, simulation cloud service support for simulation Infrastructure-as-a-Service (IaaS)/Platform-as-a-Service (PaaS)/Software-as-a-Service (SaaS)/Data-as-a-Service (DaaS)/Product-as-a-Service (PROaaS)/Capability-as-a-Service (CaaS)/CPS engineering cloud simulation service function layer, and intelligent user interface layer.

(5) Intelligent simulation system cloud service platform layer of intelligent high performance simulation cloud includes intelligent simulation application of collaboration, service, customization, flexibility, socialization, and intelligence.

(6) The human/organization layer includes cloud service providers, cloud operators, and cloud service users of intelligent high performance simulation.

(7) All levels of the system have standard specifications and safety measures.

5.2.3 CPSEO-IHPSC Technical System

As shown above, CPSEO-IHPSC has brought about transformations and innovations in the CPS-oriented modes, technical means and types of business, and will produce significant social and economic benefits (Fig. 5.6).

5.3 The Research of Cyber-Physical System Engineering Oriented Intelligent High Performance Simulation Cloud (Prototype)

This section presents the research of Cyber-Physical System Engineering Oriented Intelligent High Performance Simulation Cloud (Prototype) developed by our team. "Prototype" means that only parts of cyber-physical system engineering oriented modeling and simulation technology is adopted.

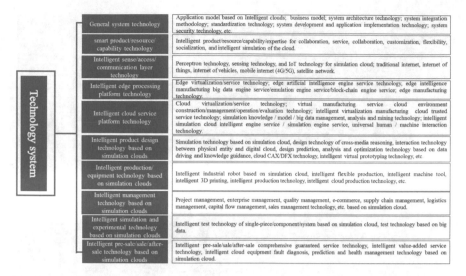

General system technology	Application model based on Intelligent clouds; business model; system architecture technology; system integration methodology; standardization technology; system development and application implementation technology; system security technology, etc.
smart product/resource/ capability technology	Intelligent product/resource/capability/expertise for collaboration, service, collaboration, customization, flexibility, socialization, and intelligent simulation of the cloud.
Intelligent sense/access/ communication layer technology	Perceptron technology, sensing technology, and IoT technology for simulation cloud; traditional internet, internet of things, internet of vehicles, mobile internet (4G/5G), satellite network.
Intelligent edge processing platform technology	Edge virtualization/service technology; edge artificial intelligence engine service technology, edge intelligence manufacturing big data engine service/emulation engine service/block-chain engine service; edge manufacturing technology.
Intelligent cloud service platform technology	Cloud virtualization/service technology; virtual manufacturing service cloud environment construction/management/operation/evaluation technology; intelligent virtualization manufacturing cloud trusted service technology; simulation knowledge / model / big data management, analysis and mining technology; intelligent simulation cloud intelligent engine service / simulation engine service, universal human / machine interaction technology.
Intelligent product design technology based on simulation clouds	Simulation technology based on simulation cloud, design technology of cross-media reasoning, interaction technology between physical entity and digital cloud, design prediction, analysis and optimization technology based on data driving and knowledge guidance, cloud CAX/DFX technology, intelligent virtual prototyping technology, etc.
Intelligent production/ equipment technology based on simulation clouds	Intelligent industrial robot based on simulation cloud, intelligent flexible production, intelligent machine tool, intelligent 3D printing, intelligent production technology, intelligent cloud production technology, etc.
Intelligent management technology based on simulation clouds	Project management, enterprise management, quality management, e-commerce, supply chain management, logistics management, capital flow management, sales management technology, etc. based on simulation cloud.
Intelligent simulation and experimental technology based on simulation clouds	Intelligent test technology of single-piece/component/system based on simulation cloud, test technology based on big data.
Intelligent pre-sale/sale/after-sale technology based on simulation clouds	Intelligent pre-sale/sale/after-sale comprehensive guaranteed service technology, intelligent value-added service technology, intelligent cloud equipment fault diagnosis, prediction and health management technology based on simulation cloud.

Fig. 5.6 CPSEO-IHPSC technological system

5.3.1 System Architecture of Cyber-Physical System Engineering Oriented Intelligent High Performance Simulation Cloud Prototype

Cyber-Physical System Engineering Oriented Intelligent High Performance Simulation Cloud (Prototype) is based on the further development and practice of the concept, technology, model, and format of the intelligent high performance simulation cloud. It has preliminarily practiced and verified the architecture and technical system of some intelligent high performance simulation clouds. Among them, "Prototype" is an integration of a part of new-generation artificial intelligence technology, a part of new simulation technology, new information and communication technology, and new simulation application technology. Its system architecture mainly includes new resource/product/capability layer, new industrial Internet of Things/access layer, new edge processing layer, new common cloud service layer, new cloud industrial SaaS layer (CMSS), application layer, and user layer (Fig. 5.7).

The explanation of each layer is as follows:

(1) New resources/products/capabilities layer: including various types of simulation equipment, simulation services, and CPS products.
(2) New industrial Internet of Things/access/edge processing layer: providing communication connectivity of simulation equipment, supporting the communication connectivity of mainstream industrial field communication protocols such as OPC-UA, MQTT, Modbus and Profinet, as well as the communication connectivity of industrial field bus, wired network and wireless network; and providing Smart IOT intelligent gateway access products and INDICS-API

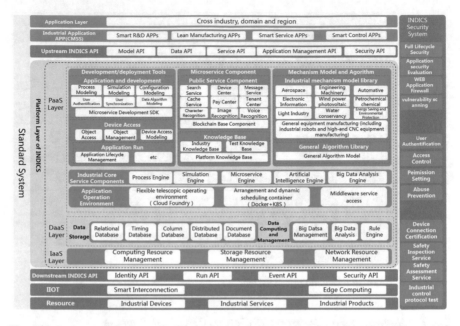

Fig. 5.7 System architecture of cyber-physical system engineering oriented intelligent high performance simulation cloud prototype

software access interface, supporting the hybrid computing mode of "Cloud Computing + Edge Computing".

(3) New cloud service common layer: including the IaaS layer that supports computing resource management, storage resource management and network resource management, the DaaS layer that provides data storage, data computing and data management, the generic PaaS platform that provides the application runtime environment, industrial core service components and other services, and the simulation PaaS platform service.

(4) New cloud simulation SaaS layer: providing simulation application APP (simulation SaaS) service. This includes intelligent business, intelligent research development, intelligent control, and the simulation application services throughout the entire manufacturing industry chain with remote monitoring, intelligent diagnosis, after-sales service, asset management as the core.

(5) Application layer: containing co-simulation application for cross-industry, cross-domain, and cross-region.

(6) User layer: involving intelligent high performance simulation cloud user.

(7) Each layer has its own standard specifications and safety specifications.

5.3.2 Key Technical Achievements of Cyber-Physical System Engineering Oriented Intelligent High Performance Simulation Cloud Prototype

(1) Intelligent High Performance Simulation's Resource/Capability/Product Layer
(1) Intelligent High Performance Simulation Computer Technology [9].

Intelligent high performance simulation computer technology is a kind of high performance simulation infrastructure integrated hardware and software as one system which mainly oriented to two kinds of users (advanced simulation users and mass user group) and three kinds of simulation (virtual, constructive and live simulation). It was aimed to build an intelligent high performance simulation computer system for optimizing the overall performance of "system modeling, simulation operation, and result analysis/processing". Single node indicators achieved by the team are as follows: The peak performance ≥ 25 trillion floating-point operations per second, the internal storage capacity ranges 2–10 TB, and the total external storage capacity ranges 40–200 TB. The system can be expanded to four nodes, up to 100 trillion floating-point operations per second. It has simulation computing units based on X86 multi-core processor and deep computing processor (DCU), simulation acceleration components based on big data processing technology, and simulation acceleration components based on artificial intelligence algorithm technology. There are multiple types of interconnection interfaces for live and virtual equipment and intelligent manufacturing systems which could support the full life cycle activities of CPS engineering based on simulation (Fig. 5.8).

(2) Intelligent Sense/Access/Communication Layer
(1) Ubiquitous Sensing, Accessing, and Integrated Processing Technology Based on New Internet [10].

Based on ubiquitous networks such as the new Internet, Internet of Things, mobile Internet, and the satellite network, the data with respect to CPS product life cycle, as well as the equipment, products, and processes in the process of simulation/manufacturing are intelligently sensed, acquired, transmitted, accessed, analyzed, and processed by means of intelligent sensing technology and IoT technology, so as to achieve network services, access to the Internet of Things, security and controllability, and collaborative integration for equipment and products. The team has made network services support GPRS, 3G, 4G, 5G, NB-IOTIoT based simulation, and other wireless functions, and met the field interaction and video acquisition requirements. The access to the Internet of Things supports industrial mainstream communication protocols such as MODBUS, OPC-UA and real-time industrial Ethernet protocol Profinet, and has realized the IoC function of access devices. As for safety and controllability, self-controlled gateway devices and edge application management software are provided for industrial edge intelligence communications. For collaborative integration, fog computing, embedded computing, multi-language edge application runtime environment, multi-industry edge model library, and heterogeneous integration are supported (Fig. 5.9).

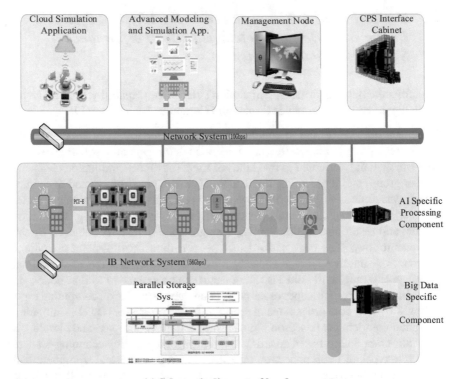

(a) Schematic diagram of hardware system

Fig. 5.8 Schematic diagram of the intelligent high performance simulation computer system **a** schematic diagram of hardware system **b** schematic diagram of the software system

(3) Intelligent High Performance Simulation's Edge Processing Platform Layer
(1) Multi-Edge Coordinated Management Platform Service Technology [11].

Multi-edge coordinated management platform service technology involves virtualization and service oriented encapsulation of simulation resources/capabilities/products at multiple edges (distributed in different enterprises, departments or operators), as well as the coordinated and optimized management and operation, which enable users to build a simulation environment and run a simulation system on demand at any time, so as to perform the whole simulation life cycle activities through the network and terminals. Our team proposed a unified virtualization and service-oriented framework, which could realize the virtualization and service oriented encapsulation of different types of resources such as computing, storage, software, model, and simulator device resources; and a unified management framework of heterogeneous resources, which could reuse the management system at multiple edges and realize coordinated and optimized management based on resource management middleware; and a multi-edge allocation optimizing model

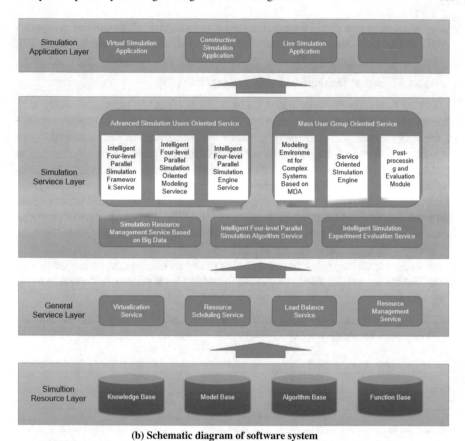

(b) Schematic diagram of software system

Fig. 5.8 (continued)

Fig. 5.9 Sensing, access and integrated processing technology application

and algorithm, which could realize dynamic build and run of the simulation system and its environment based on optimal scheduling of simulation resources (Fig. 5.10).

(4)   Intelligent High Performance Simulation's Cloud Service Platform Layer
(1)   The AI Engine Technology Based on Big Data [12].

The AI engine technology based on big data is aimed to train adequately with appropriate and optimized training algorithms on the mass data of the physic space or the cyber space, and ultimately achieves the best fitting of decision-making or decision boundary in the CPS engineering applications based on given information by 3 kinds of means including (semi-)supervised learning, unsupervised learning, and reinforcement learning. The author's team has the following research progress. In the aspect of integrating human intelligence into a machine, algorithm designers and experts in various fields could design and integrate deep neural networks, simulation environment, and evaluation system. In the aspect of machine intelligence, the designed deep neural networks would be parallel trained with a large number of parallel simulation examples. In the aspect of man-machine hybrid enhanced intelligence, the trained results would be evaluated and fed back to experts in various fields (Fig. 5.11).

(2)   New-Generation Intelligent High Performance Modeling and Simulation Language Technology [13]

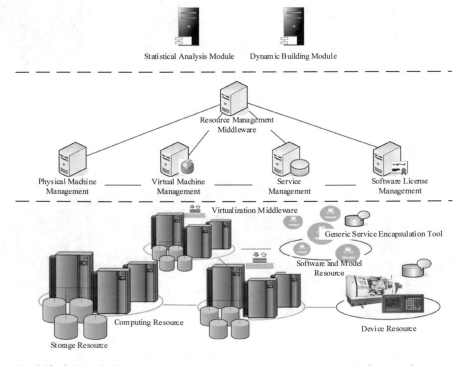

Fig. 5.10 Schematic diagram of the multi-edge coordinated management platform service

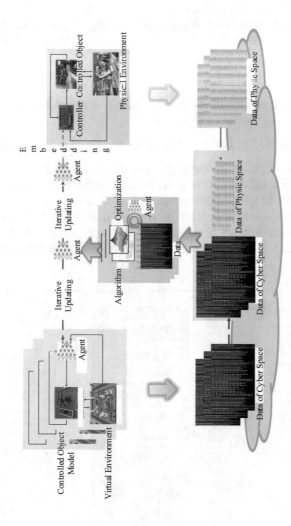

Fig. 5.11 Schematic diagram of the AI engine based on big data

The next generation intelligent high performance modeling and simulation language technology mainly provides intelligent high performance software system for complex system modeling and simulation, enabling system researchers to focus on complex system problems themselves, and thus greatly reducing software development and debugging work in modeling and simulation. Our team proposed a new specification for system description from the aspects of requirements, functions, logic, physics (continuous, discrete, qualitative, optimized, etc.), and a new approach to perform system engineering modeling through text/graphics input [13]. A compiler for parallel simulation was developed to automatically identify parallelism and generate a unified parallel simulation program framework linked with various simulation functions and algorithms. The modeling and simulation language program is seamlessly integrated with cloud simulation scheduling and experiment management which could schedule the programs to the nodes and cores on the back end for high performance execution (Fig. 5.12).

(3) Multi-level Parallel High Performance Simulation Solving Technology [9]

In order to make full use of the super parallel computing environment to accelerate simulation of CPS problems, four-level parallel high-efficiency simulation algorithms are proposed in the study as shown in Fig. 5.13 [9], including job-level parallel methods for large-scale simulation, task-level parallel methods among simulation system members, model-level parallel methods within the federate, and thread-level parallel methods based on complex model solution. The preliminary research results

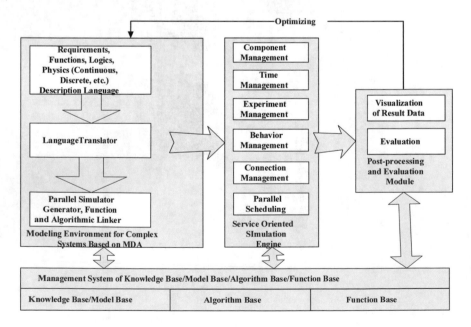

Fig. 5.12 Framework of the new-generation intelligent high performance M&S language

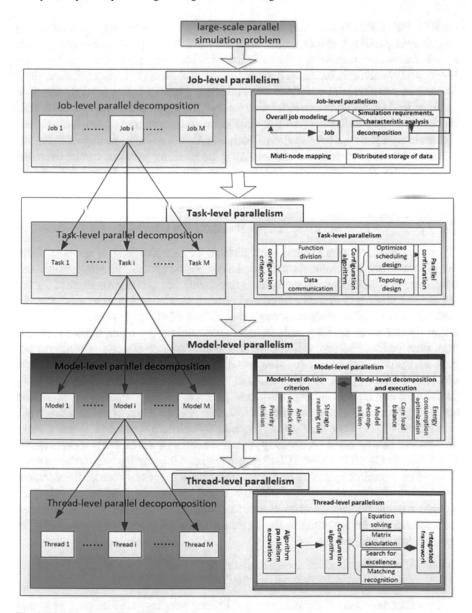

Fig. 5.13 Four-level parallel high-efficiency simulation algorithms

of the author's team consist of: (1) job-level parallel methods for large-scale simulation [14]: QMAEA (quantum multi-agent evolutionary algorithm), adaptive double chain quantum genetic algorithm, cultural genetic algorithm, and multi-population parallel differential evolution algorithm integrated with cuckoo search; (2) task-level parallel methods among federate members [15]: task-level parallel method based on RTI, task-level "hybrid simulation" parallel method based on event table; (3) model-level parallel methods within the federate member [16]: multi-core parallel simulation engine technology for complex variable structure systems; (4) thread-level parallel methods based on complex model solution [17]: parallel algorithms for ordinary differential equations of continuous systems based on right-function uniform load of equations, parallel discrete-system algorithms based on optimistic mechanism, and parallel algorithms for qualitative systems based on qualitative linear algebraic equations and optimization algorithms. Usually, a large-scale parallel simulation problem is decomposed into relatively independent jobs that can be executed in parallel for faster simulation. Similarly, as shown in Fig. 5.13, a job can be further decomposed into tasks, then models and then threads, to utilize the internal parallelism and accelerate the simulation executions.

(4)   Simulation Evaluation Technology Based on Big Data [18]

For the application of system simulation in CPS engineering, a series of functions, such as big data acquisition, management, visual analysis and processing, and intelligent evaluation, should be realized to provide comprehensive support for the application engineers in the analysis and evaluation of simulation results. The big data-based simulation evaluation method is a kind of effective methods for simulation evaluation of the CPS with uncertain mechanisms by using massive simulation, observation, and application data. The main research directions comprise big data acquisition technology for simulation experiments, big data integration and cleaning technology, big data storage and management technology, big data analysis and mining technology, big data visualization technology, intelligent simulation and evaluation technology, Benchmark technology (two types of users, three types of simulation), big data standards and quality systems, and big data security technology. For example, the preliminary research results of the author's team include: DaaS layer in CASICloud INDICS platform [19] and the intelligent simulation evaluator of the complex system (Fig. 5.14), which have functions such as the dynamic acquisition of simulation data, evaluation based on various intelligent evaluation algorithms, and playback and analysis of simulation process, so that the designed CPS system can be evaluated in the implementation of complex system modeling and simulation engineering.

(5)   Intelligent High Performance Simulation's Cloud Service Application Layer
(1)   Container-based APP Technology for CPS Engineering [20]

Container-based APP technology for CPS engineering provides an elastic and scalable environment for the development of APP through the hybrid container orchestration technology based on Cloud Foundry, Kubernetes + Docker. Container technology enables developers to implement applications in a resource-isolated process, and the components necessary for running applications are packaged into a mirror. They can be reused to support the design and construction of industrial APP based on the model and automatic machine learning, as well as the intelligent management of the full-life cycle of industrial APP, achieving the encapsulation,

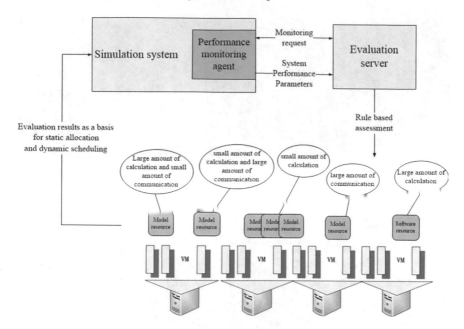

Fig. 5.14 Intelligent evaluation algorithms of CPS cloud simulation

distribution, and operation of the APP development. The container-based APP technology for CPS engineering proposed by the team integrates the product life cycle services such as intelligent R&D, lean production, intelligent service, and intelligent control horizontally, and forms application services at the four levels of interconnected enterprise layer, enterprise layer, production line layer, and equipment layer by vertically combining with various professional applications and collaborative applications. Customized business interface is adopted to provide services for enterprise users (Fig. 5.15).

(6)   Others
(1)   Standardization Technology

The standard technology system of CPSEO-IHPSC consists of the basic common standard, the general standard, and the application service standard. Among them, (a) the basic common standard is an important guiding foundation on which related standards of intelligent high performance simulation cloud for CPS engineering. The basic common standard supports of the general standard and the application standard, including the terminology definition, general requirements, architecture, testing and evaluation, management, etc. (b) The general standard is to standardize the generality, versatility, and guiding standard for intelligent high performance simulation cloud, including network and connectivity, identity resolution, edge computing, platform and data, industrial APP, and security standard. (c) The application service standard includes the standards for typical application scenarios such as intelligent production, service production, networked collaboration, and customization, as well as the

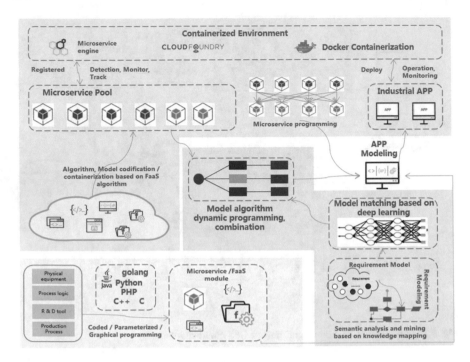

Fig. 5.15 Container-based APP technology for CPS engineering

standards based on basic common, overall and typical application also cover application guidelines, specific technical standards, and management specifications for key industries such as automotive, aerospace, petrochemical, machinery manufacturing, light industrial, appliances, electronic information, etc. (Fig. 5.16).

(2) Safety Technology

CPSEO-IHPSC security technology system is divided into six categories: equipment security, network security, platform security, application security, security management, and data security. Among them, (a) equipment security includes equipment access authentication, transmission link encryption, anomaly monitoring and other security protection technologies involved in the design, research and development, manufacture and operation process of the simulation cloud system; (b) network security mainly includes abnormal flow analysis, access control and other protection technology for factory network security that supporting industrial intelligent production applications; (c) platform security mainly includes information encryption, antivirus, security defense, intrusion prevention, and commercial security technology; (d) application security refers to security protection technologies related to business applications such as vulnerability scanning, application firewall, application security assessment; (e) security management mainly refers to the management and security prevention technologies involving security risk management, security event,

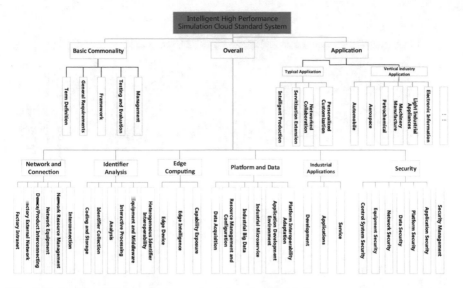

Fig. 5.16 Intelligent High performance simulation cloud standard system

and emergency response of simulation cloud; (f) data security mainly includes security storage, data encryption, and other industrial data security protection technology (Fig. 5.17).

5.3.3 Technical Novelty of Cyber-Physical System Engineering Oriented Intelligent High Performance Simulation Cloud Prototype

According to the achievements presented above, there are several technical novelties of CPSEO-IHPSC including architecture, hardware, software, and application technology which are shown as follows:

(1) Integrated simulation cloud architecture technology for two kinds of users (advanced simulation users and mass user group) and three kinds of simulation (virtual, constructive, and live simulation) collaborated intelligently between cloud and edge.

(2) Hybrid computing units technology based on X86 multi-core processor and deep computing processor (DCU); simulation acceleration component technology based on big data processing and simulation acceleration component technology based on artificial intelligence algorithm; multiple kinds of interconnection interface technology for live, virtual equipment, and intelligent manufacturing systems.

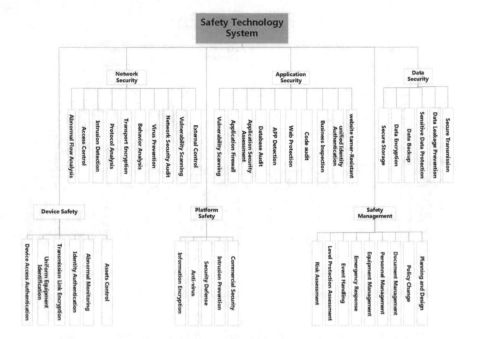

Fig. 5.17 Safety technology system

(3) Multi-level parallel high performance simulation engine technology; a large number of parallel simulation examples building and running technology; intelligent high performance simulation language technology for CPS engineering.

(4) VVA technology based on intelligent evaluation driven by big data; intelligent experiment design and result visualization technology.

5.4 Case Study of Cyber-Physical System Engineering Oriented Intelligent High Performance Simulation Cloud for CPS Engineering (Prototype)

5.4.1 Based on CPSEO-IHPSC Digital Twins Technology of Intelligent Manufacturing System Application

Virtual factory is the application of CPSEO-IHPSC digital twin technology in intelligent manufacturing system. Based on the intelligent high performance simulation cloud for CPS engineering, the digital twin technology is used to build a virtual manufacturing factory with its environment, production capacity, and process exactly corresponding to the reality, and integrate various manufacturing information (including environmental and device state data, equipment constraints, etc.).

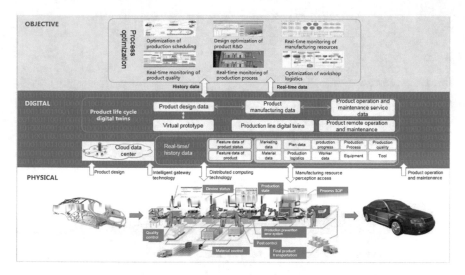

Fig. 5.18 Digital twin technology of intelligent manufacturing system application

By making full use of big data intelligence, the structured data (integrated data and simulation data) and the unstructured data are transformed into structured knowledge to support the monitoring, simulation, analysis and optimization of production line planning, workshop production, and operation services at all stages, we achieved real-time monitoring, correction and closed-loop feedback optimization control of production processes, assembly lines, and assembly stations (Fig. 5.18).

5.4.2 Based on CPSEO-IHPSC Digital Twins Technology of Smart City System Case

Digital Twin City is a smart city system based on CPSEO-IHPSC digital twin technology [21]. It is a virtual mapping object and intelligent manipulator of the physical city. It can serve as a comprehensive real-time link between the physical worlds and the digital world, and then record, analyze, and predict the full-life cycle activities of the operating object. Through comprehensive digital modeling of infrastructure, full perception, and dynamic monitoring of urban operation state, the digital twin city is formed to represent and map the precise information of the physical city. The virtual and real space is integrated and coordinated so that various traces are observable in the physical space, while the various information is retrievable in the virtual space. As software-defined simulation, the behaviors of humans, events, and objects in the physical environment are simulated in software to show urban development and evolution based on big data. Through the digital twin city simulation and extrapolation, some reasonable and feasible countermeasures may be put forth to optimize the urban planning and management and achieve intelligent intervention (Fig. 5.19).

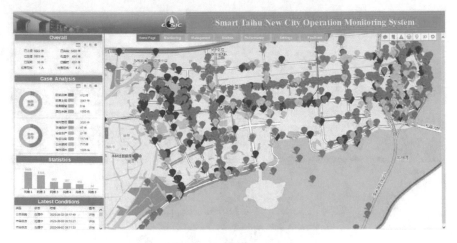

Fig. 5.19 Digital twin technology of smart city system

5.5 Suggestions on Developing Cyber-Physical System Engineering Oriented Intelligent High Performance Simulation Cloud in the New Era

5.5.1 Interpretation of "New Internet + Cloud Computing + Big Data + New Artificial Intelligence+"

The rapid development of new Internet technology, new information and communication technology, new artificial intelligence technology, and their integration with new specialized technology in new application areas, is ushering major changes in new models, new means, and types of business in various fields such as national economy, people's livelihood, and national security. It heralds the arrival of a new era featuring "New Internet + Cloud Computing + Big Data + New Artificial Intelligence".

(1) New Internet technology: Internet of Things, Internet of Vehicles, mobile Internet (5G/6G), satellite network, space-ground integrated information network, and the future internet.
(2) New information and communication technology: cloud computing, big data, 5G, high performance computing, modeling/simulation, digital twin, blockchain, quantum computing, etc.
(3) New artificial intelligence technology: data-driven deep reinforcement learning intelligence, network-based swarm intelligence, technology-oriented hybrid intelligence of human–computer and brain–computer interaction, cross-media reasoning intelligence, autonomous intelligent unmanned system, etc.
(4) New specialized technology in application fields: new specialized technology in fields such as national economy, national security, and people's livelihood.

5.5.2 Focusing on the Coordinated Development of Technology, Industry, and Application

The high performance intelligent simulation cloud for CPS engineering is an important part of the modeling and simulation technology system in the new era, and its development requires the coordinated development of technology, application, and industry. Its development should unswervingly follow the "innovation-driven approach" and deep integration of modeling and simulation technology, information and communication technology, new-generation artificial intelligence technology, and domain-application technology. Undoubtedly, its development and implementation also call for global cooperation and exchanges and meantime give full attention to the characteristics of each country, field, and system.

Acknowledgements Thanks for all the members of author's team, especially for Duzheng Qing, Han Zhang, Liqin Guo, Yingying Xiao, Chi Xing, Zhengxuan Jia of Beijing Simulation Center, Qinping Zhao, Lin Zhang, Lei Ren, Fei Tao, Ni Li, Aimin Hao of Beihang University, Jun Li, Rong Dai of Sugon Information Industry Co., Ltd, Ming Yang of Harbin Institute of Technology, Yiping Yao of National University of Defense Technology.

References

1. Helen G (2008) From vision to reality: cyber-physical systems. National science foundation November 18–20
2. Wolf W (2009) Cyber-physical systems. Computer 42(3):88–89
3. Rajkumar RR, Lee I, Sha L et al (2010) 44.1 Cyber-Physical systems: the next computing revolution. In: Proceedings of the 47th design automation conference DAC 2010, Anaheim, California, USA, 13–18 July 2010. IEEE
4. Lee EA (2008) Cyber physical systems: design challenges. object oriented real-time distributed computing (ISORC). In: 11th IEEE international symposium on IEEE. Orlando, Florida, USA, 5–7 May 2008
5. Lee J, Bagheri B, Kao HA (2015) A cyber-physical systems architecture for industry 4.0-based manufacturing systems. Manuf Lett 3:18–23
6. Zhou P et al (2017) Cyber physical system white book
7. Li BH, Zhang L, Li T et al (2017) Simulation-Based Cyber-Physical Systems and Internet-of-Things. In: Mittal S, Durak U, Ören T. (eds) Guide to Simulation-Based Disciplines. Simulation Foundations, Methods and Applications. Springer, Cham
8. Song X, Ma Y, Teng D (2015) A load balancing scheme using federate migration based on virtual machines for cloud simulation. Math Probl Eng 506432:1–12
9. Li BH, Chai XD, Li T et al (2012) Research on high efficiency simulation technology for complex systems. J Chin Acad Electron Sci 7(3)
10. Yang C, Lan SL, Shen WM et al (2017) Towards product customization and personalization in IoT-enabled cloud manufacturing. Cluster Comput 20(2):1–14
11. Lin TY, Li BH, Yang C (2013) A multi-centric model of resource and capability management in cloud simulation. In: Proceedings of the 2013 8th EUROSIM congress on modelling and simulation. Washington, DC, United States, September. IEEE
12. Li BH, Lin TY, Jia ZX et al (2019) An intelligent cloud design technology for intelligent industrial systems. Computer Integr Manuf Syst 25(12)

13. Li BH, Song X, Zhang L et al (2017) Cosmsol: complex system modeling, simulation and optimization language. Int J Model Simul Sci Comput 08(02)
14. Tao F, Zhang L, Laili YJ (2014) Configurable intelligent optimization algorithm. Des Pract Manuf 5(26):587–588
15. Zhang ZH, Li BH, Chai XD (2010) Research on high performance RTI architecture based on shared memory. In: The proceedings of ICCEE
16. Yang C, Chi P, Song X et al (2016) An efficient approach to collaborative simulation of variable structure systems on multi-core machines. Cluster Comput 19(1):29–46
17. Li BH, Chai XD, Li T et al (2012) Research on high efficiency simulation technology for complex systems. J Chin Acad Electron Sci 7(3):285–289
18. Li BH, Chai XD, Zhang L et al (2018) Preliminary study of modeling and simulation technology oriented to neo-type artificial intelligent systems. J Syst Simul
19. Li BH, Chai XD, Hou BC et al (2018) New generation artificial intelligence-driven intelligent manufacturing (NGAIIM). In: 2018 IEEE smartworld, ubiquitous intelligence and computing, advanced and trusted computing, scalable computing and communications, cloud and big data computing, internet of people and smart city innovation (SmartWorld/SCALCOM/UIC/ATC/CBDCom/IOP/SCI). Guangzhou, China, 8–12 Oct 2018. IEEE
20. Shi GQ, Lin TY, Zhang YX et al (2019) Lightweight virtualization cloud simulation technology for industrial internet. Comput Integr Manuf Syst
21. Li BH, Chai XD, Zhang L et al (2018) Preliminary study of modeling and simulation technology oriented to neo-type artificial intelligent systems. J Syst Simul 02:349–362

Bo Hu Li is an Academician of Chinese Academy of Engineering, a Ph.D. adviser and the Honorarium President of the School of Automatic Science and Electrical Engineering in Beijing University of Aeronautics and Astronautics (BUAA),the Honorarium President of Chinese Simulation Federation, Co-Chief-Editor of the journal "International Journal of Modeling, Simulation, and Scientific Computing". He was the Director of Beijing Institute of Computer Application and Simulation Technology and Beijing Simulation Center,as well as the President of the School of Automatic Science and Electrical Engineering in BUAA. He was the President of Chinese Simulation Federation and the first President of Federation of Asian Simulation Societies (ASIASIM) and a member of the council of directors of Society for Modeling and Simulation International (SCS). In addition, He was the Director of Expert Committee of Contemporary Integrated Manufacturing System (CIMS) Subject in Chinese National High Technology Research and Development Plan. In the fields of simulation and manufacturing informatization, he authored or co-authored 370 papers, 14 books and 4 translated books, and he got 1 first class scientific award and 3 second class scientific awards from China State, and 17 scientific awards from the Chinese Ministries. In 2012, He got the SCS Life time Achievement Award of Society for Modeling and Simulation International, and the "National Excellent Scientific and Technical Worker" honorary title from the China Association for Science and Technology. In 2017, He got the CCF Life time Achievement Award of the China Computer Federation. In 2019, He got the Life time Achievement Award of the China Simulation Federation.

Xudong Chai received the Ph.D. degree in navigation, guidance and control from Beihang University, in 1999. He did post-doctoral research on automation system control theory and control engineering Tsinghua University in 1999–2001. He is standing director of the Chinese Association for System Simulation, and Chairman of Technical Standards Group of China Industrial Internet Industry Alliance. His research interests include complex system modeling and simulation, design of virtual products, and integrated manufacturing systems. He has previously been awarded first and second prize in the National Defense Science and Technology Progress Awards of China, and second prize in the Science and Technology Progress Award of Ministry of Education of China. He has published 100 academic papers.

Ting Yu Lin received the B.S. degree and the Ph.D. degree in control system from School of Automation Science and Electrical Engineering in Beihang University, China, in 2007 and 2014 respectively. From 2019 to day, he was a senior engineer in State Key Laboratory of Intelligent Manufacturing System Technology of Beijing Institute of Electronic System Engineering. His research interest includes complex system modeling and simulation and cloud simulation. He authored and co-authored 40 papers, 3 books and chapters in the fields of his major, and has got a first class Scientific and Technological Progress Award of from a Ministry, China, in 2018. He is now a member of China Simulation Federation (CSF).

Chen Yang obtained his Ph.D. in Control Science and Engineering, 2014 and B. Eng. in Automatic and Information Technology, 2008, both from the School of Automation Science and Electrical Engineering and the Honors College (an elite program) of Beihang University (BUAA), Beijing, China. He has worked in the HKU-ZIRI Laboratory for Physical Internet, the University of Hong Kong as an associate research officer, and in Huawei Technologies, as a senior engineer on R&D tools. He is currently an Associate Professor with the School of Computer Science and Technology, Beijing Institute of Technology, Beijing, China. His research interests include Internet of Things, Industry 4.0, Cloud Manufacturing, modeling and simulation of complex systems, Artificial Intelligence and Big Data Analytics.

Baocun Hou Ph.D., researcher, general manager of China Aerospace Science And Industry Coperation Limited, Casicloud-Tech Co., Ltd., Beijing Aerospace Smart Manufacturing Technology Development Co., Ltd., deputy leader of the experimental platform group of Alliance of Industrial Internet, deputy leader of the platform AD Hoc Group, leader of production-study-research Cooperation and Application Group in AIIA of China. He has long been engaged in the technical research of smart cloud manufacturing/smart manufacturing, industrial internet platform, modeling and simulation of complex systems, and software system engineering. Participated in the research, development and management of China's first and the world's first industrial Internet platform-the INDICS platform (Casicloud) His related research results won 2 second-level awards of National defense science and technology progress. More than 50 papers were published in academic journals and conferences in domestic and abroad. Co-published 1 Academic work.

Yang Liu received the B.S. degree in measurement and control technology and instruments from Xi an University of technology in 2007 and M.S. degree in machinery and electronics engineering from Beijing University of technology, in 2010. From 2010 to 2016, she was a Process Designer with Beijing Aerospace Xin feng Machinery Co., Ltd. Since 2016, she was a Technical Director with Beijing Aerospace Manufacturing Technology Co., Ltd. Her research interest includes fundamental study of intelligent manufacturing, industrial internet, artificial intelligence, and machine vision. Author's awards and honors include, publishing 6 papers and applying for 2 invention patents, and participating in research result "intelligent manufacturing cloud platform led by the new generation of artificial intelligence technology" was listed as one of the top 10 international intelligent manufacturing technological advances in 2019.

Xiao Song received the B.S., Ph.D. degree in automation from Beihang University (BUAA), Beijing, China, in 1999 and 2006, respectively. He is currently an associate professor in school of automation, Beihang University (BUAA), Beijing, China. His research interest includes the artificial intelligence, manufacturing, modeling and simulation. Mr. Song's awards and honors include the best paper nominee of AsiaSim conference 2019, 2017 and 2012.

Hui Gong received the B.S. degree in Communication Engineering from North University of China, Taiyuan, China, in 2002 and the Ph.D. degree in Optical engineering at Beijing Institute of Technology, Beijing, China, in 2012. From 2002 to 2006, she was a teacher of North University of China. After receiving the Ph.D. degree in 2012, she has been working in the Second Academy of CASIC as a senior engineer. Her research interest includes the design of smart city and complicated Information systems.

Part II
Methodology

Chapter 6
Service Composition and Scheduling in Cloud-Based Simulation

Lin Zhang, Feng Li, Longfei Zhou, and Bernard Zeigler

Abstract Cloud-based simulation, as a new simulation mode under the support of cloud computing, big data, internet of things, etc., has attracted wide attention recently. A cloud-based simulation system is actually a cyber-physical system. In the cloud environment, simulation models or simulation resource can be encapsulated into simulation services. The issue of service composition and scheduling is an important concern to cloud-based simulation. Under the assumption that the steps of tasks being executed are certain, most of the work is purely dedicated to studying different scheduling schemes based on some specific priorities. However, the number and types of resource services in a practical cloud-based simulation are numerous. These services can constitute a service network and they have different granularities. These services have different granularities, which will lead to the number of composition steps is indeterminate. This chapter will give a review of the literature on related research. Then a service network-based method to implement service composition and scheduling in a simulation environment is proposed. In addition, a dynamic Data-Driven Simulation-based dynamic service scheduling method is proposed.

L. Zhang (✉)
School of Automation Science and Electrical Engineering, Beihang University, Beijing 100191, China
e-mail: zhanglin@buaa.edu.cn

F. Li
School of Computer Science and Engineering, Nanyang Technological University, Singapore, Singapore
e-mail: feng.li@ntu.edu.sg

L. Zhou
Computer Science & Artificial Intelligence Laboratory, Massachusetts Institute of Technology, Cambridge, MA 02139, USA
e-mail: longfei@mit.edu

B. Zeigler
RTSync Corp, Arizona Center for Integrative Modeling and Simulation, Sierra Vista, AZ, USA
e-mail: zeigler@rtsync.com

© Springer Nature Switzerland AG 2020   121
J. L. Risco Martín et al. (eds.), *Simulation for Cyber-Physical Systems Engineering*,
Simulation Foundations, Methods and Applications,
https://doi.org/10.1007/978-3-030-51909-4_6

6.1 Introduction

6.1.1 The Characteristics of Cloud-Based Simulation

Cloud-based Simulation (CBS) which integrates Cloud Computing, Web-based Simulation (WBS), Internet of Things (IoT), Cyber-Physical System (CPS), etc., has been paid lot of attention in recent years [1]. CBS is derived from WBS, using cloud service to manage various simulation resources and build different simulation environments [2].

A cloud-based scenario can be described as follows [3]:

(1) Simulation users can run simulation models on cloud infrastructure (Simulation as a Service).
(2) Some simulation users (especially simulation modelers) may want to have control over the models by creating models using simulation development tool hosted on cloud infrastructure (Modeling as a Service).
(3) Some simulation modelers may want to have greater control over the simulation development tool itself. Provided that the simulation development tool is reconfigurable or modular, the components of the tool can be mixed and matched. This is close to the PaaS.
(4) Simulation users may want to have control over storage (for example, in data-driven simulation) and the execution platform and middleware (for example, in high performance simulation or distributed simulation). This is similar to the Infrastructure as a Service (IaaS).

A cloud platform can be built to support modeling and simulation in a cloud (named as "simulation cloud"), which improves the capabilities of the existing networked simulation platform. It utilizes network and cloud platform to compose the simulation resources (simulation cloud) in network on demand, and to provide various simulation services to users. Generally, a cloud platform for simulation consists of 4 layers: resource layer, simulation service layer, portal, and support tools layer, and application layer (Fig. 6.1) [4].

Resource layer provides network and various simulation resources encapsulated by virtualization technology, including model resources, tools and software resources, computing resources, storage resources, model/data resources, knowledge resources, and various types of simulators, scientific instruments, and so on. Simulation service layer provides simulation related core services, including multi-user-oriented resource scheduling and management services, pervasive co-simulation services, virtual simulation resources information management services, intelligent resource discovery services, co-simulation scheduling and composition services, and so on. Application portal/support tools layer provides browser and desktop portals/tools for users to login to the platform and carry out simulation activities. Application layer includes multidisciplinary virtual prototype collaborative simulation applications, large-scale system level collaborative simulation applications, and other Collaborative Simulation Applications.

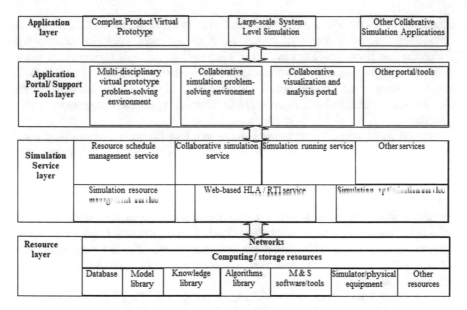

Fig. 6.1 The framework of a cloud platform for simulation

Due to a variety of services and the complexity of tasks on the simulation cloud, it is necessary to combine different services to fulfill a task requirement. Hence, the issue of service composition and scheduling is one of the most challenging research problems. Service composition and scheduling is a necessary way to realize multiple complexity simulation requirements. It can combine different services to fulfill a requirement and schedule different services for multiple requirements. By and large, researches on resources service scheduling on a simulation cloud can be divided into computing resource scheduling (such as software resource scheduling) and simulation entity resource scheduling (such as simulation devices resource scheduling).

There are many challenges for service composition and scheduling in cloud-based simulation. For example, (1) massive resources from numerous providers are aggregated in the cloud platform and encapsulated into cloud services, which result in the services of large-scale and high complexity; (2) geographical distributions of simulation resources have the features of interoperability; (3) the simulation cloud should rapidly respond to personalized requests; (4) there are some uncertainties and disturbance of tasks and services.

6.1.2   Related Works

In the cloud-based simulation environment, models and the related tools are encapsulated into simulation services. In a simulation cloud, simulation tasks, which are composed of dynamically distributed simulation models, are always difficult to be completed. The number and resource services types and tasks in practical cloud-based simulation are numerous. How to correctly incorporate multiple services together to accomplish a multidisciplinary simulation task and how to tackle multiple tasks requirements are the two important concerns not only for users, but also for the cloud center itself. Service composition and scheduling play an important role in the process.

By and large, service composition and scheduling (SCS) issues can be divided into computing resource services SCS and simulation entity resources services SCS.

In the realm of computing resource services, task scheduling has been put considerable efforts, among which, directed acyclic graph (DAG)-based scheduling is one of the most widely researched approaches. These research fruits mainly focus on different scheduling targets and different scheduling priorities. With respect to scheduling targets, maximize resource utilization [5, 6], fairness maximization, and minimize scheduling time [7], are the most common ones. For the establishment of the scheduling priority, sequential methods and round-robin method [8], etc., have attracted much attention.

As for scheduling algorithms, these can be composed of traditional methods and intelligent optimization algorithms. For traditionally methods, there are Min-Min and Max-Min algorithm [9], and the Sufferage algorithm [10], etc. For intelligent optimization algorithms, research achievements are abundant. To obtain a near-optimal solution, the authors adopted the genetic algorithm [11, 12]. Arockiam and Sasikaladevi [13] proposed the simulated annealing to get a task solution. Li et al. [14] discussed the particle swarm algorithm to cope with task problem. Li et al. [15] presented the ant colony algorithm to solve a scheduling issue which can balance the entire system load while minimizing the makespan of a given task set. Li et al. [16] built an abstract clustering network to solve service composition issues. Based on the built service network, a genetic algorithm is applied to obtain service composition solutions.

In the field of simulation, optimization algorithms also play important roles. Meketon [17] summarized the optimization algorithms for simulation. Fu et al. [18] discussed the integration of simulation with optimization. They concerned about algorithm characteristics and problem scopes that simulation optimization addressed. Tekin and Sabuncuoglu [19] reviewed and analyzed different simulation optimization approaches. Hong and Nelson [20] classified simulation optimization problems into three different kinds. i.e., a finite number of solutions, continuous decision variables, and discrete variables that are integer-ordered. Fu [21] introduced the simulation software and the optimization routines and provided a convergence theory of simulation optimization. Zhou et al. [22–24] studied dynamic scheduling problems in a cloud environment. Simulation-based methods were proposed to deal with the multi-task scheduling problem of distributed cloud services.

Although different methods perform well in their own concerns, most of them assumed that task executive steps are determined in advance. The granularities and functional diversity of services that result in uncertainties of service composition processes for completing the task in a cloud environment are not well considered, which is an important issue for practical cloud-based simulation scheduling.

6.2   A Service Network Model for Simulation Entity Service Composition and Scheduling

In a simulation cloud, all kinds of resources are encapsulated into services. For the purpose of service composition and scheduling, these services should be described in a uniform way and the relationship among them should be built. Inspired by the method of establishing an abstract cluster network, in this section, we describe a new model structure of service composition and scheduling based on a concrete service network [25]. First, the service description model is discussed followed by a task description model. After that, service network-based service composition and scheduling is built with the support of service and task description models.

6.2.1   Service and Task Description Models

(1)   A service description model

Considering that there are many kinds of simulation units registered into the simulation cloud, each simulation unit can provide different number and different kinds of services. A service represents a specific function of the corresponding simulation unit. The granularities of services registered in the platform provided by different simulation units are different and they are represented by $s = \{s_1, s_2, \ldots s_m, \ldots s_M\}$, M reveals the number of services. As been described, services come from different simulation units and have different granularities. Hence, it is necessary to establish a uniform service model to describe different services. Service description includes functional description and nonfunctional description. For functional description, we only concentrate on input information and output information, which are denoted by S_m^i, S_m^o for service m, respectively. The nonfunctional description is denoted by QoS_m. The unified model of service description can be expressed as follows:

$$S_m = \langle S_m^i, S_m^o, QoS_m \rangle,$$

$$QoS_m = \langle Pro_m, Cost_m, Rel_m \rangle$$

where Pro_m denotes productivity of one day for simulation unit m and $Cost_m$ is one day's cost, while Rel_m means reliability of simulation unit m. The initial values of the three parameters are constants which are given by simulation units. In the process of scheduling, some of them will change based on the initial values.

(2) A task description model

In this chapter, let $\text{Task}(K) = \langle T_1, T_2, \ldots T_K \rangle$ denote K tasks to be scheduled, in which T_k means the K_{th} task and is represented by unified description model, namely $T_k = \langle t_i, t_o, t_{load} \rangle$. Among them, t_i expresses the input parameters of task and t_o signifies the output parameters of it. Besides, t_{load} means the requirement number of the task, which denotes the amount of this kind of task. Every task T_k is a multi-resource-requirement task and can be completed by several kinds of services. The number of needed services is uncertain, and it is dependent on the functionalities of the chosen services. In other words, the granularities of services are different. Some of them can only perform one atomic function while others contain large granular functions. Hence, a complicated task can be completed by different number of services in composition schemes.

6.2.2   Service Network-Based Service Composition and Scheduling Model

Based on the descriptions of the service and task model, a relationship network among services to services and services to tasks is established. Based on the built service network, we can obtain a service composition and scheduling result for a certain task requirement. The framework of this process is shown in Fig. 6.2. In this figure, resources and requirements are encapsulated into services and tasks, respectively. The relationship between services and tasks can be built based on the descriptions of services and tasks. Then, the results of service composition and scheduling can be obtained based on some kinds of scheduling algorithms in the relationship network. It can be seen that two processes, i.e., establishment of service network and composition scheduling based on the built network, are two important steps. Some more details will be described in the following.

In our model, services are denoted by nodes. Composition relationships among services which are determined by functional information of services are signified by edges. Specifically, if the output information of a service i can match with the input information of a service j, then there is a directed edge from i to j. The edge weight consists of QoS characteristics of the latter service and the two services' distance which is determined by the actual distance of two resources that the services represent. For instance, the nonfunctional information of S_i and S_j are $S_i = \langle Pro_i, Cost_i, Rel_i \rangle$, $S_j = \langle Pro_j, Cost_j, Rel_j \rangle$, respectively. The relationship between them is $S_i \rightarrow S_j$, then the edge weight calculation is expressed by the following formula (6.1)

$$W_{i,j} = w_1 \left(Cost_j / Pro_j \right) + w_2 * (1 - Rel_j) + w_3 * Dis(i, j) \qquad (6.1)$$

in which, $Cost_j / Pro_j$ means cost per unit productivity of the latter service that the edge connects. Rel_j denotes the reliability of service j. $Dis(i, j)$ expresses the

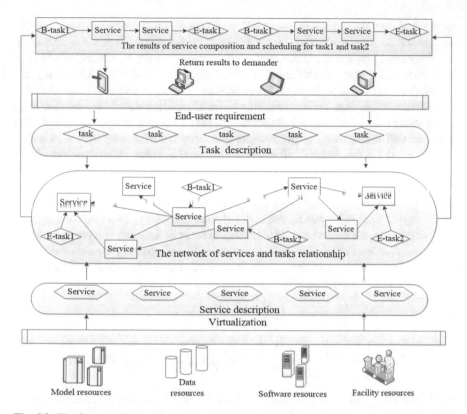

Fig. 6.2 The framework of service composition and scheduling based on service network

physical distance index between service i and service j. Only when there is physical entity transportation between the two services, the value of $Dis(i, j)$ is meaningful. Inversely, the value of $Dis(i, j)$ equals to 0. Every index has been normalized in Eq. (6.1). w_1, w_2, w_3 are weights of indexes which reveal the importance of different indexes. Based on the above-mentioned method, the network of service relationship can be established.

Based on the built network, requirements of tasks can be fulfilled. To this end, the first thing being done is to find a service node whose input information is the same as the task' input information and to find a service node whose output information can fulfill the requirement of the task. In most cases, one task cannot be completed by only one service. As a result, service composition is needed. There is more than one kind of composition strategies since the number of services is big and granularities and functions of services are different. At the same time, if one service is used by different tasks, scheduling on them is essential.

It is worth noting that $W_{i,j}$ mentioned above is a relatively fixed static edge weight without considering multiple tasks conflict. In fact, at a certain time, there exist a set number of tasks in the simulation cloud. They can be scheduled in a sequential way. When they are scheduled, every task is denoted by two nodes and joins into the service relationship graph. One task forms two nodes, of which one is a start node and the other is an end node. The input information t_i of the task is included in the start node and output information t_o is contained in the end node. The start node establishes connection with the nodes in service network whose input functionality can match with the requirement of the start node. The end node performs the same step as the start node. Based on the built service network, service composition searching methods will be executed to get composition results. When one task scheduling is finished, the nonfunctional information values of the chosen services should be changed. One reason is that, if a service is chosen to execute a task, the original processing time may be changed when it processes new tasks because of the waiting time. For the other reason, from the perspective of the balance of service utilization, the change of QoS value can decrease the chosen rate of some services being used, while increasing the chosen rate of other services not being used. Hence, if service j is used to complete the previous task, the productivity of service j should be renewed according to Eq. (6.2), and the new weight of edge that takes service j as an end node can be expressed as Eq. (6.3)

$$proj_{\_new} = proj_{\_old} - \Delta proj \qquad (6.2)$$

$$W_{i,j\_new} = w_1\left(\frac{Cost_j}{Pro_{j_{new}}}\right) + w_2 \times \left(1 - Rel_j\right) + w_3 \times Dis(i, j) \qquad (6.3)$$

in which $Proj_{\_old}$ is the old value of productivity that a service processes in normal condition. $Proj_{\_new}$ means a new value of productivity considering service reutilization. $\Delta Proj$ is the change of the productivity and it can be designed according to different situations.

The purpose of service composition and scheduling for multiple tasks is to find an optimal solution with minimum total edge weight which is shown in Eq. (6.4). In this equation, j belongs to service network and the number of j is not a constant. In other words, the composition steps for a certain task are uncertain which are determined by services relationship and the fixed weight value and the changed weight value.

$$\min \sum_{j \in SN} W_j \qquad (6.4)$$

6.2.3   Case Study

Taking the scheduling problem in manufacturing as an example, task requests and resource functions are various, which mean that the types of inputs and outputs are manifold. For the simulation environment of manufacturing enterprises in the cloud, there are many kinds of manufacturing resource models. These models come from different level of granularities of manufacturing resources, such as a machine tool, a machine center, or a production line. When they are encapsulated into manufacturing service, the services are managed and scheduled in the cloud platform, while the corresponding manufacturing resource in the real world will act according to the results of management or scheduling in the cloud. In this sense, the system that consists of cloud services and manufacturing resources is a typical cyber-physical system.

These services mentioned above could have different manufacturing capabilities, which can be denoted by different types of input and output information. These services can be combined together to complete a complicated task. Due to the aggregation of services in a cloud environment and the diversity of service granularities, the steps and number of the services to complete the request are uncertain. The description information of tasks and simulation unit services are shown in Table 6.1 and Table 6.2, respectively. Only 3 tasks and 15 simulation units are listed due to the limitation of space. Based on the given data and the proposed model, a network built based on services and tasks is shown in Fig. 6.3. In this figure, three big ovals in the bottom line denote requirements whereas other circular nodes mean services. The

Table 6.1 Tasks descriptions

Task number	Input type	Output type	Task load
t_1	1	6	80
t_2	2	8	96
t_3	3	6	70

Table 6.2 Simulation unit services descriptions

Simulation unit number	Input type	Output type	QoS
S_1	1	2	$\langle 10, 15, 0.8 \rangle$
S_2	4	5	$\langle 9, 14, 0.7 \rangle$
S_3	3	5	$\langle 7, 25, 0.9 \rangle$
S_4	4	5	$\langle 10, 16, 0.8 \rangle$
S_5	2	3	$\langle 8, 17, 0.8 \rangle$
S_6	6	7	$\langle 11, 16, 0.7 \rangle$
S_7	2	3	$\langle 9, 14, 0.9 \rangle$
S_8	3	4	$\langle 9, 12, 0.8 \rangle$
S_9	1	3	$\langle 8, 21, 0.8 \rangle$
S_{10}	3	4	$\langle 10, 16, 0.7 \rangle$
S_{11}	5	6	$\langle 11, 17, 0.8 \rangle$
S_{12}	5	7	$\langle 10, 23, 0.9 \rangle$
S_{13}	7	8	$\langle 9, 14, 0.8 \rangle$
S_{14}	5	6	$\langle 7, 13, 0.8 \rangle$
S_{15}	6	8	$\langle 6, 28, 0.9 \rangle$

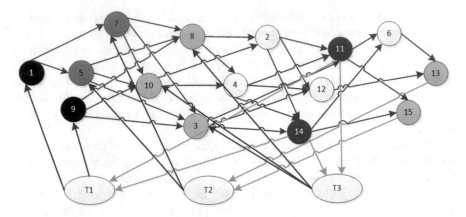

Fig. 6.3 A network of services and tasks

service network has been built based on their relationship. Simulations conducted in the following section are based on the network. From this figure, we can see that one task can be fulfilled by different number of steps. For example, task 1 can be finished by five steps: 1 -> 5 -> 10 -> 4 -> 14 or four steps: 1 -> 5 -> 3 -> 14. Composition and scheduling must be considered as a whole to decide which services are better in performance.

Due to the space limitation, the distance matrix between different services is not given. Based on the above information, the calculation of edge weight is as follows. In this chapter, the values of w_1, w_2, w_3 are chosen as 0.7, 0.2, 0.1.

(1) Calculating $Cost_j/Pro_j$
(2) Normalization $Cost_j/Pro_j$ and Dis_j

$$\frac{Cost_{nor}}{Pro_{nor}} = \frac{(Cost_j / Pro_j - Cost_{min} / Pro_{max})}{(Cost_{max} / Pro_{min} - Cost_{min} / Pro_{max})}$$

$$Dis_{nor} = \frac{Dis_j - Dis_{min}}{Dis_{max} - Dis_{min}}$$

(3) Calculating $W_{i,j}$

$$W_{i,j} = 0.7 * (Cost_{nor} / Pro_{nor}) + 0.2 * (1 - Rel_{nor}) + 0.1 * Dis_{nor}$$

6.3   DDDS-Based Dynamic Service Scheduling in a Cloud Environment

The advantages of simulation technologies in dynamic scheduling are reflected in predicting future scheduling performance and eventualities to guide the current scheduling decision-making. Traditional simulations are largely decoupled from real simulation systems due to the limited availability of real-time data of simulation systems in the past. With the recent development of sensors and IoT-based technologies [26]. Kuck et al. [27] proposed a DDDS-based optimization heuristic for the dynamic shop floor scheduling problem. Keller and Hu [28] presented a data-driven simulation modeling method for mobile agent-based systems.

To solve the dynamic simulation scheduling problem in a cloud environment, a Dynamic Data-Driven Simulation (DDDS)-based dynamic service scheduling method (DSS) is proposed to address the randomly arrived tasks and unpredicted service time in the dynamic environment [29]. In this section, the DSS method is introduced from the aspects of system framework, DDDS strategy, and scheduling rules. Like Sect. 6.2.3, we also use the manufacturing industry as an example to illustrate the method.

6.3.1   System Framework

The system framework of DSS is presented in Fig. 6.4. There are three main roles in the framework: service demanders, service providers, and simulation platform. Service demanders submit task requirements to the simulation platform, and the simulation platform applies a service scheduling system to generate task schedules based on real-time task information, real-time service information, scheduling rulesScheduling rules, and optimization objectives. Service providers receive allocated subtasks and execute these subtasks at a specific time according to task schedules. The completed products/parts are then delivered from selected service providers to service demanders through logistics.

(1)   Task processing

Demanders' fuzzy and qualitative requirements are transformed into determinate subtask sequences and quantitative requirements by task processing modules. The information of tasks is then stored in the task Database (task DB) and matched to services through service scheduling system.

(2)   Service scheduling

The service scheduling system generates real-time subtask schedules based on the input information and specific scheduling rules. The input information of the service scheduling system includes real-time subtask information and real-time service information. There are two types of scheduling rules in the service scheduling system:

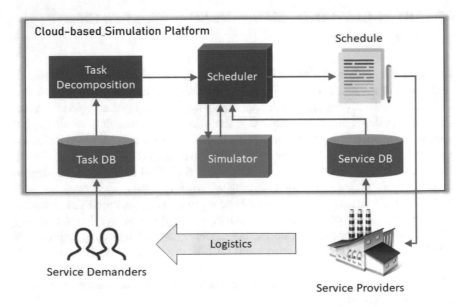

Fig. 6.4 System framework of the proposed DSS method

subtask ranking rules and service selection rules. The generated real-time schedule stipulates the selected providers, services, and start times of all triggered subtasks.

(3) Simulator call

Due to the stochasticity of service time, it is hard to guarantee that the service time in the preset simulation model is equal to the actual service time in real simulation systems. But the probabilistic model can be applied to describe the service time to make the generated service time in the simulation process statistically closer to the actual situation. The expected value of service time is obtained based on historical data of service time. Normal distribution is used to describe the simulation model of service time.

On one hand, when the deviation between the actual service time and the expected service time is greater than a certain threshold, the DDDS simulator is invoked to simulate different scheduling rules. On the other hand, the platform operator can switch among different scheduling optimization objectives. When the scheduling optimization objective is changed, the DDDS simulator is also triggered to test the scheduling performance of different scheduling strategies for the current optimization objective. Therefore, the scheduling strategies in the service scheduling system are optimized through the DDDS simulation, which is based on the real-time system states and the current optimization objective.

6.3.2   Scheduling Rules

There are two kinds of scheduling rules in the proposed method including subtask priority rules and service selection rules. In the subtask priority rules, task arrival time, task due date, and subtask serial number are considered to rank the simultaneously triggered subtasks. When several subtasks compete for a single service, the task arrival times of these triggered subtasks are compared and the scheduling program selects subtasks with the earliest task arrival time. If the task arrival times of these subtasks are equal, the scheduling program compares the due dates and select subtasks with the earliest due date. If the task due dates of these subtasks are equal, the scheduling program compares the subtask serial numbers and select subtasks with minimum serial numbers. If the task arrival times, task due dates, and subtask serial numbers of these subtasks are equal, the program randomly selects a triggered subtask.

Function values of different scheduling rules are calculated based on service time, logistics time, and queue time of candidate services. Three single rules MS (Minimum Service Time), ML (Minimum Logistics Time), and MQ (Minimum Queue Time), and three combined rules MLS (Minimum Logistics and Service Time), MMQL (Minimum Maximum Queue and Logistics Time), and MSQL (Minimum Service, Queue and Logistics Time) are proposed as service selection strategies. In MS, only service time is considered as the optimization criterion, and the function value of $S_{i,j}$ is $e_{i,j,k}$ where k is the type of subtask.

Figure 6.5 presents an example of these scheduling rules in DSS. There are three service providers P_1, P_2, and P_3. P_1 provides services $S_{1,1}$ and $S_{1,2}$. P_2 provides services $S_{2,1}$ and $S_{2,2}$. P_3 only provides service $S_{3,1}$. As shown in Fig. 6.5a), subtasks $I_{1,1}$, $I_{2,2}$, and $I_{3,1}$ are completed at time t_1. Subtasks $I_{1,2}$, $I_{2,3}$, and $I_{3,2}$ are, therefore, triggered at time t_1. At this moment, the current subtask ranking rule is applied to

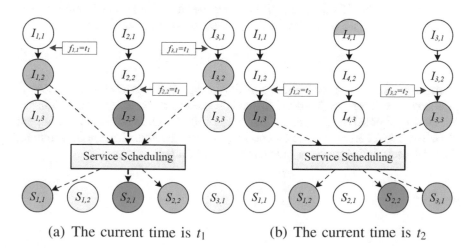

(a) The current time is t_1 (b) The current time is t_2

Fig. 6.5 An example of the scheduling rules in DSS method

rank these triggered subtasks. Task arrival time a_i, task due date c_i, and task priority p_i are considered in the subtask ranking rules of DSS. Assume that $I_{2,3}$ is selected as the next subtask to be scheduled and the candidate services of $I_{2,3}$ are $S_{1,1}$, $S_{2,1}$, and $S_{2,2}$. Then the current service selection rule is applied to select an optimal service from these candidate services to execute $I_{2,3}$. As shown in Fig. 6.5b, subtasks $I_{1,2}$ and $I_{3,2}$ are completed at time t_2, and subtasks $I_{1,3}$ and $I_{3,3}$ are triggered at this moment. And then subtasks $I_{1,3}$ and $S_{2,2}$ are selected by subtask ranking rules and service selection rules, respectively.

The MS rule selects the service with the minimum service time to execute the current subtask $I_{2,3}$. The ML rule selects the service with minimum logistics time from service $S_{1,2}$. The total remaining service time of service $S_{i,j}$ is considered by the MQ rule. The total remaining service time of service $S_{i,j}$ is equal to the sum of the total service time of the subtasks in the queue of $S_{i,j}$ and the remaining service time of service $S_{i,j}$ at time t_1. The MQL rule takes both the logistics time of the candidate service and the service time of the triggered subtask into account. The subtask queue of service $S_{i,j}$ continues executing on $S_{i,j}$ after $I_{2,3}$ selects $S_{i,j}$ and moves by logistics. Therefore, the MMQL rule selects the candidate service with minimum max(logistics time, queue time). In MSQL, the service time, logistics time, and queue time of candidate services are considered to select the optimal service. Compared with the above five service selection rules, MSQL is of higher computational complexity.

6.3.3 DEVS Modeling

The DEVS formalism provides a sound and practical foundation for the architecture of model engineering and simulation environment [30]. A DEVS atomic model specification M is presented as Eq. (6.5), the four functions δ_{ext}, δ_{int}, λ, and ta specify the behavior of the DEVS model, also defines the state of all possible input values X, the set of all possible output values Y, and the set of all possible states S.

$$M = \langle X, Y, S, \delta_{ext}, \delta_{int}, \lambda, ta \rangle \qquad (6.5)$$

An example of a Task-Service scheduling problem modeled with DEVS is presented here. There are 3 tasks published by a user. Each task needs to be finished with one type of service. There are 2 services with the same function but different QoS. The 3 tasks are needed to be coupled to finish a composite task in minimum time. This scheduling problem can be modeled as follows:

Service Model

$$service = < X, Y, S, \delta_{ext}, \delta_{in}, \lambda, ta >$$
$$X = Task$$
$$Y = Task$$
$$S = \{"idle", "sendJob"\} \times R_0^+ \times Task$$
$$\delta_{ext}\{"idle", \delta , ifdo, e, storedJob\}=$$

$$\left\{ \begin{array}{ll} ("idle", \infty, storedJob) & !ifdo||ability! = storedJob.typeNow \\ ("sendJob", storedJob.serviceTime, storedJob) & !ifdo||ability == storedJob.typeNow \end{array} \right.$$

$$\delta_{int}\{"sendJob", \delta, ifdo, storedJob\}=("idle", \infty, storedJob)$$
$$\delta_{ext}\{"sendJob", \delta, ifdo, e, storedJob\}=("sendJob", \delta - e, storedJob)$$
$$\lambda("sendJob", \delta, ifdo, storedJob)=storedJob$$
$$ta(phase, \delta, storedJob)=\delta$$

where Task is the data type of a task, *stroedJob* denotes the received task, *ifdo* is a Boolean variable that denotes if the service is available or not, *ability* denotes the service type. δ is the total duration of the current state, e is the elapsed time of the current state.

Task Model

$$TaskGenerate = < X, Y, S, \delta_{ext}, \delta_{in}, \lambda, ta >$$
$$X = R$$
$$Y = Task$$
$$S = \{"Passive", "Generate"\} \times R_0^+ \times R$$
$$\delta_{int}\{"Generate", \delta, task, count\}=("Generate", \delta, task, count + 1)$$
$$\lambda("Generate", \delta, task, count)=task$$
$$ta(phase)=\delta$$

where *count* denotes the number of tasks.

Scheduling Model

$Scheduling = < X, Y, S, \delta_{ext}, \delta_{in}, \lambda, ta >$

$X = Task$

$Y = Task$

$S = \{"wait", "send"\} \times R_0^+ \times Task$

$\delta_{ext}\{"wait", \delta, storedTask, e, clock\}=("wait", \delta, storedTask, clock +e)$

$\delta_{int}\{"send", \delta, storedTask, clock\}=("wait", \infty, storedJob, clock)$

$\lambda("send", \delta, storedTask, clock)=storedTask$

$ta(phase)=\delta$

Provider Model

$Provider = < X, Y, D, \{M_d | d \in D\}, EIC, EOC, IC >$

$InPorts = \{"inJob"\}$

$X_{in} = Task$

$X = \{("inJob", v) | v \in Task\}$

$OutPorts = \{"outJob"\}$

$Y_{in} = Task$

$Y = \{("outJob", v) | v \in Task\}$

$D = \{Service1, Service2, Service3\}$

   $Service1 = Service2 = Service3=Service$

EIC

$= \{((N, inJOb), (Service1, inJob)), ((N, inJOb), (Service2, inJob)), ((N, inJOb), (Service3, inJob))\}$

   $EOC = \{((Service1, outJOb), (N, outJob)), ((Service2, outJOb), (N, outJob)),$
   $((Service3, outJOb), (N, outJob))\}$

      $IC = \{\}$

where Service is the Atomic model.

System Model

$$\text{System} = \ < X, Y, D, \{M_d | d \in D\}, EIC, EOC, IC >$$
$$\text{InPorts} = \{\}$$
$$X_{in} = \ \phi$$
$$X = \{("inJob", v) | v \in Task\}$$
$$\text{OutPorts} = \{\}$$
$$Y_{in} = \ \phi$$
$$Y = \{("outJob", v) | v \in Task\}$$
$$D = \{taskGenerate, scheduling, provider1, provider2\}$$

taskGenerate = TaskGenerate, scheduling = Scheduling, provider1 = provider2
         = Provider

$$EIC = \{\}$$
$$EOC = \{\}$$
IC =
{((taskGenerate, outTaskque),(scheduling,inTaskque)),((scheduling,outJob),(provider1,inJob))
,((provider1,outJob),(scheduling,inTaskque)),
((scheduling,outJob),(provider2,inJob)),((provider2,outJob),(scheduling,inTaskque))}

where TaskGenerate is an atomic model, *taskgenerate* is the instance of TaskGenerate. Scheduling is an atomic model, *scheduling* is the instance of Scheduling. Provider is an atomic model, *provider1* and *provider2* are instances of Provider.

The simulation results with MS4 Me is shown in Fig. 6.6.

This is just one example of a discrete system that can be modeled using the DEVS formalism. By incorporating time into every state transition, DEVS can be used to represent nearly any time-varying system. Due to its generality, DEVS can be considered foundational to modeling and simulation in much the same way that state machines are foundational to conventional software development.

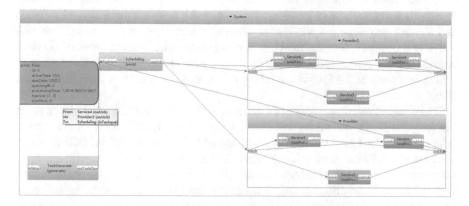

Fig. 6.6 Simulation result of the Task-Service scheduling problem with MS4 Me

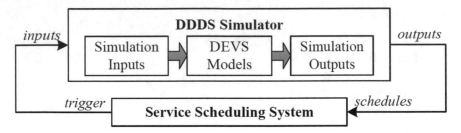

Fig. 6.7 The real-time interaction between the DDDS simulator and the service scheduling system

6.3.4 DDDS Strategies

The interaction between the DDDS simulator and the service scheduling system in DSS is a loop, as shown in Fig. 6.7. When trigger events occur, the service scheduling system sends the real-time service information to the DDDS simulator and makes a simulation call. The DDDS simulator runs the DEVS simulations and then sends the evaluation results of different scheduling strategies to the service scheduling system.

The input information of the DDDS simulator includes real-time task information, real-time service information, the optimization objective, and the simulation period. The real-time information of the task includes the arrival time, due date, task priority, subtask numbers, completion time of completed subtasks, and the serial number of the subtask under execution. The real-time information of service includes the service busy/idle status and the remaining service time. The optimization objective is given by the platform operator to reflect the target of the current platform. The simulation period is set to be the total simulation time of the DEVS simulations.

The DEVS modeling method is applied in the DDDS simulator to build the model of the task scheduling environment. The task scheduling processes of multiple distributed service providers are then simulated by the DDDS simulator. A DEVS simulation model includes tasks, service providers, services, and scheduling rules. The attributes of tasks include task arrival time, task types, task due date, task priority, and subtask sequence length. The attributes of service providers include service provider number N, service numbers, and logistics time. The attributes of services include service types, service capacity, and service time. There are two types of scheduling rules in the DEVS models: subtask ranking rules and service selection rules.

The detailed running process of DEVS models is shown in Fig. 6.8. After being invoked, the DDDS simulator will initialize the DEVS model based on the current system status. Different scheduling rules are added to this initialized DEVS model to construct different simulation models. These simulation models with different scheduling rules are then separately simulated for the same length of time. The simulation results of different DEVS models under the current scheduling objective are then compared after these simulations. Finally, the optimal scheduling rules under the current system status are obtained and sent to the service scheduling system.

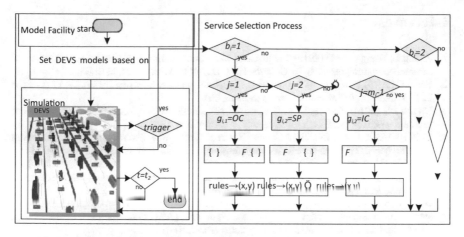

Fig. 6.8 Detail running process of DEVS models

Initial results in some examples were encouraging, nevertheless, further research needs to be done to confirm the feasibility and advantages of the proposed method.

6.4 Conclusions

In this work, we investigated the problem of service composition and scheduling in the cloud-based simulation environment. A new service network-based service composition and scheduling model is proposed. Then a dynamic service scheduling method based on dynamic data-driven simulation is proposed to select better scheduling strategies based on real-time information of service status and task execution.

Composition and scheduling of simulation services in cloud environments are very challenging. There are many research topics need to be studied, such as credibility evaluation of composite services, adaptivity to dynamic changes and uncertainty of services, composition and scheduling with more constraints, composition and scheduling considering time synchronization of services.

References

1. Zhang L, Wang F, Li F (2019) Cloud-based simulation, summer of simulation—50 years of seminal computing research. In: Sokolowski J, Durak U, Mustafee N, Talk A (eds). Springer
2. Erdal C (2013) Modeling and simulation as a cloud service: a survey. In: Proceedings of the 2013 Winter simulation conference, Savannah, GA
3. Onggo BS (2014) The need for cloud-based simulation from the perspective of simulation practitioners. In: Operational research society simulation workshop
4. Li BH, Chia XD, Hou BC, Tan L, Zhang YB et al (2009) Networked modeling & simulation platform based on concept of cloud computing—cloud simulation platform. J Syst Simul
5. Arabnejad H, Barbosa J (2012) Fairness resource sharing for dynamic workflow scheduling on heterogeneous systems. In: 2012 IEEE 10th international symposium on parallel and distributed processing with applications (ISPA), pp 633–639
6. Hönig U, Schiffmann W (2006) A meta-algorithm for scheduling multiple dags in homogeneous system environments. In: Proceedings of the eighteenth IASTED international conference on parallel and distributed computing and systems (PDCS'06)
7. Barbosa JG, Moreira B (2011) Dynamic scheduling of a batch of parallel task jobs on heterogeneous clusters. Parallel Comput 37:428–438
8. Bittencourt LF, Madeira ER (2010) Towards the scheduling of multiple workflows on computational grids. J Grid Comput 8:419–441
9. Ibarra OH, Kim CE (1977) Heuristic algorithms for scheduling independent tasks on nonidentical processors. J Assoc Comput Mach 24(2):280–289
10. Maheswaran M, Ali S, Siegel HJ, Hensgen D, Freund RF (1999) Dynamic mapping of a class of independent tasks onto heterogeneous computing systems. J Parallel Distrib Comput 59(2):107–131
11. Yao W, Li B, You J (2002) Genetic scheduling on minimal processing elements in the grid. In: Australian joint conference on artificial intelligence, pp 465–476
12. Martino VD, Mililotti M (2002) Scheduling in a grid computing environment using genetic algorithms. In: IPDPS, 0235
13. Arockiam L, Sasikaladevi N (2012) Simulated annealing versus genetic based service selection algorithms. Int J u- e-Serv Sci Technol 5:35–50
14. Li H, Wang L, Liu J (2010) Task scheduling of computational grid based on particle swarm algorithm. In: 2010 third international joint conference on computational science and optimization (CSO), pp 332–336
15. Li K, Xu G, Zhao G, Dong Y, Wang D (2011) Cloud task scheduling based on load balancing ant colony optimization. In: 2011 Sixth annual Chinagrid conference (ChinaGrid), pp 3–9
16. Li F, Zhang L, Liu YK, Laili YJ, Tao F (2017) A clustering network-based approach to service composition in cloud manufacturing. Int J Comput Integr Manuf 1–12
17. Meketon MS (1987) Optimization in simulation: a survey of recent results. In: Proceedings of the 19th conference on Winter simulation, pp 58–67
18. Fu MC, Andradóttir S, Carson JS, Glover F, Harrell CR, Ho Y-C et al (2000) Integrating optimization and simulation: research and practice. In: Proceedings of the 32nd conference on Winter simulation, pp 610–616
19. Tekin E, Sabuncuoglu I (2004) Simulation optimization: a comprehensive review on theory and applications. IIE Trans 36:1067–1081
20. Hong LJ, Nelson BL (2009) A brief introduction to optimization via simulation. In: Proceedings of the 2009 Winter simulation conference (WSC), pp 75–85
21. Fu MC (2002) Optimization for simulation: theory vs. practice. Inform J Comput 14:192–215
22. Zhou LF, Zhang L, Sarkerc BR, Laili YJ, Ren L (2018) An event-triggered dynamic scheduling method for randomly arriving tasks in cloud manufacturing. Int J Comput Integr Manuf 31(3)
23. Zhou LF, Zhang L, Zhao C, Laili YJ, Xu LD (2018) Diverse task scheduling for individualized requirements in cloud manufacturing. Enterp Inf Syst 12(3)
24. Zhou LF, Zhang L, Laili YJ, Zhao C, Xiao YY (2018) Multi-task scheduling of distributed 3D printing services in cloud manufacturing. Int J Adv Manuf Technol 96(9–12):3003–3017

25. Li F, LaiLi YJ, Zhang L, Hu XL, Zeigler BP (2018) Service composition and scheduling in cloud-based simulation environment. In: SpringSim 2018, 15–18 April, Maryland, Baltimore, USA
26. Cai H, Xu L, Xu B, Xie C, Qin S, Jiang L (2014) IoT-based configurable information service platform for product lifecycle management. IEEE Trans Industr Inf 10(2):1558–1567
27. Kuck M, Ehm J, Hildebrandt T, Freitag M, Frazzon EM (2016) Potential of data-driven simulation-based optimization for adaptive scheduling and control of dynamic manufacturing systems. In: 2016 Winter simulation conference, Washington, DC, pp 2820–2831
28. Keller N, Hu X (2016) Data driven simulation modeling for mobile agent-based systems. In: 2016 symposium on theory of modeling and simulation, Pasadena, CA, pp 1–8
29. Zhou LF, Zhang L, Ren L, Wang J (2019) Real-time scheduling of cloud manufacturing services based on dynamic data-driven simulation. IEEE Trans Industr Inf. https://doi.org/10.1109/TII.2019.2894111 (online published)
30. Zeigler RP Zhang L (2018) Service-oriented model engineering and simulation for system of systems engineering. In: Yilmaz L (ed) Concepts and methodologies for modeling and simulation: a tribute to Tuncer Oren. Springer, pp 19–44

Lin Zhang is a professor of Beihang Unversity, China. He received the B.S. degree in 1986 from Nankai University, China, the M.S. and the Ph.D. degree in 1989 and 1992 from Tsinghua University, China. His research interests include service-oriented modeling and simulation, cloud manufacturing and simulation, model engineering for simulation. He served as the President of the Society for Modeling and Simulation International (SCS), the executive vice president of China Simulation Federation (CSF). He is currently the president of Asian Simulation Federation (ASIASIM), a Fellow of SCS, ASIASIM and CSF, a chief scientist of the National High-Tech R&D Program (863) and National Key R&D Program of China, and associate Editor-in-Chief and associate editors of 6 peer-reviewed international journals. He authored and co-authored more than 200 papers, 10 books and chapters. He received the National Award for Excellent Science and Technology Books in 1999, the Outstanding Individual Award of National High-Tech R&D Program, 2001, the National Excellent Scientific and Technological Workers Awards in 2014.

Feng Li received the Ph.D degree in Automation Science and Electrical Engineering from Beihang University, Beijing, China, in 2019. She is currently a Research Fellow in Computer Science and Engineering, Nanyang Technological University, Singapore. Her research interests include service-oriented manufacturing resource scheduling, cloud simulation, simulation-based optimization.

Longfei Zhou received the B.S. degree in automation engineering from Harbin Engineering University, Harbin, China, in 2012, and received the Ph.D. degree in manufacturing engineering from Beihang University, Beijing, China, in 2018. He is currently a postdoc associate in the Computer Science and Artificial Intelligence Laboratory at Massachusetts Institute of Technology. His research interests include intelligent manufacturing, real-time scheduling, discrete system modeling and simulation, computer vision for transportation and medicine. At MIT, he develops machine learning and computer vision algorithms for multiple object tracking and medicine image analysis. Dr. Zhou serves as one of the chairs of the MIT Postdoc Association (MIT PDA) from 2019 to 2020, and serves in the editorial board of the International Journal of Modeling, Simulation, and Scientific Computing (IJMSSC), and serves as a reviewer for more than 20 academic journals. His awards and honors include the Beijing Municipal Education Commission Outstanding Graduate Award in 2018, and the China Simulation Society Outstanding Doctoral Dissertation Award in 2019.

Bernard Zeigler, RTSync Corp. and Arizona Center for Integrative Modeling and Simulation, Tucson. AZ, is best known for his theoretical work concerning modeling and simulation based on systems theory. Zeigler has received much recognition for his various scholarly publications, achievements, and professional service. His 1984 book, Multifaceted Modelling and Discrete Event Simulation, was published by Academic Press and received the outstanding Publication Award in 1988 from The Institute of Management Sciences (TIMS) College on Simulation. Zeigler is a Fellow of the Institute of Electrical and Electronics Engineers (IEEE) as well as The Society for Modeling and Simulation International (SCS) which he served as President (2002–2004). He is a member of the SCS Hall of Fame. In 2013 he received the Institute for Operations Research and Management Sciences (INFORMS) Simulation Society Distinguished Service Award. In 2015 he received the INFORMSLifetime Professional Achievement Award, which is the highest honor given by the Institute for Operations Research and the Management Sciences' Simulation Society.

Chapter 7
Agent-Directed Simulation and Nature-Inspired Modeling for Cyber-Physical Systems Engineering

Tuncer Ören

Abstract The definition of cyber-physical systems is revised from the point of view of the evolution of physical tools. The desirable synergies and contributions of the following disciplines are elaborated to tackle the complexity of cyber-physical systems: simulation, agent-directed simulation, and systems engineering, including simulation-based cyber-physical systems engineering. Over 80 types of systems engineering are also listed. The richness of paradigms offered by nature-inspired modeling and computing are elaborated on and sources of information, as well as over 60 possibilities offered by nature-inspired modeling, for simulation-based cyber-physical systems engineering are pointed out. Some other important possibilities, such as cloud computation, big data analytics, cyber security, and ethical issues are treated in other chapters of the book and are not elaborated in this chapter.

7.1 Introduction to Cyber-Physical Systems (CPS)

Cyber-Physical Systems (CPS) are physical systems augmented with software to allow them the ability to process knowledge, communicate, and control. The abilities of communication and control justify the term "cyber." However, sometimes the term cyber is used to denote any software component to allow knowledge processing in general and not for communication or control. The physical components can receive inputs mostly through several types of sensors, and in the case of advanced intelligent systems, inputs can also be endogenous, i.e., can be generated internally within the knowledge processing component. The components of CPSs can communicate with their environments, other systems, and/or with humans.

Cyber-physical systems can also be considered as the fourth level of the evolution of physical tools, as shown in Fig. 7.1.

T. Ören (✉)
School of Electrical Engineering and Computer Science, University of Ottawa, Ottawa, Ontario, Canada
e-mail: oren.tuncer@sympatico.ca

© Springer Nature Switzerland AG 2020   143
J. L. Risco Martín et al. (eds.), *Simulation for Cyber-Physical Systems Engineering*,
Simulation Foundations, Methods and Applications,
https://doi.org/10.1007/978-3-030-51909-4_7

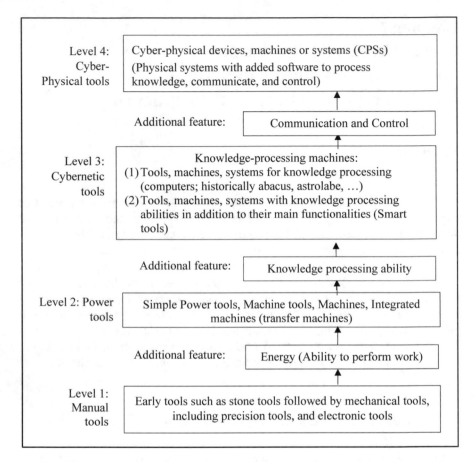

Fig. 7.1 Cyber-physical systems within the evolution of physical tools. Adopted from [44]

As shown in Fig. 7.1, at the first level of tools, there are manual tools which consist of early tools, including different types of stone tools and mechanical tools, including precision tools [27].

By adding an additional feature to the manual tools, such as energy, i.e., the ability to perform work, one passes to the second level of tools or power tools which include simple power tools, machine tools, machines, and integrated machines.

To pass to the third level of tools, namely to cybernetic tools, the needed additional feature is knowledge processing ability. Cybernetic tools are distinguished by being knowledge processing machines or knowledge processing systems. There are two types of knowledge processing machines: (1) Tools, devices, machines, or systems for knowledge processing (currently computers; historically there were other tools for knowledge processing such as the abacus, astrolabe, punched card machines, etc. [42]). (2) Tools, machines, devices, or systems with knowledge processing abilities in addition to their main functionalities. They are smart tools such as smart devices, smart machines, and smart systems.

The addition of communication and control abilities to cybernetic tools define cyber-physical tools or cyber-physical systems. Cyber-physical systems are already widely used for civilian and military applications and they are becoming ubiquitous. A survey article outlines CPS applications in: Smart manufacturing, emergency response, air transportation, critical infrastructure, health care and medicine, intelligent transportation, and robotic for service [17]. The body of knowledge of cyber-physical systems is recently published [56]. When groups of humans are involved, cyber-physical systems are called cyber-social-physical systems.

In the sequel, the following topics are elaborated since they can contribute to the study of cyber physical systems: Simulation, Agent-directed simulation, Nature-inspired modeling and computing, and Systems engineering and cyber-physical systems engineering. In each section, their relevance is pointed out.

7.2 Simulation and Its Increasing Importance

Simulation has two major aspects: experimentation and experience. From the point of view of experimentation, simulation is performing experimentations with dynamic models of a system. From the point of view of experience, simulation is gaining/enhancing three types of skills through the use of models or representations of real systems. These skills and corresponding simulations are: (1) motor skills (virtual simulation, since virtual equipment is used), (2) communication and decision-making skills (constructive simulation such as war simulation, peace simulation), and (3) operational skills (live simulations). Simulation has been maturing. Nine aspects of the evolution of simulation are documented by Durak et al. [12]. Already several disciplines benefit from simulation-based approaches [32, 46]. Several aspects of the simulation are well documented. For example, for its scientific basis see Zeigler et al. [80] and Ören and Zeigler [51]. A historic view of simulation is well documented by Sokolowski et al. [65]. Tolk covers engineering principles of combat modeling and distributed simulation [67].

Cyber-physical system studies can benefit from simulation in various ways. Simulated experiments can allow detection of any type of flaw in cyber-physical systems. Since extensive simulated experiments can be performed under various conditions including extreme conditions. Some complex cyber-physical systems may require human operators. Simulation would allow future operators to gaining/enhancing skills to operate such complex systems. For autonomic control of CPSs, simulation can be used for extensive tests of control software.

In this section, two aspects of simulation, namely inputs and coupling of component models are mentioned, due to their relevance to the theme of this chapter.

Table 7.1 Externally generated (exogenous) inputs (Adopted from [43, 47])

Exogenous Inputs	
Mode	Type
Imposed input (Passive acceptance of exogenous input)	**Access** to input: – Direct input, input from coupling, argument passing, knowledge in a common area (blackboard), message passing, broadcasting (to all, to a fixed or varying group, to an entity) **Nature** of input: – **Information**: Data, facts, events, goals – **Sensation** (from transducers) (Converted sensory data (Table 7.3) from analog to digital—single or multi sensor—sensor fusion)
Perceived input (**Active** perception of exogenous input)	**Perception process** includes Noticing, recognition, decoding, selection (filtering), regulation **Nature** of input: – **Interpreted sensory input** data (Table 7.3) and selected events – **Infochemicals** (Table 7.4) (chemical messages/chemical messengers for chemical communication) • Sources: animate, inanimate – **Infotraces**—traces of information transactions among: • Interconnected infohabitants of Internet of things Users of media and search engines

7.2.1 Inputs

There are two categories of inputs: externally generated (or exogenous) inputs and internally generated (or endogenous) inputs. Internally generated inputs and evaluated externally generated inputs are especially important for intelligent systems and can be useful for the autonomic management of complex cyber-physical systems. This concept was first introduced in 2001 [43]. Tables 7.1 and 7.2 (adopted from [43, 47]) outline externally generated and internally generated inputs.

7.2.2 Coupling

Coupling allows model composition by specifying input/output relationships of component models. The concept of coupling is already well-documented both for DEVS (Discrete Event Modeling System Specification) formalism and for GEST (General System Theory implementor) language. For example, Zeigler et al. [81] and Seck and Honig [61], clarified different aspects of coupling of DEVS models.

The concept of simulation model coupling was introduced with the GEST language [39, 41, 49, 51]. In an early article, time-varying coupling was also introduced where either input/output relations and/or some component models (or coupled

Table 7.2 Internally generated (endogenous) inputs (Adopted from [43, 47])

Endogenous (Internally-generated) Inputs (for intelligent/smart systems)	
Mode of input	Type of input
Perceived endogenous input	**Introspection**
	Perceived (cogitated) internal facts, events; or realization of lack of them
Anticipated/deliberated endogenous input	**Anticipation**: **Anticipated facts** and/or **events** (behaviorally anticipatory systems) **Deliberation** of past facts and/or events (deliberative systems) **Generation** of goals, questions and hypotheses by· Expectation-driven reasoning (Forward reasoning, or Bottom up reasoning, or Data-driven reasoning) – Model-driven reasoning

Table 7.3 Types of sensations from sensors and transducers (Adopted from [49])

Type of stimulus	Type of perception
Light	– **Vision** (visual perception): visible light vision, ultraviolet vision, infrared vision and vision at other wavelengths
Sound	– **Hearing** (auditory sensing): audible/infrasonic/ultrasonic sound (medical ultrasonography, fathometry: sonic depth finding)
Chemical	– (**Gas** sensing/detection): smell (smoke/CO_2), humidity sensor – (**Solid**, **fluid** sensing): taste, microanalysis
Heat	– **Heat** sensing (thermal image input)
Magnetism	– **Magnetism** sensing: geomagnetism, thermo-magnetism sensing, electrical field sensing, radio frequency identification
Touch	– Sensing **surface characteristics**
Motion	**Acceleration** sensing (fall detection)
Vibration	– **Vibration** sensing: seismic sensor
Thought	– **Brain**-controlled technology

models) may change during runtime of simulation studies [40]. Currently, agent-monitored simulation aspect of agent-directed simulation can allow the implementation of this powerful concept. Nested couplings, where some component models in a coupling can be resultant of other couplings, are useful for the composition of systems of systems. Ören and Yilmaz [49] elaborated on nature-inspired modeling in model coupling and listed over 90 types of model couplings.

Several aspects of model coupling can be very useful for the studies of cyber-physical systems: *Couplings* allow model composition. In the *hierarchical model coupling*, some of the component models are already coupled models. Hierarchical model couplings allow modeling of systems of systems. *Time-varying couplings* allow modeling of varying input/output relationships, as well as dynamically changing component models.

Table 7.4 Types of infochemicals for chemical communication (Adopted from [49])

Infochemicals (*Chemical messengers* for chemical communication)

Nature		*Hormones* (Interactions are within a living organism between different organs or tissues)		
		Interactions are:	Intra-specific (**same species**)	*Pheromones* (ant-communication) (In early literature *Ectohormones*)
Animate	Semio-chemicals		Inter-specific (**different species**)	*Allelochemics* (allomones, antimones, kairomones, synomones) (Table 7.5)
	Inanimate	*Apneumones*		

Table 7.5 Types of some allelochemicals, based on their affection characteristics (Adopted from [3, 49])

Signal to the benefit of		Receiver	
		Yes	No
Sender	Yes	Synomones (e.g., floral sent, pollinator)	Allomones (defense secretion, repellant; e.g., venom of snake for a person)
	No	Kairomones (e.g., a parasite seeking a host)	Antimones (e.g., chemicals of a pathogene/host)

7.3   Agent-Directed Simulation (ADS)

Agent-directed simulation considers two categories of synergies of simulation and software agents. In addition to agent simulation or agent-based simulation (i.e., simulation of systems modeled by agents), the power of agent-supported simulation and agent-monitored simulation are very important possibilities for advanced simulation environments [77–79]. Table 7.6 outlines the relationships between three types of agent-related simulation.

Agent-monitored simulation [45], where agents can monitor simulation runs is appropriate for implementing time-varying couplings where some component models and/or their input/output relationships may change under specified conditions.

Table 7.6 Agent-Directed Simulation—(ADS) (Synergies of simulation and software agents) [44]

	Contribution of:		Types of simulation
Agent-Directed Simulation—(ADS) (Synergies of Simulation and Software Agents)	Contribution of simulation to agents		**Agent simulation** (Commonly called agent-based simulation) – Simulation of agent systems or simulation with agent-based models
	Contribution of agents to simulation	Agents as support facilities	**Agent-supported simulation** – Agent support for user/system interfaces (both front-end for formulation of the problem; and back-end for presentation and explaining the results)
		Agents as run-time monitoring facilities	**Agent-monitored simulation** – Model behavior generation – Agent monitoring during run-time (including time-varying coupling)

All three aspects of agent-directed simulation are appropriate for complex CPSs: It would be appropriate to model some components of a CPS by an agent. Agent-monitored simulation may allow agent monitoring some components or the performance of the whole system. Furthermore, agents may be used in modeling both front-end and back-end simulation interfaces.

7.4 Nature-Inspired Modeling and Computing

Nature-inspired modeling (NIM) provides a very rich paradigm for simulation modeling and computation, especially, to empower tools for simulation-based cyber-physical systems engineering which deals with complex problems of cyber-physical systems. As an everyday example of nature-inspired modeling, let's consider the light which travels on a straight line between two points A and B. However, as shown in Fig. 7.2, if the points A and B are on different media, light travels from A to C and from C to B, in such a way that, after the refraction, time to travel from A to B is kept to a minimum; as light travels longer distance in a medium where it's speed is faster. This model can be emulated, for example, to minimize total time or total risk in robot displacement.

The sources for nature-inspired models can be from inanimate or animate entities. The latter includes, humans, animals, and plants as outlined in Fig. 7.3.

Two related terms with nature-inspired models and nature-inspired modeling are biomimicry and biomimetics. *Biomimicry* is the "imitation of natural biological designs or processes in engineering or invention". *Biomimetics* is "the study of the formation, structure, or function of biologically produced substances and materials (such as enzymes or silk) and biological mechanisms and processes (such as protein synthesis or photosynthesis) especially for the purpose of synthesizing similar products by artificial mechanisms which mimic natural ones—called also biomimicry [29]."

Fig. 7.2 Path of light between two points (One medium: A–B, two media: A–C–B)

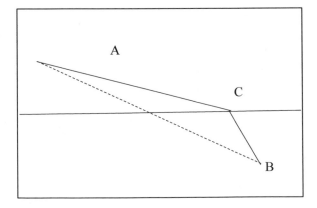

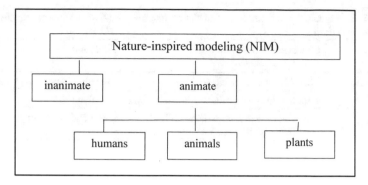

Fig. 7.3 Sources for nature-inspired models

7.4.1 Sources of Information

Several categories of nature-inspired models and nature-inspired computing are already widely available. There are several sources of information about nature-inspired models and nature-inspired computing as outlined in Tables 7.7 and 7.8. Table 7.7 displays some specialized disciplines and centers/institutes and Table 7.8 displays some conferences, lecture series, publications, and history.

7.4.2 Categories of Nature-Inspired Models

In the sequel, over 60 types of nature-inspired engineered product are outlined in Tables 7.9, 7.10, 7.11, 7.12, 7.13, and 7.14. They are in the following 15 categories: Agriculture, Architecture, Construction, and Cities, Communication, Decision-making, Energy, Locomotion/Transport, Materials, Optics, Optimization, Resource management, Security/cyber security, Sensors, Surfaces, Tools, and Tracking. Additional focus points are possible. For example, Wyss Institute for biologically-inspired engineering have 9 major focus areas: (1) Adaptive Material Technologies, (2) Bioinspired Soft Robotics, (3) Bioinspired Therapeutics and Diagnostics, (4) Diagnostics Accelerator, (5) Immuno-Materials, (6) Living Cellular Devices, (7) Molecular Robotics, (8) 3D Organ Engineering, and (9) Synthetic Biology [75]. Due to the richness of the concepts, other focus areas can be found in other resources (and especially from nature).

Nature-inspired modeling and nature-inspired computing allow very rich paradigms already for many disciplines such as NIE (Nature-inspired engineering), BIE (Biologically-inspired engineering), NISE (Nature-inspired surface engineering), and NICE (Nature-Inspired Chemical Engineering). As outlined in this section, there are already many good sources of information about NIM and NIC. Exploration of NIM and NIC for complex CPSs appear to be very promising.

Table 7.7 Some sources of information about Nature-Inspired (NI) Models: *Specialized Disciplines and Centers/Institutes*

	Specialized Disciplines and Centers/Institutes	References
Specialized disciplines	*Art*—Nature-Inspired Art	Painting, Sculpture, Literature
	Architecture—Biomimicry in Architecture	Pawlyn [53]
	Artificial Intelligence—Bio-Inspired Artificial Intelligence	Floreano and Mattiussi [14]
	Biomedical Engineering—Nature-Inspired Intelligent Techniques for Solving Biomedical Engineering Problems	Kose, U., G. Emre, and O. Deperlioglu (eds.) [21]. https://www.igi-global.com/book/nature-inspired-intelligent-techniques-solving/185481
	Chemical Engineering—Nature-Inspired Chemical Engineering (NICE)	[54] https://royalsocietypublishing.org/doi/10.1098/rsta.2018.0268
	Engineering	BIE—Biologically-Inspired Engineering [6] https://wyss.harvard.edu/
		NIE—Nature-Inspired Engineering
	Surface Engineering—NISE—Nature-Inspired Surface Engineering	http://ameriscience.org/nise-2019/
	Transportation—Biomimicry and Transportation	https://www.theexplorationplace.com/galleries-3/galleries/2019-winter-disheveling-a-hair-raising-exhibition
Centers/Institutes	BIO MIMICRY IBERIA	https://biomimicryiberia.com/en/about-us/
	BIOMIMICRY Institute	https://biomimicry.org/
	NIFTI—(USA) The Air Force Center of Excellence on Nature-Inspired Flight Technologies and Ideas [35]	http://nifti.washington.edu/
	Wyss—Institute for Biologically Inspired Engineering	https://wyss.harvard.edu/

In agent-directed simulation for cyber-physical systems engineering [25], NIM and NIC can be used to: (1) model some component models, (2) specify inputs to the systems, and (3) specify the input/output relationships, or couplings of component models. Furthermore, agent-monitored simulation can allow run-time monitoring of simulation models to allow time-varying couplings of component models.

Table 7.8 Some sources of information about Nature-inspired (NI) Models: *Conferences, Lecture Series, Publications, and History*

	Sources of information	References
Conferences, Lecture Series	Biomimicry Lecture Series	Alberta, Canada and Argentina https://biomimicryalberta.com/
	BioSTAR 2018—3rd Workshop on Bio-inspired Security, Trust, Assurance and resilience	2018 May 24 San Francisco, CA http://biostar.cybersecurity.bio/
	NIE—Nature-Inspired Engineering [34]	2019 Sept. 8–13, Calabria, Italy
	NISE-2019, The 1st International Conference on Nature Inspired Surface Engineering	2019 June 11–14, The Stevens Institute of Technology, New Jersey, USA http://ameriscience.org/nise-2019/
Books	Nature-Inspired Cyber Security and Resiliency	El-Alfy et al. [13]
	A Comparison of Bio-Inspired Approaches for the Cluster-Head Selection Problem in WSN (wireless sensor network)	Miranda et al. [30]
	Nature-Inspired Cyber Security and Resiliency	Olariu and Zomaya [38]
	Advances in Nature-Inspired Computing and Applications	Shandilya et al. [63]
Journals	IJBIC—International Journal of Bio-Inspired Computing (in 2019, vol, 14)	https://www.inderscience.com/jhome.php?jcode=ijbic
Publication list	CfNIE—Centre for Nature Inspired Engineering	https://www.natureinspiredenginee ring.org.uk/publications
History	Biomimicry: A History By W. Schreiner [59]	Dept. of History, OSU https://ehistory.osu.edu/exhibitions/biomimicry-a-history

7.5 Systems Engineering and Cyber-Physical Systems Engineering

The Systems Engineering Body of Knowledge [60] provides the following definitions:

A **system** *is a collection of elements and a collection of inter-relationships amongst the elements such that they can be viewed as a bounded whole relative to the elements around them. Open Systems exist in an environment described by related systems, with which they may interact and conditions to which they may respond. While there are many definitions of the word "system," the SEBoK authors believe that this definition encompasses most of those which are relevant to SE [Systems Engineering].*

Table 7.9 A list of Nature-Inspired Models for: *Agriculture, Architecture, Construction, and Cities*

Inspiration		Engineered product	References
Type	From		
Agriculture	Prairies	To grow food in resilient ways	Biomimicry Inst. [7]
Architecture, Construction, and Cities	Bird's nest	Beijing national stadium	Perera and Coppens [54]
	Flower, bird-of-paradise	Hingeless mechanism: Flectofin	Lienhard et al. [24], Wikipedia-b [73]
	Goldfish	Olympic pavilion in Barcelona	Perera and Coppens [54]
	Leaves	Gathering energy	Benyus [4]
	Lotus Flower	Lotus temple in India	Perera and Coppens [54]
	Phyllotaxy (arrangement of leaves)	Better solar power collection	Wikipedia-b [73]
	Plant, carnivorous	Hingeless mechanism: Flectofold	Wikipedia-b [73]
	Self-diagnosis and self-repair	Tensegrity modules (for bridges)	Rhode-Barbarigos [57], Wikipedia-b [73]
	Termites	Eastgate development in Zimbabwe	Perera and Coppens [54]
		Natural ventilation system	Adams [1]; Biomimicry Inst. [7]
	Biotic analogies	Self-organizing cities	Narraway et al. [33]; CNIE (news) [8]

> **An engineered system** *is an open system of technical or sociotechnical elements that exhibits emergent properties not exhibited by its individual elements. It is created by and for people; has a purpose, with multiple views; satisfies key stakeholders' value propositions; has a life cycle and evolution dynamics; has a boundary and an external environment; and is part of a system-of-interest hierarchy.*

> **Systems engineering** *is "an interdisciplinary approach and means to enable the realization of successful (engineered) systems". It focuses on holistically and concurrently understanding stakeholder needs; exploring opportunities; documenting requirements; and synthesizing, verifying, validating, and evolving solutions while considering the complete problem, from system concept exploration through system disposal.*

Systems engineering [19] which became model-based systems engineering soon after its inception has been specialized in over 80 areas, including cyber-physical systems engineering [16, 58], and can benefit from a shift of paradigm from model-based systems engineering to simulation-based systems engineering [68].

Due to its universality, there are several types of systems engineering as shown in Tables 7.15 and 7.16. In Table 7.15, over 60 types of systems engineering as are listed by application are. Table 7.16 includes 20 domain-independent types of systems engineering.

Cyber-physical systems can benefit from the power of systems engineering. Otherwise, complex CPSs may fail. (See the last chapter of this book).

Table 7.10 A list of Nature-Inspired Models for: *Communications; Decision-making; and Energy*

Inspiration		Engineered product	Reference
Type	From		
Communication	Biology-based algorithms	Wireless networks	Hatti and Sutagundar [18]
	Dolphins	to send underwater signals (for tsunami early warning systems)	Biomimicry Inst. [7]
	Swarm intelligence	Swarms of robots	Dorigo and Birattari [11]
	Swarm theory	Self managed air traffic flow management	Turuu [69]
Decision-making	Neurons	Artificial neural networks	Dash et al. [10]
Energy	Butterflies	Solar power	Alexander [2]
	Lung	Fuel cell	Bethapudi et al. [5] CNIE (news) [8]
	Moth's eye	To reduce the reflectivity of solar panels	Wikipedia-b [73]
	Whale fins	Wing turbine blades	Smithers [64] Biomimicry Inst. [7]

Table 7.11 A list of Nature-Inspired Models for: *Locomotion/Transportation*

Inspiration from	Engineered product	References
Bat echolocation	SONAR (Sound Navigation and Ranging)	Smithers [64]
Gecko's toes	Climbing materials for humans	Alexander [2]
Kingfisher bills	Shinkansen bullet train	Smithers [64]; Biomimicry Inst. [7]
Legs (2, 4, 6, 8)	Legged locomotion	
Snake	Locomotion: Snake robot	Koopaee et al. [20]
Wings, tips of eagle	Curved tips on the ends of aeroplane wings	Smithers [64]

7.6 Conclusions

As of 2019, references of nature-inspired simulation is extremely limited. A very few, though important references are:

Table 7.12 A list of Nature-Inspired Models for: *Materials* (including adaptive or biomimetic materials, and materials for biomedical applications)

Inspiration from	Engineered product	References
Biomineralization (process by which living organisms produce minerals)	Biomorphic mineralization (technique that produces materials with morphologies and structures resembling those of natural living organisms)	Wikipedia-b [73]
Helicoidal structures of stomatopod clubs	Carbon fiber epoxy composites	Wikipedia-b [73]
Honeycob	Honeycomb structure	Wikipedia-b [73]
Human skin and other biological materials	Autonomic healing	Wikipedia-b [73]
Leg attachment pads of several animals	Climbing robots	Wikipedia-b [27]
Living tissues	Artificial composite material	Wikipedia-b [73]
Nacre	Tissue regeneration, implants and support materials	Perera and Coppens [54]
Natural layered structures	Freeze casting	Wikipedia-b [73]
Protein folding	Self-assembled functional nanostructures	Wikipedia-b [73]
Spiders	Bird-safe glass	Alexander [2]
	Fiber weaving	Benyus [4]
Tree frogs, mussels	– Adhesives (glues) – Nanolithography	Wikipedia-b [73]
Woodpecker	Shock absorbers: – Shock-resistant flight recorders – Micrometeorite-resistant spacecrafts	Adams [1]

(1) A recent Ph.D. thesis titled: "A bioinspired simulation-based optimization framework for multi-objective optimization" by Leung [22].

(2) Another Ph.D. thesis titled:"Intelligence artificielle et robotique bio-inspirée: Modélisation de fonctions d'apprentissage par réseaux de neurones à impulsions" par Cyr [9].

(3) An article titled: "Bioinspired design of lightweight and protective structures" [28], and

(4) A recent article titled: "Modeling evacuation of high-rise buildings based on intelligence decision P system" by Niu et al. [37].

The fact that some of the existing references on nature-inspired simulation studies are at an advanced research level is very promising.

Advancements in simulation as a model-based knowledge-processing activity [50], initiated model-based activities, including model-based Systems Engineering [72]. Currently, the importance of the shift of paradigm, from model-based approach

Table 7.13 A list of Nature-Inspired Models for: *Optics, Optimization, Resource management, and Security/cyber security*

Inspiration		Engineered product	References
Type	From		
Optics	Bioluminescence	Engineered bioluminescence	Wikipedia-bl [74]
	Cat's eyes	Cat's eyes used in the roads	Smithers [64]
	Cephalopod camouflage	Thermochromatic "skin" for camouflage	Adams [1]
	Polarization-based vision in animals	Polarization vision for robotics (camouflage breaking, and signal detection and discrimination, …)	Shabayek et al. [62]
Optimization	Decentralized and self-organized behavior (such as artificial bee colony algorithm, particle swarm optimization, ant colony optimization, bat algorithm, firefly algorithm, glow worm swarm optimization)	Evolutionary algorithms Path planning Training neural networks Feature selection Image processing Computational fluid dynamics Hand gesture detection Data clustering Optimal nonlinear feedback control design Machine learning Photonics	Mane and Gaikwad [26], Wagner et al. [71], Mirjalili et al. [31]
Resource management	Stenocara beetle	Harvesting water from the air	Adams [1]
Security/cyber security	Diverse nature-inspired techniques	Cyber security and resiliency	El-Alfy et al. [13]
	Fingerprint	Security of wireless devices	ur Rehman et al. [70]

to simulation-based approach for several disciplines, is well documented [32, 46]. Yilmaz [55] elaborated on system engineering for agent-directed simulation. And already there are developments along "modeling and simulation-based systems engineering" [15].

Due to the richness and power of the paradigms offered by nature-inspired models and nature-inspired computing, it is expected that they will be part of the advanced agent-directed simulation tools and environments including agent-directed simulation-based systems engineering. This way, managing already ubiquitous and rapidly wide-spreading complex, cyber-physical systems and cyber-social-physical systems can be based on powerful scientific concepts.

Table 7.14 A list of Nature-Inspired models for: *Sensors, Surfaces, Tools, and Tracking*

Inspiration		Engineered product	References
Type	From		
Sensors	Eyes of jumping spiders	Metalens depth sensor for microrobotics, augmented reality, wearable devices	Qi [55]
	Skin	Electronic glove for robots	Stanford [66]
Surfaces	Burr on plants	Velcro	Smithers [64]
	Lotus	Superhydrophobic material	Alexander [2]
	Nature-inspired surface engineering		NISE [36]
	Skin, Shark	– Swimming costumes – Hospital surfaces	Smithers [64]
	Surface tension biomimetics (Hydrophobic and hydrophilic coatings)		Wikipedia-b [73]
Tools	Mosquito proboscis	Painless injection needle	Smithers [64]; Biomimicry Inst [7].
Tracking	Pheromones	To track internet users	
		To guide swarms of robots	Li et al. [23]
		To coordinate unmanned vehicles	Parunak et al. [52]
		Traffic lights	Zou and Yilmaz [82]

Table 7.15 64 types of systems engineering *by application areas* (adopted from [48]—"S. Eng." represents: systems engineering)

A
Aerospace S. Eng.
Air transport S. Eng.
Aircraft S. Eng.
Automation S. Eng.
B
Biological S. Eng.
Biomedical S. Eng.
Boiler S. Eng.
Building S. Eng.
Business S. Eng.
C
Civil S. Eng.
Cognitive S. Eng.
Communication S. Eng.
Computer S. Eng.
Control S. Eng.
Creative S. Eng.
Cyber-physical S. Eng.
Cyber-social-physical S. Eng.
Cyber-social S. Eng.
E
Earth S. Eng.
Electronic S. Eng.
Embedded S. Eng.
Energy S. Eng.
Enterprise S. Eng.
Environmental S. Eng.
F
Fuel Cell S. Eng.
H
Healthcare S. Eng.
Homeland Security S. Eng.
Human S. Eng.
I
Image S. Eng.
Industrial S. Eng.
Information S. Eng.

(continued)

Table 7.15 (continued)

M
Manufacturing S. Eng.
Marine S. Eng.
Marine Autonomous S. Eng.
Mechanical S. Eng.
Mechatronic S. Eng.
Molecular S. Eng.
N
Naval S. Eng.
Naval Combat S. Eng.
Nuclear S. Eng.
Nuclear Systems Reliability Engineering
P
Petroleum S. Eng.
Photovoltaic S. Eng.
Pipeline S. Eng.
Pneumatic S. Eng.
Political S. Eng.
Power S. Eng.
Q
Quality S. Eng.
R
Railway S. Eng.
Robotics S. Eng.
S
Satellite S. Eng.
Satellite Communication S. Eng.
Service S. Eng.
Simulation-based cyber-physical S. Eng.
Simulation-based cyber-social-physical S. Eng.
Social S. Eng.
Software S. Eng.
Space S. Eng.
Spacecraft S. Eng.
Structural S. Eng.
System of S. Eng.
T
Transportation S. Eng.

(continued)

Table 7.15   (continued)

W
Web information S. Eng.
Wireless S. Eng.

Table 7.16   Domain-independent 20 Types of Systems Engineering (adopted from [48])

	Systems Engineering
Agent-based	Systems Engineering
Cognitive	Systems Engineering
Digital	Systems Engineering
Distributed	Systems Engineering
Ethical	Systems Engineering
Intelligent	Systems Engineering
Interactive	Systems Engineering
Model-based	Systems Engineering
Predictive	Systems Engineering
Process	Systems Engineering
Product	Systems Engineering
Quality	Systems Engineering
Reliable	Systems Engineering
Reliability	Systems Engineering
Secure	Systems Engineering
Simulation	Systems Engineering
Simulation-based	Systems Engineering
System of	Systems Engineering
Systems Theory-based	Systems Engineering

References

1. Adams D (2017) The best of biomimicry: here's 7 brilliant examples of nature-inspired design. https://www.digitaltrends.com/cool-tech/biomimicry-examples/. Accessed 22 Dec 2019
2. Alexander D (2018). Biomimicry: 9 ways engineers have been 'inspired' by nature. https://interestingengineering.com/biomimicry-9-ways-engineers-have-been-inspired-by-nature. Accessed 22 Dec 2019
3. Barrows EM (2017) Animal behavior desk reference: a dictionary of animal behavior, ecology, and evolution. CRC Press, Taylor & Francis Group. Third Edition, Boca Raton, FL
4. Benyus JM (2009). Biomimicry-innovation inspired by nature. New York, NY: Harper Perennial (First published in 1997)

5. Bethapudi VS, Hack J, Trogadas P, Cho JIS, Rasha L, Hinds G, Shearing PR, Brett DJL, Coppens M-O (2019) A lung-inspired printed circuit board polymer electrolyte fuel cell. Energy Convers Manag 202:112198

6. BIE—Biologically-inspired engineering. https://wyss.harvard.edu/. Accessed 22 Nov 2019

7. Biomimicry Institute. https://biomimicry.org/biomimicry-examples/. Accessed 22 Nov 2019

8. CNIE—Centre for Nature Inspired Engineering. https://www.natureinspiredengineering. org.uk/. Accessed 22 Nov 2019

9. Cyr A (2016) Intelligence artificielle et robotique bio-inspirée: Modélisation de fonctions d'apprentissage par réseaux de neurones à impulsions. PhD thesis, Université du Québec à Montréal, Montréal, PQ, Canada

10. Dash N, Priyadarshini R, Mishra BK, Misra R (2017) Bio-inspired computing through artificial neural network. In: Sangaiah AK, Gao XZ, Abraham A (eds) Chapter 11—Handbook of research on fuzzy and rough set theory in organizational decision making. IGI Global, Hershey, PA, pp 246–274

11. Dorigo N, Birattari M (2014) Swarm intelligence. Scholarpedia. http://www.scholarpedia.org/ article/Swarm_intelligence#Cooperative_Behavior_in_Swarms_of_Robots. Accessed Nov 22 2019

12. Durak U, Ören T, Tolk A (2017) An index to the body of knowledge of simulation systems engineering. In: Tolk A, Ören T (eds) Chapter 2—The profession of modeling and simulation. Wiley, Hoboken, NJ, pp 11–33

13. El-Alfy EM, Eltoweissy M, Fulp EW, Mazurcyk W (eds) (2019) Nature-inspired cyber security and resiliency: fundamentals, techniques and applications. The Institute of Engineering and Technology (IET). ISBN: 978-1-78561-638-9

14. Floreano D, Mattiussi C (2008) Bio-inspired artificial intelligence: theories, methods, and technologies. MIT Press, Cambridge, MA

15. Gianni D, D'Ambrogio A, Tolk A (eds) (2014) Modeling and simulation-based systems engineering handbook, 1 edn. CRC Press, Boca Raton, FL. ISBN 9781466571457

16. GCPSE. Guide to Cyber-Physical Systems Engineering. Road to CPS. An EU project. http:// road2cps.eu/events/wp-content/uploads/2016/03/D3.2_Guide-to-Cyber-Physical-Systems-Egineering_v1.0–1.pdf. Accessed 22 Nov 2019

17. Gunes V, Peter S, Givargis T, Vahid F (2014) A survey on concepts, applications, and challenges in cyber-physical systems. KSII Trans Internet Inf Syst 8:4242–4268. https://doi.org/10.3837/ tiis.2014.12.001

18. Hatti DI, Sutagundar A (2019) Nature inspired computing for wireless networks applications: a survey. Int J Appl Evolut Comput (IJAEC) 10(1):1–29. https://doi.org/10.4018/IJAEC.201 9010101

19. INCOSE (2015) Systems engineering handbook—a guide for system life cycle processes and activities. Wiley, Hoboken, NJ

20. Koopaee MJ, Gilani C, Scott C, Chen X (2018) Bio-inspired snake robots: design, modelling, and control. In: Habib M (ed) Chapter 11—Handbook of research on biomimetics and biomedical robots. IGI Global, Hershey, PA, pp 246–275

21. Kose U, Emre G, Deperlioglu O (eds) (2018) Nature-inspired intelligent techniques for solving biomedical engineering problems. IGI Global, Hershey, PA

22. Leung CSK (2018) A bio-inspired simulation-based optimization framework for multi-objective optimization. PhD thesis, University of Hong Kong

23. Li G, Chen C, Geng C, Li M, Xu H, Lin Y (2019) A pheromone-inspired monitoring strategy using a swarm of underwater robots. Sensors 19(19) https://doi.org/10.3390/s19 194089. Accessed: 22 Nov 2019

24. Lienhard J, Schleicher S, Poppinga S, Masselter T, Milwich M, Speck T, Knippers J (2011) Flectofin: a hingeless flapping mechanism inspired by nature. Bioinspir Biomim 6(4):045001

25. Mäkiö-Marusik E (2017) Current trends in teaching cyber physical systems engineering: a literature review. In: Proceedings of the 2017 IEEE 15th international conference on industrial informatics (INDIN), 24–26 July 2017, Emden, Germany, pp 518–525. https://ieeexplore.ieee. org/document/8104826. Accessed 22 Nov 2019

26. Mane SU, Gaikwad PG (2014) Nature inspired techniques for data clustering. In: 2014 International conference on circuits, systems, communication and information technology applications (CSCITA), 4–5 April 2014. IEEE
27. McGeough JA, Hartenberg RS (2019). Hand tool (*Encyclopedia Britannica*). https://www.britannica.com/technology/hand-tool/The-Mousterian-flake-tools (Revised and updated by JP Rafferty, 2019 Jan 16). Accessed 22 Nov 2019
28. Mehta P, Ocampo JS, Tovar A, Chaudhari P (2016) Bio-inspired design of lightweight and protective structures, SAE Technical Paper 2016-01-0396, 2016. https://doi.org/10.4271/2016-01-0396. Accessed 22 Nov 2019
29. Merriam-Webster. https://www.merriam-webster.com/dictionary/biomimicry. Accessed Nov 22 2019
30. Miranda K, Zapotecas-Martínez S, López-Jaimes A, García-Nájera A (2019) A comparison of bio-inspired approaches for the cluster-head selection problem in WSN, In: Shandilya S, Nagar A (eds) Advances in nature inspired computing and applications. EAI/Springer innovations in communication and computing. Springer, New York
31. Mirjalili S, Dong JS, Lewis A (2020) Nature-inspired optimizers—theories, literature reviews and applications. Springer, New York
32. Mittal S, Durak U, Ören T (eds) (2017) Guide to simulation-based disciplines: advancing our computational future. Springer, New York
33. Narraway C, Davis O, Lowell S, Lythgoe K, Turner JS, Marshall S (2019) Biotic analogies for self-organising cities. In: Environment and planning B: urban analytics and city science. https://doi.org/10.1177/2399808319882730
34. NIE—Nature-inspired engineering. http://www.engconf.org/conferences/civil-and-environmental-engineering/nature-inspired-engineering/. Accessed 22 Nov 2019
35. NIFTI—Nature-inspired flight technologies + ideas. The Air Force Center of Excellence on Nature-Inspired Flight Technologies and Ideas. Research: Sensory Modalities. http://nifti.washington.edu/research/sensor-modalities/. Accessed 22 Nov 2019
36. NISE (2019) Abstracts of the 1st international conference on nature inspired surface engineering, 11–14 June 2019, Stevens Institute of Technology, Hoboken, NJ
37. Niu Y, Zhang J, Zhang Y, Xiao J (2019) Modeling evacuation of high-rise buildings based on intelligence decision P system. Sustain—Open Access J. Licensee MDPI, Basel, Switzerland. This article is an open access article distributed under the terms and conditions of the Creative Commons Attribution (CC BY) license. http://creativecommons.org/licenses/by/4.0/. Accessed 22 Nov 2019
38. Olariu S, Zomaya AY (eds) (2019) Handbook of bioinspired algorithms and application. Chapman and Hall/CRC, Boca Raton, FL
39. Ören TI (1971a) GEST: General system theory implementor, a combined digital simulation language. PhD dissertation, 265 p. University of Arizona, Tucson, AZ: https://dl.acm.org/citation.cfm?id=905488&dl=GUIDE&coll=GUIDE. Accessed 22 Nov 2019
40. Ören TI (1971b) GEST: a combined digital simulation language for large-scale systems. In: Proceedings of the Tokyo 1971 AICA (Association Internationale pour le Calcul Analogique) symposium on simulation of complex systems, 3–7 September. Tokyo, Japan, pp B-1/1–B-1/4)
41. Ören TI (1984) GEST—A modelling and simulation language based on system theoretic concepts. In: Ören TI, Zeigler BP, Elzas MS (eds) Simulation and model-based methodologies: an integrative view. Springer Heidelberg, Germany, pp 281–335. http://www.site.uottawa.ca/~oren/pubs-pres/1984/pub-1984-03_GEST_NATO_ASI.pdf. Accessed 22 Nov 2019
42. Ören TI (1990) A paradigm for artificial intelligence in software engineering. In: Ören TI (ed) Advances in artificial intelligence in software engineering, vol 1. JAI Press, Greenwich, Connecticut, pp 1–55
43. Ören TI (2001–Invited Paper) Software agents for experimental design in advanced simulation environments. In: Ermakov SM, Kashtanov YuN, Melas V (eds) Proceedimgs of the 4th St. Petersburg workshop on simulation, 18–23 June 2001, pp 89–95
44. Ören TI (2009) Modeling and simulation: a comprehensive and integrative view. In: Yilmaz L, Ören TI (eds) Agent-directed simulation and systems engineering. Wiley series in systems engineering and management. Wiley, Berlin, Germany, pp 3–36

45. Ören T (2018) Powerful higher-order synergies of cybernetics, systems thinking, and agent-directed simulation for cyber-physical systems. Acta Syst 18(2):1–5 (July 2018)
46. Ören T, Mittal S, Durak U (2017) The evolution of simulation and its contributions to many disciplines. In: Mittal S, Durak U, Ören T (eds) Guide to simulation-based disciplines: advancing our computational future. Springer, New York, pp 3–24
47. Ören T, Yilmaz L (2004) Behavioral anticipation in agent simulation. In: Proceedings of WSC 2004—winter simulation conference, 5–8 December 2004, Washington, D.C, pp 801–806
48. Ören T, Yilmaz L (2012) Synergies of simulation, agents, and systems engineering. Expert Syst Appl 39(1):81–88
49. Ören T, Yilmaz L (2015, Invited article) Awareness-based couplings of intelligent agents and other advanced coupling concepts for M&S. In: Proceedings of the 5th international conference on simulation and modeling methodologies, technologies and applications (SIMULTECH'15), 21–23 July 2015, Colmar, France, pp 3–12
50. Ören T, Zeigler BP (1979) Concepts for advanced simulation methodologies. Simulation 32(3):69–82
51. Ören T, Zeigler BP (2012) System theoretic foundations of modeling and simulation: a historic perspective and the legacy of A. Wayne Wymore. Special Issue of Simulation—The Transactions of SCS 88(9):1033–1046
52. Parunak H, Van D, Bruekner SA, Sauter J (2004) Digital pheromones for coordination of unmanned vehicles. In: Weyns D, Parunak H, Van D, Michel F (eds) Environments for multi-agent systems. E4MAS 2004. Lecture notes in computer science, vol 3374. Springer, New York
53. Pawlyn M (2016) Biomimicry in architecture. Riba Publishing, New castle upon Tyme, England
54. Perera AS, Coppens MO (2018) Re-designing materials for biomedical applications: from biomimicry to nature-inspired chemical engineering. Philo Trans Roy Soc A: Math, Phys Eng Sci. Published by the Royal Society under the terms of the Creative Commons Attribution License. http://creativecommons.org/licenses/by/4.0/. Accessed 17 Nov 2019
55. Qi G, Shi Z, Huang Y-W, Alexander E, Qiu C-W, Capasso F, Zickler T (2019) Compact single-shot metalens depth sensors inspired by eyes of jumping spiders. In: Proceedings of the national academy of sciences of the United States of America, 28 October 2019. https://www.pnas.org/content/early/2019/10/22/1912154116. Accessed 22 Nov 2019
56. Rashid A et al (eds) (2019) The cyber security body of knowledge. CyBOK Version 1.0 © Crown Copyright. The National Cyber Security Centre 2019. https://www.cybok.org/media/downloads/CyBOK-version_1.0.pdf. Accessed: 22 Nov 2019. Licensed under the Open Government Licence. http://www.nationalarchives.gov.uk/doc/open-government-licence/version/3/. Accessed 22 Nov 2019
57. Rhode-Barbarigos L, Hadj Ali NB, Motro R, Smith IFC (2010) Designing tensegrity modules for pedestrian bridges. Eng Struct 32(4):1158–1167
58. Romanovsky A, Ishikawa F (2016) Trustworthy cyber-physical systems engineering. CRC Press, Boca Raton, Florida
59. Schreiner WB (2019) A history. Department of History, OSU. https://ehistory.osu.edu/exhibitions/biomimicry-a-history. Accessed 22 Nov 2019
60. SEBoK (2019) Guide to the systems engineering body of knowledge (SEBoK), version 2.2. https://www.sebokwiki.org/wiki/Guide_to_the_Systems_Engineering_Body_of_Knowledge_(SEBoK. Accessed 22 Nov 2019
61. Seck MD, Honig HJ (2012) Multi-perspective modelling of complex phenomena. Comput Math Organ Theory 8:128–144. https://doi.org/10.1007/s10588-012-9119-9
62. Shabayek ElR, Morel O, Fofi D (2018) Bio-inspired polarization vision techniques for robotics applications. In: Chapter 17—Computer Vision: concepts, methodologies, tools, and applications. IGI Global, Hershey, PA, pp 421–457
63. Shandilla SK, Shandilla S, Nagar AK (eds) (2019) Advances in nature-inspired computing and applications. Springer, New York
64. Smithers B (2016) Looking to nature for engineering inspiration: science made simple. http://www.sciencemadesimple.co.uk/curriculum-blogs/biology-blogs/looking-to-nature-for-engineering-inspiration. Accessed 22 Nov 2019

65. Sokolowski J, Durak U, Mustafee N, Tolk A (eds) (2019) Summer of simulation—50 years of seminal computer simulation research. Springer, New York
66. Stanford (2018) Stanford develops an electronic glove that gives robots a sense of touch. https://news.stanford.edu/2018/11/21/stanford-develops-electronic-glove-gives-robots-sense-touch/. Accessed 22 Nov 2019
67. Tolk A (ed) (2012) Engineering principles of combat modeling and distributed simulation. Wiley, Hoboken, NJ
68. Tolk A, Glazner CG, Pitsco R (2017) Simulation-based systems engineering. In: Mittal S, Durak U, Ören T (eds) Guide to simulation-based disciplines: advancing our computational future. Springer, New York, pp 75–102
69. Torres S (2011) Swarm theory applied to air traffic flow management. Procedia Comput Sci 12:463–470. https://www.researchgate.net/publication/257719407_Swarm_Theory_App lied_To_Air_Traffic_Flow_Management. Accessed 22 Nov 2019
70. ur Rehman S, Alam S, Ardolinni IT (2014) Security of wireless devices using biological-inspired RF fingerprinting technique. In: Alam S, Dobbi G, Koh Y, ur Rehman S (eds) Chapter 15—Biologically-inspired techniques for knowledge discovery and data mining, edn. 1. IGI Global, Hershey, PA
71. Wagner N, Şahin CŞ, Pena J, Streilein WW (2019) Automatic generation of cyber architectures optimized for security, cost, and mission performance: a nature-inspired approach. In: Shandilla SK, Shandilla S, Nagar AK (eds) Advances in nature-inspired computing and applications. Springer, New York, pp 1–25
72. Wymore AW (1993) Model-based systems engineering. CRC Press, Boca Raton, FL
73. Wikipedia-b (2019). https://en.wikipedia.org/wiki/Biomimetics. Accessed 22 Nov 2019
74. Wikipedia-bl (2019). https://en.wikipedia.org/wiki/Bioluminescence. Accessed 22 Nov 2019
75. WYSS Institute—The WYSS institute for biologically inspired engineering. https://wyss.harvard.edu. Accessed 22 Nov 2019
76. Yilmaz L (2009) Toward systems engineering for agent-directed simulation. In: Yilmaz L, Ören T (2009a). Chapter 8—Agent-directed simulation and systems engineering. Wiley, Berlin, Germany, pp 219–235
77. Yilmaz T (2015) Toward agent-supported and agent-monitored model-driven simulation engineering. In: Yilmaz L (ed) Concepts and methodologies for modeling and simulation: a tribute to Tuncer Ören. Springer, New York, pp 3–18
78. Yilmaz L, Ören T (2009) Agent-directed simulation and systems engineering. Wiley, Berlin, Germany
79. Yilmaz L, Ören T (2009b) Agent-directed simulation (ADS). In Yilmaz L, Ören T (eds) Agent-directed simulation and systems engineering. Wiley Series in systems engineering and management. Wiley, Berlin, Germany, pp 111–143
80. Zeigler BP, Muzy A, Kofman E (2019) Theory of modeling and simulation—discrete event and iterative system computational foundations. Academic Press, London, UK
81. Zeigler BP, Praehofer H, Kim TG (2000) Theory of modeling and simulation. Academic Press, San Diego, CA
82. Zou G, Yilmaz L (2019) Self-organization models of urban traffic lights on digital infochemicals. Simulation 95(3):271–285

Tuncer Ören is a professor emeritus of computer science at the School of Electrical Engineering and Computer Science of the University of Ottawa, Canada. He has been involved with simulation since 1965. His Ph.D. is in Systems Engineering from the University of Arizona, Tucson, AZ (1971). His basic education is from Galatasaray Lisesi, a high school founded in his native Istanbul in 1481 and in Mechanical Engineering at the Technical University of Istanbul (1960). His **research interests** include: advancing methodologies for modeling and simulation; agent-directed simulation; agents for cognitive and emotive simulations (including representations of human personality, understanding, misunderstanding, emotions, and anger mechanisms); computational awareness; reliability, QA, failure avoidance, ethics; as well as body of knowledge and terminology of modelling and simulation. He has over 550 **publications**, including 56 books and proceedings (+2 in press and in preparation). He has contributed to over 500 **conferences and seminars** held in 40 countries. Dr. Ören **has been honored** in several countries: **USA**: He is a Fellow of SCS (2016), an inductee to SCS Modeling and Simulation Hall of Fame–Lifetime Achievement Award (2011), and received SCS McLeod Founder's Award for Distinguished Service to the Profession (2017). **Canada**: Dr. Ören has been recognized, by IBM Canada (2005), as a pioneer of computing in Canada. He received the Golden Award of Excellence from the International Institute for Advanced Studies in Systems Research and Cybernetics (2018). **Turkey**: He received "Information Age Award" from the Turkish Ministry of Culture (1991), an Honor Award from the Language Association of Turkey (2012), and Lifetime service award from the Turkish Informatics Society and Turkish Association of Information Technology (2019). A book was edited by Prof. Levent Yilmaz: Concepts and Methodologies for Modeling and Simulation: A Tribute to Tuncer Ören. Springer (2015).

Chapter 8
Composing Cyber-Physical Simulation Services in the Cloud via the DEVS Distributed Modeling Framework

Rob Kewley

Abstract Systems engineering and simulation of cyber-physical systems require the aggregation of disparate models from the component cyber and physical domains in order to understand the whole system. Military multi-domain operations employ emerging technologies such as unmanned sensors, cyber, and electronic warfare. The Discrete Event System—Distributed Modeling Framework (DEVS-DMF) is a simulation technology that enables composition of multiple models via the actor model of computation, parallel and asynchronous messaging, and location transparency. Using a system of systems engineering approach, we compose models of military operations, unmanned systems, and electronic warfare technologies to analyze mission performance using different advanced equipment sets. Important performance metrics span the physical (sensor performance), cyber (electronic attack), human factors (soldier load), and military (mission success) domains. Simulation services are allocated to each domain, and the simulation's microservice architecture allows for independently deployable services that own their internal state. Containerization and cloud deployment allow geographically distributed users to manipulate simulation inputs, conduct large-scale experiments, and analyze simulation output using browser and web tools. The resulting ensemble enables system of systems engineering and analysis of cyber and electronic systems in support of small tactical operations.

8.1 Introduction

Systems engineering for modern cyber-physical systems presents a tremendous challenge. Increasing complexity in both the cyber and physical domains leads to interactions between various sub-systems and the operating environment. These interactions often drive system performance. Using simulation to analyze the interactions is difficult because the simulation models written to analyze each domain cannot be easily integrated into one simulation environment. While disparate simulations

R. Kewley (✉)
simlytics.cloud LLC., 55 Madison Avenue, Suite 400, Morristown, NJ 07960, USA
e-mail: rob@simlytics.cloud

© Springer Nature Switzerland AG 2020   167
J. L. Risco Martín et al. (eds.), *Simulation for Cyber-Physical Systems Engineering*,
Simulation Foundations, Methods and Applications,
https://doi.org/10.1007/978-3-030-51909-4_8

can be integrated using the High Level Architecture (HLA) [12], this is a technology with some challenges in modern cloud-based computing environments [5, 6]. Cloud-based simulations are better when re-engineered for cloud deployment by composing individual domain models [9, 18]. This paper works a modest example of using distributed simulation to support systems engineering of cyber-physical systems for military multi-domain operations in the cloud using the DEVS Distributed Modeling Framework (DEVS-DMF) [17]. This technical approach enables the logical and physical separation of simulated entities and their state into separate models independently developed and deployed as microservices to cloud computing nodes. The approach also uses advances in Modeling and Simulation as a Service (MSaaS) [24] and its associated engineering process [23] as drivers for simulation design.

The intent of this chapter is to provide a road map for simulation engineers and systems engineering integrators who would like to employ this approach in their work in order to ease integration and to improve performance in the cloud. It begins with overviews of multi-domain operations and the technology advances that support their analysis in distributed simulation. It then introduces a multi-domain military scenario involving dismounted soldiers, unmanned aircraft sensors, and electronic attack. As it follows the MSaaS engineering process, models in 3 different domains—dismouted movement, sensor detection, and radio networks—are separately deployed to the cloud as microservices, then integrated to produce simulation results. The MSaaS cloud-based approach using the DEVS Distributed Modeling Framework enables discovery and integration of simulation services in different domains that can be deployed on-demand to support tradespace analysis between competing system designs. Note that this engineering is an evolution of the process described in "Federated Simulation for Systems Engineering" [15] where the current approach moves from HLA integration of simulations running on workstations to DEVS-DMF integration of models running as services in the cloud.

8.2 Systems Engineering for Multi-Domain Operations

Multi-domain operations is a common theme in emerging military doctrine across the globe [25, 29, 33]. The common thread among the implementations is the ability to integrate effects from space, air, sea, land, information, human, electronic, and cyber domains in a single place and at a single echelon. This differs from traditional military organizations where different military services or large organizations had primary responsibility for a single domain, and integration occurred across large spaces and organizations, vice small spaces and single echelons. This includes operations at the lowest tactical level [28], where leaders will employ automated and semi-automated systems that compete in the ground, air, cyber, electronic warfare, and information domains.

This poses a particular challenge for systems engineering to integrate these capability packages at each echelon. For example, a cyber or electronic system may disrupt enemy communications, but its interaction with the physical domain is also important. How will it impact situation awareness and maneuver? While it is possible to analyze each of these questions in isolation using domain models, it is often the dynamic interactions that have the greatest impact on mission performance. In order to properly assess design tradeoffs, the systems engineer must integrate multiple domain models into a single simulation scenario. This requires a new technical approach.

8.3 Emerging Technologies to Support Cloud-Based Modeling and Simulation

The widespread adoption of cloud and integration technologies in the commercial sector sheds light on new opportunities for composition and deployment of simulation models. The same cloud and web-based technologies used by businesses to improve integration, continuously deploy services, and drive analytics are available to the simulation sector.

8.3.1 Cloud-Based Modeling and Simulation

Despite the success of the cloud in the commercial sector, the modeling and simulation community's cloud adoption has been inhibited by some managerial and technical challenges. Cybersecurity and information assurance policies inhibit the movement of applications and data to the cloud [9]. Additionally, cloud deployments must compete with the requirement to continuously maintain and deliver the existing capability. Until a cloud-based system is fully deployed, accredited, and tested, the legacy system consumes resources. Several efforts to move simulation programs to the cloud have virtualized the legacy applications and moved them as is, resulting in modest success and modest savings [20, 26].

Despite the paucity of implementations, there is a growing consensus in the research community that the best practice for cloud-based modeling and simulation is to decompose the simulations into a set of discrete models, each representing a single concept. [9, 14]. One emerging architectural model proposes a cloud-native simulation stack with the following layers [18]:

Layer 1–Infrastructure provisioning layer: The cloud-based infrastructure onto which simulation components are deployed.
Layer 2–Elastic simulation platform: Performs coordinated deployment of simulation components on Layer 1 for simulation execution.
Layer 3–Simulation service composing: Performs composition of stateful and stateless simulation models into one running simulation.
Layer 4–Simulation: Executes the composed models over time to produce results.

Mittal and Tolk propose mobile propertied agents as components of a concept-driven architecture [32]. Each agent "encapsulates a semantic concept, its associated properties (by way of syntactic data elements) and provides interfaces to manipulate the properties by external services." Each of these architectures builds on the underlying notion that cloud-based simulations should be composed of individual models that encapsulate single concept. Note the distinction between models, purposeful abstractions of a concept, and simulations, coordinated execution of models over time. Models should be offered as services and semantically aligned in a composed simulation service. This idea is an emerging best practice for cloud-based simulation. The following technical approaches underpin this paper's implementation of that best practice.

8.3.2   Modeling and Simulation as a Service

The NATO Science and Technology Office has a focused technology development program for Modeling and Simulation as a Service (MSaaS). NATO Modeling and Simulation Group (MSG) 131 "Modelling and Simulation as a Service: New concepts and Service Oriented Architectures" completed work in 2014. The follow-on group, MSG-136 "Modelling and Simulation as a Service: Rapid Deployment of Interoperable and Credible Simulation Environments," released its final report in 2018.[1] NATO's MSaaS vision is that "M&S products, data and processes are conveniently accessible and available on-demand to all users in order to enhance operational effectiveness." This is achieved via the following goals [24]:

1. To provide a framework that enables credible and effective M&S services by providing a common, consistent, seamless, and fit for purpose M&S capability that is reusable and scalable in a distributed environment.
2. To make M&S services available on-demand to a large number of users through scheduling and computing management. Users can dynamically provision computing resources, such as server time and network storage, as needed, without requiring human interaction. Quick deployment of the solution is possible since the desired services are already installed, configured and, online.
3. To make M&S services available in an efficient and cost-effective way, convenient short setup time and low maintenance costs for the community of users will be available and to increase efficiency by automating efforts.
4. To provide the required level of agility to enable convenient and rapid integration of capabilities, MSaaS offers the ability to evolve systems by rapid provisioning of resources, configuration management, deployment, and migration of legacy systems. It is also tied to business dynamics of M&S that allow for the discovery and use of new services beyond the users' current configuration.

---

[1] Both MSG-131 and MSG-136 received earned the NATO Scientific Achievement award for their repective contributions.

The operational concept calls for an MSaaS implementation to enable three processes for users—service discovery, service composition, and service execution. The group also published a technical reference architecture [22], a service discovery and metadata standard [21], and an engineering process [23]. The MSaaS engineering process is an overlay to the IEEE Recommended Practice for Distributed Simulation Engineering and Execution Process (DSEEP) [13] that brings in engineering considerations for cloud-based M&S services. This chapter follows the MSaaS engineering process in Sect. 8.4 to develop a composed simulation service for multi-domain operations. NATO's current effort, MSG-164 "Modelling ans Simulation as a Service Phase 2," is following and extending this process to prototype an MSaaS reference implementation and a set of services.

8.3.3 Microservices and Domain-Driven Design

Another architectural concept that supports cloud-based M&S is a microservice architecture [17, 18]. It "advocates creating a system from a collection of small, isolated services, each of which owns their data, and is independently isolated, scalable and resilient to failure [2]." Reactive microservices should

- Fail in isolation without cascading failure to other services
- Act autonomously only making promises about their own behavior
- Do one thing well
- Own their own state exclusively
- Embrace asynchronous message passing
- Stay mobile, but addressable.

Extending the microservices idea to the modeling and simulation concept driven architecture [32], the microservice is the independent unit of employment for a model of a concept, offered as a service.

Domain-driven design is a technique that has been applied to develop a reactive microservices architecture [34]. It offers a methodology for decomposing a system into multiple independent subdomains that consist of concepts, actions, and rules about a particular part of the system. These all share a common ubiquitous langauge and are isolated from the rest of the system in a bounded context. Communication with other domains is through an abstract interface, and care is taken during development to ensure isolation of state and components within a single context. Domain-driven design can be applied to cloud-based M&S as a way to separate and isolate different conceptual models as building blocks for the composed simulation.

8.3.4 Discrete Event System (DEVS) Distributed Modeling Framework (DMF)

DEVS-DMF is a microservices-based simulation framework that is designed for the cloud by applying the emerging technologies highlighted in previous sections [16, 17]. The Parallel DEVS framework [4] allows independent models to be specified as atomic models that maintain an internal state. This encapsulation of state is a powerful abstraction that relieves the larger simulation architecture from state management. Instead, external models accept messages which contain state information. The software implementation makes use of the actor model of computation [1] as implemented by the Akka Framework [19] resulting in a distributed simulation system that incorporates location transparency, follows reactive principles, and employs a microservices architecture. Parallel DEVS manages scheduling via its model coordinators and simulators. It achieves separation of concerns and partitioning by isolating state, event management and output within DEVS atomic models. Its implementation as Akka actors enables location transparency, which allows flexible scaling within a single machine, to other containers in the cloud, or across cloud infrastructures. Its asynchronous message-based framework provides a pathway for integrating existing microservices.

The DEVS-DMF infrastructure:

- Allows easy discovery, understanding, and integration of existing conceptual models.
- Supports the coupling of models with entities, functions, or behaviors in the systems architecture.
- Exposes input data so that it can easily be manipulated in the development of complex scenarios.
- Exposes the trajectory of state data so that it can be used in output analysis.
- Enables efficient design and execution of experiments.
- Allows parallel computation for execution of large-scale experiments.

Figure 8.1 shows the key components in a DEVS-DMF model. The simulation itself is the outermost component and has responsibility to initialize and run the overall simulation. It has to manage independent parallel random number streams and logging for subordinate components. All DEVS models are contained in a parent coordinator responsible for their execution. The workhorse component of a DEVS simulation is the DEVS model, shown in Fig. 8.2. It contains the static properties and dynamic state variables for a simulation entity wrapped in an actor. In response to the advance of time, internally scheduled events, and external events, it runs internal and external state transition functions to modify internal state and produce output.

These state transition functions are the smallest unit of composability, and they can easily be shared across simulation implementations because they contain no internal state. Instead, they take current state and time advance as input, perform calculations and produce new state as output. A key characteristic of a function is that it should always produce the same output for any given input. In stochastic simulations, random

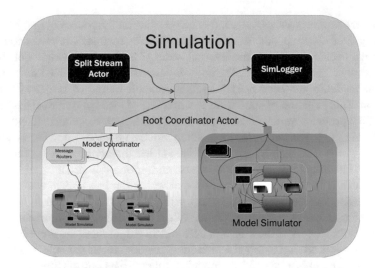

Fig. 8.1   DEVS-DMF simulation components

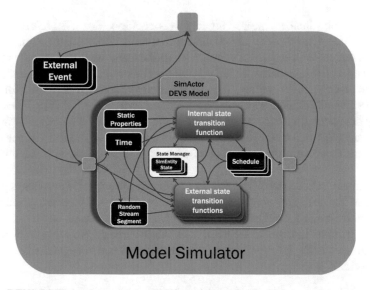

Fig. 8.2   DEVS-DMF simulator and the DEVS model contained withng

numbers should be passed in to state transition functions, not generated internally. For example, a dead reckoning transition function, used to move an entity to a new location based on its internal position, velocity and acceleration, could be used by multiple entities in the simulation, or by entirely different simulations. Following these rules enables composability and re-use of these functions.

The DEVS model is the next higher unit of composition for DEVS. They are portable across simulations as long as those simulations pass the same events and consume the same output messages. It manages state and produces output by calling its internal and external state transition functions in response to time advance and to external events. However, the internal state of a DEVS model is not directly accessible by other components. The only way for external entities to know the state is for the model to publish state data in its immutable output messages after a state transition. This isolation of state from external threads is also a characteristic of an actor, and it provides tremendous advantages with respect to concurrent and parallel programming [1]. For example, the soldier movement model updates a soldier's location at each time step by advancing the soldier the appropriate distance along its route via its internal state transition function. Important internal variables are the soldier's location, planned route, and time at which the last update was made. An external event, such as a change in route, will also generate an update in position via the external state transition function. When these functions complete, the new position values are sent to the state manager and an output message is published with the new location, giving external entities information about the soldier's position. Each DEVS Model is wrapped by a model simulator that controls time advance and passes messages into and out of the DEVS model in accordance with the simulation clock.

A model coordinator is the next higher level of composition. All DEVS simulators and subordinate coordinators are contained in a model coordinator. A model coordinator interacts with its parent coordinator in the same way as a model simulator. In this way, entire hierarchies of DEVS models can be wrapped by a coordinator as if it were just one DEVS model. This hierarchical aggregation is a powerful feature of DEVS because models are also portable across different simulations, constrained only by the messaging into and out of the models. The model coordinator manages the time advances and message routing for its internal models. It knows which models are imminent and controls their state transitions in accordance with the simulation clock.

Is summary, DEVS-DMF provides a mechanism to exploit emerging cloud and simulation technologies. It's simulation engine is engineered from the ground up for the cloud. Its hierarchical DEVS structure allows us to model a single concept and to isolate its state in a single DEVS model. These models are services, which can be integrated via DEVS coordinators into distributed and independently deployed microservices.

8.4 Simulation for Multi-Domain Operations

In Sect. 8.2, we saw that systems engineering multi-domain operations called for dynamic composition of models from different domains, and in Sect. 8.3 we got an overview of the emerging simulation technologies that will support us. In this section, we will follow the MSaaS Engineering Process to build a hierarchical DEVS-DMF simulation that integrates independent models of soldier movement, sensor target acquisition, and radio network communications in the face of jamming.

8.4.1 DSEEP Phase 1—Define Composed Simulation Service Objectives

During phase 1 of the engineering process, it is most important to understand the engineering objectives we wish to accomplish with the simulation system. This requires interaction with the simulation study sponsor and a strong understanding of the engineering domains to be modeled. This understanding is best built with the simulation engineering team meeting collaboratively with the systems engineering, domain engineering, project management teams, and customer for the system under design.

For this example problem, the sponsor is trying to gain an airborne intelligence, surveillance, and reconnaissance advantage against an advanced enemy. They are considering different alternatives for including a small unmanned aircraft, such as a quad-copter and a counter-UAV (Unmanned Aerial Vehicle) jammer in the squad's equipment package. Sensor models support analysis of the capabilities provided by different sensors on the UAV. Radio network models support analysis of jamming capabilities against enemy UAV radios. In addition, the sponsor is concerned that the additional weight of this equipment will unduly hinder soldier performance. It is up to the lead systems engineer to provide analytical data to support these decisions. We will assist with modeling and simulation.

Based on the sponsor needs, the high-level simulation objective is to support a tradespace analysis between situation awareness, degrading enemy situation awareness, and soldier load. The decision alternatives include various configurations of the UAV and jammer with their associated carry weights. Another goal is for the systems engineering team to have access to the simulation in order to run excursions as needed. These objectives give us enough information to begin conceptual analysis of the operations to be modeled.

8.4.2 DSEEP Phase 2—Perform Conceptual Analysis

In performing conceptual analysis, simulation engineers are focused on the problem definition phase of the systems engineering process. Their primary goal during this phase is to gain an understanding, documented as a set of systems engineering models, of the system of systems under design. Note that the views in this section do not define the simulation, but the system of systems to be modeled. Simulation engineers should understand system users, subsystetms, and the value derived from system operation.

8.4.2.1 Develop Scenario

In the operational scenario, a friendly team of four soldiers first performs an eight kilometer road march over mountainous terrain, shown in Fig. 8.3, so they would like to minimize the weight they carry as much as possible.

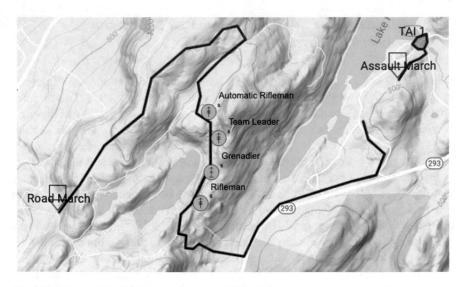

Fig. 8.3 Dismounted soldier scenario road march route

Upon arrival in the objective area, shown in Fig. 8.4, the team will begin an assault march toward the assault position. Prior to this march, they would like as much information as possible about the potential enemy in Named Area of Interest

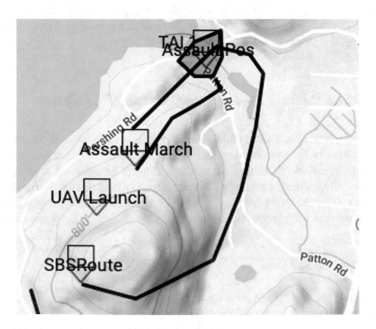

Fig. 8.4 Objective area for the dismounted scenario

1 (NAI 1). If the team is equipped with a UAV, they will employ it along the shown UAV Route to attempt to identify possible threats. In addition, the enemy forces will employ their own UAV to overwatch the roach march approach into their area. The friendly team would like to avoid detection. If they are equipped with a jammer, they will employ the jammer to disrupt the communications between the enemy UAV and its controller, causing it to land automatically.

8.4.2.2 Develop Conceptual Model

An important component of the conceptual model is the structure of the system under design. For our scenario, this system is shown in Fig. 8.5. At the top of the hierarchy is the team, which in our case, contains four soldiers. Important soldier properties for our scenario, modeled as constraint properties, are the soldier weight and soldier fitness. For all of the equipment carried, weight is an important property because of

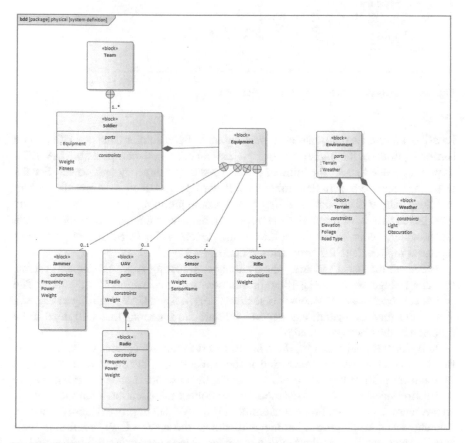

Fig. 8.5 Block definition diagram for the soldier system under design

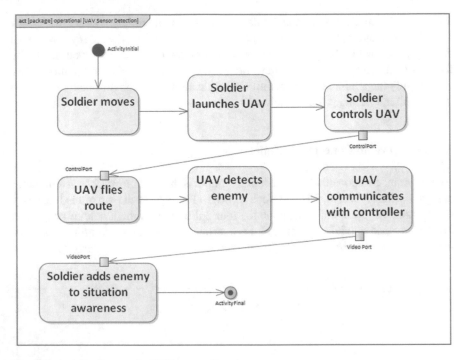

Fig. 8.6 Activity diagram for UAV sensor detection

its effect on soldier road marching performance. Each soldier carries a rifle, and a sensor. Optionally, the soldier may carry a jammer or a small man-packable UAV. The important properties for the jammer are its operating frequency and power. For the UAV, the sensor type will determine its ability to detect targets, and its radio power and frequency will impact its ability to overcome jamming. Finally, the engineers have included a set of properties that represent the environmental conditions, terrain and weather, under which the soldier team performs. These constraint properties represent input data to the simulation models.

Based on the listed scenario, the systems engineering team developed systems modeling language (SysML) [8] representations of the battlefield activities. The activities modeled are UAV sensor detection in Fig. 8.6 and UAV jamming in Fig. 8.7. With these models, operational experts validate the functions that will need to be modeled in the tradespace analysis.

In the operational scenario, there are two performance measures of interest. The first is situation awareness, measured as the percentage of the threat force identified by the team prior to beginning of the assault. Because our assault team would like to remain undetected, it is also important to collect the situation awareness of the enemy team. The expectation is that employing a UAV will improve friendly situation awareness and employing a jammer will defeat the enemy UAV, degrading their situation awareness. The other important performance measure is soldier exhaustion,

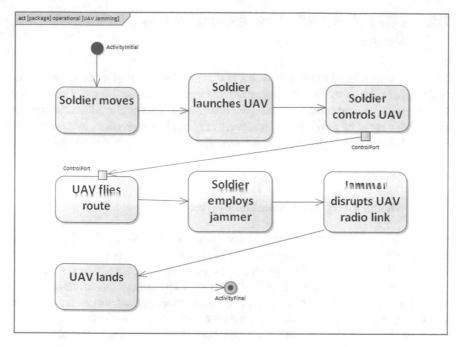

Fig. 8.7 Activity diagram for UAV jamming

measured as a percentage of available energy expended during the road march and prior to the assault. The team wishes to have sufficient energy reserves to move quickly during the combat portion of the assault.

8.4.2.3 Develop Simulation and Service Requirements

In addition to the operational requirements, there are a number of implementation and infrastructure requirements for this project. First, for ubiquitous access to the simulation from distributed workplaces, the most practical approach is to deploy the simulation environment to the cloud. Running a microservices environment in the cloud without locking into a single cloud provider closely aligns with the Docker [7], Kubernetes [31], and Helm [30] technology stack. We will deploy a DEVS-DMF simulation onto this stack. Finally, input and output data sets will be stored in cheap, efficient, cloud-based object storage. These technology choices align well with the problem set, and they offer a cost-efficient solution, because the user does not have to purchase and maintain a lab of computers. Compute resources are deployed to the cloud only during simulation runs and are shut down otherwise. This also allows the user to scale up a very large set of runs on a large number of temporarily deployed computers for short periods.

8.4.3 DSEEP Phase 3—Design Composed Simulation Service

Given our understanding of operational and infrastructure requirements, it is time to flow these down to an independently deployable composed simulation service that meets the defined requirements. We continue to follow the MSaaS engineering process steps to develop the top-level functional and physical architecture of the simulation system, consistent with the principles and technologies laid out in Sect. 8.3.

8.4.3.1 Design Composed Simulation Service

With the defined scenario and performance measures from Sect. 8.4.2, the modeling and simulation team developed the high-level simulation structure shown in Fig. 8.8. The simulation data collection system aggregates entity status during movement to determine soldier exhaustion over time, and it aggregates sensor detections to determine situation awareness over time. There are three separate simulation services. The soldier movement simulation executes the scenario and emits the entity status, to include location and exhaustion, of every entity in the scenario. Based on these

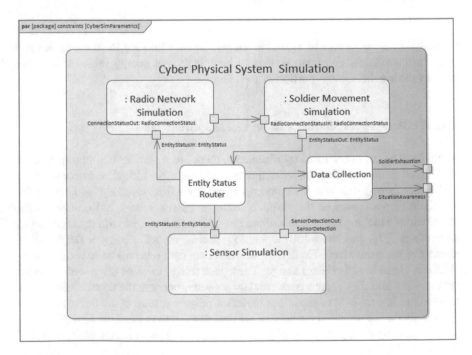

Fig. 8.8 Parametric diagram for the cyber-physical simulation

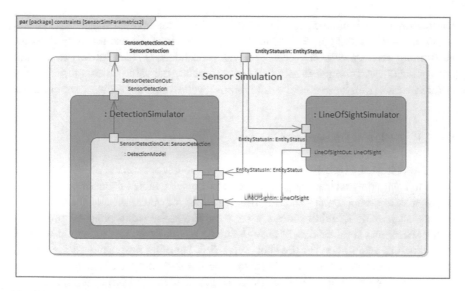

Fig. 8.9 Parametric diagram for the sensor simulation

locations, the radio network simulation tracks the connection status of all radios, to include the UAV radio connections during jamming. If a connection is broken, it updates the solider movement model so that it can disable and land the UAV. The movement model then reports this disabled entity status, along with all other entities, to the sensor simulation. Based on the status and locations of all entities, the sensor simulation emits sensor detections as they occur. Note the dynamic interplay of all the models. The movement and radio connection status of all entities influences the pattern of sensor detections over time. This type of dynamic interplay between models can only be determined using simulation. Simply executing the sensor simulation soldier movement simulation, and radio network simulation in isolation would only yield the static impacts of system properties, but not the dynamic impact resulting from their interactions.

Going one level into the hierarchy, Fig. 8.9 shows the internal structure of the sensor simulation. The line of sight simulator keeps track of enemy locations and reports which entities have line of sight across the terrain to each other. The detection simulator computes when one entity detects another. These detection reports are sent externally to the cyber-physical simulation.

8.4.3.2   Discover and Select M&S Services

One of the key characteristics of a microservices architecture is an enhanced ability to re-use services. This holds true for our analysis scenario. The engineering team selected an existing detection model and an existing terrain model to use in the simulation. In addition, the planning team developed the operational scenario using a scenario service that passed it along to the simulations. To use these models, the consuming services will only have to develop an interface, also known as an anti-corruption layer in domain-driven design, to translate the value objects representing domain events from these domains into internal value objects for their own domains.

8.4.3.3   Design M&S Services

At the lowest level, Fig. 8.10 shows the detection model for sensors. This is a standard engineering model that uses the sensor name to consult a table of sensor performance parameters to compute the detection time and level of detection under certain conditions, such as light and obscuration data. Note that the static physical properties of the sensor and weather do not change during a simulation run (constant weather is a modeling assumption for the short scenario), but other properties such as entity locations, will need to change. These are internal state properties of the detection model which must be updated over time. Target distance and target type, such as a

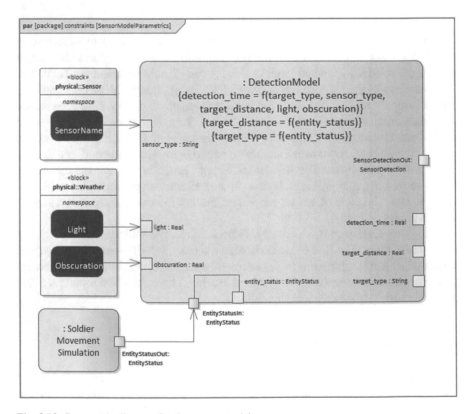

Fig. 8.10 Parametric diagram for the sensor model

crouching person or standing person, are computed based on the entity status information coming into the model from the soldier movement simulation. If the detection model computes a detection, it will emit a sensor detection report.

Corresponding diagrams like Figs. 8.9 and 8.10 were also done for the other models and simulations in the cyber-physical simulation. Take notice of the correspondence between the diagrams in Sects. 8.4.2 and 8.4.3 and the key terms in domain-driven design. The movement simulation, radio network simulation, and sensor simulation each represent a different domain, with domain experts, a sphere of knowledge, and a ubiquitous language. Therefore, these three domains should be implemented as isolated microservices. The block definition diagram in Fig. 8.5 has a set of domain entities. For example, the sensor entity and soldier entity (targets for detection) have important properties in the sensor simulation domain, and these are domain aggregates. The sensor model takes in one domain event, an entity status, which gives the location and posture of all entities in the simulation. These events allow the detection model (Fig. 8.10) to update its internal state and perform detection calculations, emitting a sensor detection as another domain event. Data corresponding to system properties, system state, and domain events is communicated with value objects. Finally, the communications ports on the simulation models represent an anti-corruption layer, so that the sensor model, for example, can operate independently, regardless of other models in the simulation. It is the job of the anti-corruption layer to translate domain events from external domains into the domain model of the sensor domain.

One should also take notice of how these conceptual models map to DEVS-DMF. Each simulation in Fig. 8.8 is a model coordinator synchronizing a number of internal coordinators or simulators. Each simulator in Fig. 8.9 is a model simulator, controlling execution of an internal DEVS model. The detection model in Fig. 8.10 is a DEVS model. Its static properties include sensor name, light, and weather data. Its dynamic internal state is the latest entity status for all entities in the simulation. For its internal state transition, the detector model invokes stateless functions to determine target distance and type, which are in turn served as input to the stateless detection model, which computes sensor detections. If a detection occurs, the DEVS model produces it as an output event.

8.4.3.4   Prepare Detailed Plan

Prior to beginning development, the project lead must prepare a detailed execution plan that ensures all requirements are met, tasks are properly allocated, work is phased, and resources are available. One team each should be assigned to the 3 independent simulation microservices under development. The team needs access to cloud infrastructure, permissions to deploy resources, and access to development and test tools to support the process.

8.4.4 DSEEP Phase 4—Develop Composed Simulation Service

The next steps call for a coordinated development of the individual services in the architecture. One advantage of isolated microservices is that they allow each team to develop and deploy independently, only concerned with their interfaces to other domains via the anti-corruption layer. However, just because their code runs independently, it does not mean they should not regularly communicate. They have to ensure a good semantic understanding of the events coming from other domains, and they have to coordinate their efforts at the top-level to ensure that the final simulation produces a semantically understood, consistent, and valid model of the operational environment. In addition, detailed development often leads to some changes in the top-level architecture and its interfaces via a collaborative change management process. Each service in the architecture, whether it is developed internally or re-used, should be independently tested and validated on the production cloud architecture prior to integration and testing of the entire cyber-physical composed service.

8.4.4.1 Develop Simulation Exchange Data Model

Unlike High Level Architecture, which passes all information via a global object model through a run time infrastructure, DEVS-DMF takes advantage of open, web-based standards and protocols. To support fast messaging, DEVS-DMF passes simulation coordination messages in accordance with a defined Google Protocol Buffers information model [11]. The sensor simulation model uses a standardized sensor domain information model, and the terrain client uses a terrain information model, both also defined in a Protocol Buffers format. For coordination between microservices, each service will have to semantically and syntactically define and standardize their output value objects, so they can be properly consumed by the other services. The information data model and standards used internal to a single service are design choices by those services. Consistent with the isolation property of microservices, they should not be exposed outside of those domains. This simplifies development within a domain because it removes constraints, and it simplifies development outside of a domain because only the externally passed data structures need to be coordinated.

8.4.4.2 Establish Service Agreements

In addition to the common standards and agreements already discussed, there are a few more service agreements for the simulation. Each service must deploy its application components to a specified Docker container accessible from within the cloud environment. They must also prepare a deployment and service specification as a Helm chart so that each service can be independently deployed via the chart and discovered in accordance with the naming service also defined in the chart. Further, coordinated storage locations for scenario data, performance data, and output data enable modification of properties during the tradespace analysis.

8.4.4.3 Implement M&S Services

The project implemented the following services:

Scenario Service   This web interface allowed operational users to define the simulation scenario (Figs. 8.3 and 8.4). The simulation services could then pull the scenario from this service via HTTP call.

Terrain Service   This stateless service accepted Protocol Buffer terrain requests via a message bus. It supported elevation change, route planning, and terrain type calls from the soldier movement service and line of sight calls from the sensor service.

Detection   This stateless service accepted calls from the sensor service via a HTTP interface in order to calculate sensor detections.

Radio Propagation   This stateless model was executed within the Radio Network Simulator, and it calculated factor of free space radio propagation loss between radios or jammers and their receivers.

Radio Network Model   Every few seconds, this DEVS model updated its internal map of entity locations based on received reports. It then calculated the pairwise strength between radios, jammers, and receivers to update the connection status of the network. If a connection status changed, it reported the change.

Jammer Model   This DEVS model contained a single jammer that broadcast at a certain frequency and power based on its properties and jamming instructions assigned as part of scenario orders.

Line of Sight Model   This DEVS model consumed entity status reports and regularly reported the line of sight status between entities in the scenario.

Sensor Model   This DEVS model, also shown in Fig. 8.10, consumed entity status reports and line of sight reports to keep a running status of entities that could possibly acquire each other. For those entities, it ran the detection model to assess and report detections.

Soldier Movement Service   This composed simulation service took in task organization and locations from the scenario, then moved dismounted entities and UAV's in accordance with the operations order defined in the scenario. It emitted status reports containing the location and posture of all entities. For UAV's, if it received a message that the radio connection had been disrupted, it landed and disabled the UAV, preventing target acquisition.

Senor Simulation Service   This composed simulation service coordinated the execution of the sensor model and line of sight model in order to calculate sensor detections for the scenario.

Radio Network Service   This composed simulation service coordinated the execution of the radio network model and jammer model to calculate the connection status of UAV radios in the scenario.

8.4.4.4   Deploy to MSaaS Infrastructure

Once the services were developed, deployment first consisted of pushing the service containers to a Docker repository. The team then initiated a Kubernetes cluster when they wanted to do simulation runs. Once the cluster was running, they deployed each service onto the cluster using Helm charts. The order of deployment is important, and Helm charts included dependencies that supported synchronized deployment. Once all supporting services were deployed, a simulation root controller was deployed to execute the simulation and store simulation results.

8.4.4.5   Isolate and Test

The testing of an isolated microservice is a continuous process, because the ideal way to debug and test evolves during the development life cycle. Unit testing of individual classes and methods is best done in the native development environment with local tools. This type of testing is much more difficult after the application has been deployed to the cloud. Integration testing, however, requires interaction with other components, which may not be available until much later in the life cycle. In this case, behavior-based testing can be achieved with individual microservices. A recommended approach includes

1. Rigorous unit testing of individual classes and methods
2. As functionality is integrated into DEVS-DMF using actors, test the expected behavior of each actor using Akka asynchronous testing
3. Once all actors have been integrated into a single microservice, test the microservice behavior locally. If the service is a DEVS-DMF component, again use Akka asynchronous testing to test expected behavior.
4. Deploy the microservice into a Docker container and test again locally
5. Containerize your test into a Docker container. Deploy both your service and the containerized test kit, and test the behavior of your deployed service. This type of testing requires the availability of interacting services, or their test stubs, for the other service in the behavior sequence. Depending on the number and availability of dependencies, complex versions of this type of testing may be deferred to integration testing, covered in Sect. 8.4.5.

Following this approach allows the developer to fall back to different stages and re-test if bugs appear. If, for example, the deployed service behaves unexpectedly, the developer may suspect the problem is a single Akka actor within a microservice and fall back to Akka asynchronous testing to make the tests more robust and capture the error behavior for debugging. Once complete, the developer would re-run microservice behavior testing at each level before redeploying to ensure that the erroneous behavior does not reappear. In addition to testing frameworks, skillful application of logging frameworks facilitates testing and debugging of microservices.

8.4.5   DSEEP Phase 5—Integrate and Test Composed Simulation Service

Testing simulation services has some advantages over testing microservices in other production environments. When we run a simulation, we have full control over the scenario. Our production environment consists of a discrete set of runs. Unlike a web retailer, our production environment can be started, executed, and shut down without having to worry about lost sales and user experiences. If we have tested individual services as discussed in Sect. 8.4.4.4, then we should embrace prudent integration testing in our production environment. Prudent means that we integrate individual services and increase scenario complexity incrementally, instead of all at once.

8.4.5.1   Plan Execution

Integration testing in the cloud has some prerequisites:

1. Isolated testing of individual microservices, as discussed in Sect. 8.4.4.4, is complete
2. Individual services are containerized and can by dynamically started and stopped, in our case using Helm charts
3. Developers of individual microservices are available so that they can debug and redeploy their services as errors arise in integration testing
4. Individual services are equipped with robust logging capabilities so that they can be monitored during simulation execution.

The next step of integration testing is to gradually increase the complexity of compositions and scenarios under test.

8.4.5.2   Integrate and Test Composed Simulation Service

Consider the cyber-physical simulation Fig. 8.8. Let's integrate the soldier movement service with the sensor simulation service as a first step, leaving the radio network service out. First, consider a simple scenario. We have one friendly sensor with 3 possible targets:

1. Target 1 is right in front of the sensor and easily identified.
2. Target 2 is behind a hill.
3. Target 3 is within line of sight but well beyond the visual range of the sensor.

Looking at the sequence of interactions in Fig. 8.8, we should expect the following to occur:

1. The entity status router should see location updates for our sensor and all 3 targets and pass them to the sensor simulation.

2. The sensor simulation should emit a sensor detection for target 1 and not targets 2 and 3 to the data collection service.
3. The situation awareness for the senor should include only target 1.

A good test is to ensure that logging captures all of these observables, then to run the scenario and evaluate output data to ensure the observables are correct. Then increase the complexity of the scenario and re-run.

When we are satisfied, the soldier movement service and sensor simulation service work well together, follow similar steps to integrate the radio network service into the composition with increasingly complex scenarios while checking for correct observables on the entire composition. Upon completion, we should have good observables across our services from multiple runs of the full scenario from Sect. 8.4.2.1. We are ready for execution.

8.4.6 DSEEP Phase 6—Execute Composed Simulation Service

For our tradespace analysis, we will use a 2×2 factorial design with the factors being the presence of a UAV and the presence of a jammer in the squad equipment.

For each execution, we capture soldier exhaustion, friendly situation awareness, and enemy situation awareness at the time the team begins the assault march.

8.4.6.1 Execute Composed Simulation Service

Four different scenario files were created, one for each cell of the design matrix, and automated execution of each scenario proceeded. The first step was to create the Kubernetes cluster. Note that compute billing for the cloud provider begins with the cluster creation. In our case, a small cluster with a master and three compute nodes is sufficient. In addition, an object storage system loaded the configuration and system performance data. This included, for example, soldier fitness, sensor performance, and radio network configuration data. For each iteration, a helm chart deployed the necessary services in sequence, executed the simulation, and pushed the data to object storage. When all iterations were done, the cluster was shut down.

8.4.6.2 Prepare Simulation Outputs

DEVS-DMF offers a very powerful, flexible, and reusable way to produce simulation outputs. Every DEVS model in the simulation, at any time, can send a message to a global simulation logger. The default logger simply logs messages to a text file, but the simulation team customized the logger to produce the output data needed for this scenario. The specific products include

- A text-based information log of generic events.
- A comma-separated value file of physical exertion data for each soldier throughout the scenario.
- A comma-separated value file of all sensor detections throughout the scenario.
- A text-based log file of radio connectivity.
- A CesiumJS [3] visualization file to support playback and visualization of soldier movement.
- A Keyhole Markup Language (KML) [10] visualization file to support visualization of sensor detections throughout the scenario.

8.4.7   DSEEP Phase 7—Analyze Data and Evaluate Results

With simulation data files available in cloud-based object storage in standardized formats such as comma-separated values and JSON logs, the large universe of cloud-based data tools is available to support analysis.

8.4.7.1   Analyze Data

Figure 8.11 shows the CesiumJS visualization of the team on its assault march toward enemy forces. This is right after the time that exhaustion and situation awareness data have been collected. For this march, it is important to have energy reserves, to know as much about the enemy as possible, and to be undetected by enemy forces.

Fig. 8.11 Soldier movement visualization

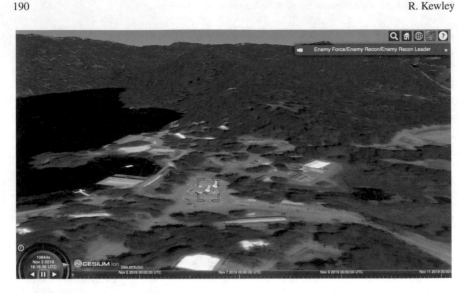

Fig. 8.12 UAV detection visualization

Table 8.1 Simulation results for soldier remaining energy with each of the factors

		Jammer	
		− (%)	+ (%)
UAV	−	42	35
	+	38	35

Figure 8.12 shows a visualization of the KML sensor detections by the UAV, supporting situation awareness. Note that the UAV has detected both the enemy soldiers prior to the assault march. The detection analysis showed that the friendly UAV was able to give 100% situation awareness, while the jammer was able to defeat the enemy UAV and degrade their situation awareness to zero, allowing the team to achieve surprise in the assault march.

For data analysis, the team pulled the comma-separated value outputs into cloud-based statistics tool, RStudio Cloud [27], in order to create the data in Table 8.1, showing the remaining energy of the team carrying the different equipment. This remaining energy impacts how far and fast the team can move during the assault phase. Note that the heavier jammer gives a greater reduction in energy, but that the combination of both does not further reduce team energy because the jammer and UAV are carried by different soldiers.

8.4.7.2 Evaluate and Feedback Results

Returning to the system design in Fig. 8.5, the simulation analysis provides valuable tradespace data. Given the UAV design and jammer design under consideration, with specific properties, the soldier team was able to begin the assault march at a 100–0% situation awareness advantage with a 7% reduction in available energy. The lead systems engineer takes this into consideration for designing the team's equipment set. In addition, they can do further engineering and analysis to try to reduce the weight of the equipment set, mitigating the negative impacts on soldier energy.

8.5 Conclusions and Follow-On Research

This chapter has shown the engineering development of a multi-domain simulation scenario using the DEVS-DMF simulation architecture while following the MSaaS engineering process. These techniques follow best practices for distributed modeling and simulation in the cloud and for a microservices architecture. By integrating independent models from each domain, instead of integrating simulations from each domain, the task is simplified because the integration engineers only need to deal with inputs and outputs to the models, not coordination of entire simulations. The domain-based separation of concerns also maintains responsibility for domain models by domain experts, supporting model validation.

Follow-on work will focus on further developing the DEVS-DMF environment and enriching it with tools. Planned advances include the development of a graphical user interface to enable code generation of model simulator and model coordinators, development of a user interface for the information exchange model, and adding a more robust and queryable logging system to support integration tests.

References

1. Agha G, Hewitt C (1985) Concurrent programming using actors: exploiting large-scale paral-lelism
2. Bonèr J (2016) Reactive microservices architecture. O'Reilly Media, Inc
3. Cesium. Cesiumjs—open source 3d mapping
4. Chow ACH, Zeigler BP (1994) Parllel DEVS: a parallel, hierarchical, modular modeling for-malism. In: Proceedings of winter simulation conference. IEEE
5. Dahmann JS (1999) The high level architecture and beyond: technology challenges. In: Proceedings thirteenth workshop on parallel and distributed simulation. PADS 99. (Cat. No.PR00155). IEEE Computer Society
6. D'Angelo G, Marzolla M (2014) New trends in parallel and distributed simulation: from many-cores to cloud computing. Simul Model Pract Theory 49:320–335
7. Docker Inc (2019) Empowering app development for developers I docker
8. Friedenthal S, Moore A, Steiner R (2014) A practical guide to SysML. Elsevier LTD, Oxford

9. Gallant S, McGroarty C, Kewley R, Diemunsch J, McDonnell J, Snively K (2018) Cloud-based modeling and simulation study group. In: Proceedings of the 2018 interservice/industry training, simulation, and education conference
10. Google. Keyhole markup language
11. Google (2019) Protocol buffers
12. IEEE (2010) IEEE standard for modeling and simulation (M&S) high level architecture (HLA)–framework and rules
13. IEEE (2011) IEEE recommended practice for distributed simulation engineering and execution process (DSEEP)
14. Johnson HE, Tolk A (2013) Evaluating the applicability of cloud computing enterprises in support of the next generation of modeling and simulation architectures. In: Proceedings of the military modeling & simulation symposium, MMS '13. San Diego, CA, USA, pp 4:1–4:8. Society for Computer Simulation International
15. Kewley R (2012) Federated simulation for system of systems engineering. Wiley-Blackwell
16. Kewley R (2019) Discrete event specification (DEVS) distributed modeling framework
17. Kewley R, Kester N, McDonnell J (2016) DEVS distributed modeling framework: a parallel DEVS implementation via microservices. In: Proceedings of the symposium on theory of modeling & simulation, TMS-DEVS '16. San Diego, CA, USA, pp 22:1–22:8. Society for Computer Simulation International
18. Kratzke N, Siegfried R (2019) Towards cloud-native simulations—lessons learned from the front-line of cloud computing. To Be Publ J Def Model Simul
19. Lightbend (2019) akka
20. Marrou L, Carr G, Nielsen K (2018) Exploring cloud-based terrain generation services. In: Proceedings of the 2018 interservice/industry training, simulation, and education conference
21. NATO Science and Technology Organization. Modelling and simulation as a service -volume 2: MSaaS discovery service and metadata
22. NATO Science and Technology Organization (2018) Modelling and simulation as a service -vol 1: MSaaS reference architecture
23. NATO Science and Technology Organization (2018) Modelling and simulation as a service -vol 3: MSaaS engineering process
24. NATO Science and Technology Organization (2019) Modelling and simulation as a service (MSaaS)–rapid deployment of interoperable and credible simulation environments
25. North Atlantic Treaty Organization (2018) NATO's joint air power strategy
26. Rieger LA (2014) Simulations in the cloud–a manager's challenge. In: Proceedings of the 2014 interservice/industry training, simulation, and education conference
27. RStudio, Inc (2019) Rstudio cloud
28. Todd South (2019) This 3-star army general explains what multi-domain operations mean for you
29. Sprang R. Russia in Ukraine 2013–2016: the application of new type warfare maximizing the exploitation of cyber, IO, and media
30. The Cloud Native Computing Foundation. Helm—the package manager for kubernetes
31. The Linux Foundation. kubernetes
32. Tolk A, Mittal S (2014) A necessary paradigm change to enable composable cloud-based M&S services. In: Proceedings of the 2014 winter simulation conference, WSC '14. IEEE Press, Piscataway, NJ, USA, pp 356–366
33. US Army Training and Doctrine Command (2018) The U.S. army in multi-domain operations 2028
34. Wade Waldron (2018) Reactive architecture: domain driven design

Rob Kewley as the Director of simlytics.cloud, has performed consulting and development on a series of projects in the military domain, to include soldier systems, radar systems, cyber analytics, and sensor systems. He has served as the head of the West Point Department of Systems Engineering and as the Chief Systems Engineer for the Assistant Secretary of the Army for Acquisition, Logistics, and Technology. His research background is in the development of simulation methodologies and engineering processes to support the application of simulation to complex system of systems problems. He has served as the co-chair of the Simulation Interoperability Standards Organization's Cloud Based Modeling and Simulation Study Group. He is also working as a member of NATO Modeling and Simulation Group 164—Modeling and Simulation as a Service.

Chapter 9
Anticipative, Incursive and Hyperincursive Discrete Equations for Simulation-Based Cyber-Physical System Studies

Daniel M. Dubois

Abstract This chapter will present algorithms for simulation of discrete space-time partial differential equations in classical physics and relativistic quantum mechanics. In simulation-based cyber-physical system studies, the main properties of the algorithms must meet the following conditions. The algorithms must be numerically stable and must be as compact as possible to be embedded in cyber-physical systems. Moreover the algorithms must be executed in real-time as quickly as possible without too much access to the memory. The presented algorithms in this paper meet these constraints. As a first example, we present the second-order hyperincursive discrete harmonic oscillator that shows the conservation of energy. This recursive discrete harmonic oscillator is separable to two incursive discrete oscillators with the conservation of the constant of motion. The incursive discrete oscillators are related to forward and backward time derivatives and show anticipative properties. The incursive discrete oscillators are not recursive but time inverse of each other and are executed in series without the need of a work memory. Then, we present the second-order hyperincursive discrete Klein–Gordon equation given by space-time second-order partial differential equations for the simulation of the quantum Majorana real 4-spinors equations and of the relativistic quantum Dirac complex 4-spinors equations. One very important characteristic of these algorithms is the fact that they are space-time symmetric, so the algorithms are fully invertible (reversible) in time and space. The development of simulation-based cyber-physical systems indeed evolves to quantum computing. So the presented computing tools are well adapted to these future requirements.

D. M. Dubois (✉)
Centre for Hyperincursion and Anticipation in Ordered Systems (CHAOS), CHAOS ASBL, Institute of Mathematics B37, University of Liège, Sart-Tilman, 4000 Liège, Belgium
e-mail: Daniel.Dubois@ulg.ac.be; Daniel.Dubois@uliege.be; ddubois.chaos@gmail.com
URL: http://www.sia.hec.ulg.ac.be

HEC Liège, Université de Liège, 14 Rue Louvrex, 4000 Liège, Belgium

© Springer Nature Switzerland AG 2020
J. L. Risco Martín et al. (eds.), *Simulation for Cyber-Physical Systems Engineering*,
Simulation Foundations, Methods and Applications,
https://doi.org/10.1007/978-3-030-51909-4_9

195

9.1   Introduction

This chapter begins with a presentation step by step of the second-order hyperincursive discrete equation of the position of the harmonic oscillator. We show that the second-order hyperincursive discrete harmonic oscillator is represented by the equations of the position and velocity of the hyperincursive discrete harmonic oscillator that is separable into two incursive discrete harmonic oscillators. We demonstrate that these incursive discrete equations of the position and velocity of the harmonic oscillators can be described by a constant of motion. After that, we give a numerical simulation of the two incursive discrete harmonic oscillators. The numerical values correspond exactly to the analytical solutions. Then we present the hyperincursive discrete harmonic oscillator. And we give also a numerical simulation of the hyperincursive discrete harmonic oscillator. The numerical values correspond also to the analytical solutions. After that, we demonstrate that a rotation on the position and velocity variables transforms the incursive discrete harmonic oscillators to recursive discrete harmonic oscillators.

Then, this chapter presents the second-order hyperincursive discrete Klein–Gordon equation.

This discrete Klein–Gordon equation bifurcates to the hyperincursive discrete Majorana equations which tend to the real 4-spinors Majorana first-order partial differential equations for the intervals of time and space tending to zero.

After that, we demonstrate that the Majorana equations bifurcate to the 8 real Dirac first-order partial differential equations that are transformed to the original Dirac 4-spinors equations. The 4 hyperincursive discrete Dirac 4-spinors equations are then presented.

Finally, we show that there are 16 discrete functions associated with the space and time symmetric discrete Klein–Gordon equation. This is in agreement with the Proca thesis on the 16 components of the Dirac wave function in 4 groups of 4 equations.

In this chapter, we restricted our derivation of the Majorana and Dirac equations to the first group of 4 equations depending on 4 functions.

This chapter is based on my papers in this field.

The paper [1] concerns the hyperincursive algorithms of classical harmonic oscillator applied to quantum harmonic oscillator separable into incursive oscillators. The paper [2] deals with a unified discrete mechanics given by the bifurcation of the hyperincursive discrete harmonic oscillator, the hyperincursive discrete Schrödinger quantum equation, the hyperincursive discrete Klein–Gordon equation and the Dirac quantum relativist equations. In this paper [2], I have demonstrated that the second-order hyperincursive discrete Klein–Gordon equation bifurcates to the 4 Dirac first-order equations, in one space dimension.

An introduction to incursion and hyperincursion is given in the following series of papers on the total incursive control of linear, non-linear and chaotic systems [3], on computing anticipatory systems with incursion and hyperincursion [4], on the computational derivation of quantum and relativist systems with forward–backward space-time shifts [5], on a review of incursive, hyperincursive and anticipatory systems, with the foundation of anticipation in electromagnetism [6], then, on the precision and stability analysis of Euler, Runge–Kutta and incursive algorithms for the harmonic oscillator [7], and finally, on the new concept of deterministic anticipation in natural and artificial systems [8].

I wrote a series of theoretical papers on the discrete physics with Adel Antippa on the harmonic oscillator via the discrete path approach [9], on anticipation, orbital stability, and energy conservation in discrete harmonic oscillators [10], on the dual incursive system of the discrete harmonic oscillator [11], on the superposed hyperincursive system of the discrete harmonic oscillator [12], on the incursive discretization, system bifurcation, and energy conservation [13], on the hyperincursive discrete harmonic oscillator [14], on the synchronous discrete harmonic oscillator [15], on the discrete harmonic oscillator, a short compendium of formulas [16], on the time-symmetric discretization of the harmonic oscillator [17], and finally, on the discrete harmonic oscillator, evolution of notation and cumulative erratum [18]. This discrete physics is based on the fundamental mathematical development of the hyperincursive and incursive discrete harmonic oscillator.

An important purpose of this chapter deals with the bifurcation of the second-order hyperincursive discrete Klein–Gordon equation firstly to the 4 hyperincursive discrete real 4-spinors Majorana equations, secondly to the 8 hyperincursive discrete real 8-spinors Dirac equations that can be rewritten as the 4 hyperincursive discrete complex 4-spinors Dirac equations.

In 1926, Klein [19] and Gordon [20] presented independently what is called the Klein–Gordon equation. In 1928, Dirac [21] introduced the relativist quantum mechanics based on this Klein–Gordon equation. His fundamental equation is based on 4-spinors and is given by 4 first-order complex partial differential equations. All the work of Dirac is well explained in his book [22].

In 1930 and 1932, Proca [23, 24] proposed a generalization of the Dirac theory with the introduction of 4 groups of 4-spinors, and with 16 first-order complex partial differential equations.

In 1937, Majorana [25] proposed a real 4-spinors Dirac equation, given by 4 first-order real partial differential equations. Ettore Majorana disappears just after having written this fundamental paper. Pessa [26] presented a very interesting paper on the Majorana oscillator based on the 4 first-order real partial differential equations.

An excellent introduction to quantum mechanics is given in the books of Messiah [27].

This chapter is essentially based on my following recent papers.

The paper [28] deals with deduction of the Majorana real 4-Spinors generic Dirac equation from the computable hyperincursive discrete Klein–Gordon equation. Then the paper [29] shows that the hyperincursive discrete Klein–Gordon Equation is the algorithm for computing the Majorana real 4-spinors equation and the real 8-spinors Dirac equation. In fact, this corresponds to bifurcation of the hyperincursive discrete Klein–Gordon equation to real 4-spinors Dirac equation related to the Majorana Equation [30].

Then the next paper is a continuation of the paper on the unified discrete mechanics [2], dealing with the bifurcation of hyperincursive discrete harmonic oscillator, Schrödinger's quantum oscillator, Klein–Gordon's equation and Dirac's quantum relativist equations. Indeed this next paper on the unified discrete mechanics II [31] deals with the space and time-symmetric hyperincursive discrete Klein–Gordon equation that bifurcates to the 4 incursive discrete Majorana real 4-spinors equations. Then the paper on the unified discrete mechanics III [32] deals with the hyperincursive discrete Klein–Gordon equation that bifurcates to the 4 incursive discrete Majorana and Dirac equations and to the 16 Proca equations.

The review paper [33], as an update of my paper [3], deals with the time-symmetric hyperincursive discrete harmonic oscillator separable into two incursive harmonic oscillators with the conservation of the constant of motion. As a novelty, we present the transformation of the incursive discrete equations to recursive discrete equations by a rotation of the position and velocity variables of the harmonic oscillator [33], as described in this chapter.

More developments are given in the paper [34] on the rotation of the two incursive discrete harmonic oscillators to recursive discrete harmonic oscillators with the Hadamard matrix. Then, a continuation deals with the rotation of the relativistic quantum Majorana equation with the Hadamard matrix and Unitary matrix U [35]. Finally in this chapter, we give the analytical solution of the quantum Dirac equation for a particle at rest following our last paper on the relations between the Majorana and Dirac quantum equations [36].

This chapter is organized as follows.

Section 9.2 deals with a presentation step by step of the second-order hyperincursive discrete harmonic oscillator.

Section 9.3 develops the 4 incursive discrete equations of the hyperincursive discrete harmonic oscillator. Then Sect. 9.4 presents the constants of motion of the two incursive discrete harmonic oscillators. In Sect. 9.5, we give numerical simulations of the two incursive discrete harmonic oscillators.

Section 9.6 presents the hyperincursive discrete harmonic oscillator. Section 9.7 gives numerical simulations of the hyperincursive discrete harmonic oscillator.

Section 9.8 deals with a rotation of the position and velocity variables of the incursive discrete equations of the harmonic oscillator which are transformed to recursive discrete equations. This result is fundamental because it gives an explanation of the anticipative effect of the discretization of the time in discrete physics. The information obtained from the hyperincursive discrete equations is richer than obtained by continuous physics.

In Sect. 9.9, we present the Klein–Gordon partial differential equation and the space and time-symmetric second-order hyperincursive discrete Klein–Gordon equation that bifurcates to the relativistic quantum Majorana and Dirac equations.

Then, in Sect. 9.10, we present the hyperincursive discrete relativistic quantum Majorana equations. For intervals of time and space tending to zero, these discrete equations tend to the 4 first-order partial differential Majorana equations.

In Sect. 9.11, next, in defining the Majorana functions by 2-spinors real functions, after some mathematical manipulations, we demonstrate that the Majorana real 4-spinors equations bifurcate into the 8 real equations. These 8 real first-order partial differential equations represent the Dirac real 8-spinors equations that are transformed to the original Dirac complex 4-spinors equations.

Then Sect. 9.12 presents the 4 hyperincursive discrete Dirac 4-spinors equations depending on 4 complex discrete Dirac wave functions.

In Sect. 9.13, we show that there are 16 complex functions associated with this second-order hyperincursive discrete Klein–Gordon equation. This is in agreement with the Proca thesis, for which the Dirac function has 16 components and divided into 4 groups of 4 functions with 4 equations. In this chapter, we restricted our derivation of the Majorana and Dirac equations to the first group of 4 equations depending to 4 functions.

Finally Sect. 9.14 deals with numerical simulations of the hyperincursive discrete Majorana and Dirac wave equations depending on time and one spatial dimension (1D) and with a null mass.

9.2 Presentation Step by Step of the Second-Order Hyperincursive Discrete Harmonic Oscillator

The harmonic oscillator is represented by the second-order temporal ordinary differential equations

$$d^2x(t)/dt^2 = -\omega^2 x(t) \tag{9.2.1a}$$

with the velocity given by

$$v(t) = dx(t)/dt \tag{9.2.1b}$$

where $x(t)$ is the position and $v(t)$ the velocity as functions of the time t and where the pulsation ω is related to the spring constant k and the oscillating m by

$$\omega^2 = k/m \tag{9.2.1c}$$

The harmonic oscillator can be represented by the two ordinary differential equations:

$$dx(t)/dt = v(t)$$
$$dv(t)/dt = -\omega^2 x(t) \tag{9.2.2a, b}$$

The solution is given by

$$x(t) = x(0)\cos(\omega t) + [v(0)/\omega]\sin(\omega t)$$
$$v(t) = -\omega x(0)\sin(\omega t) + v(0)\cos(\omega t) \tag{9.2.2c, d}$$

with the initial conditions $x(0)$ and $v(0)$.

In the phase space, given by $x(t), v(t)$, the solutions are given by closed curves (orbital stability).

The period of oscillations is given by $T = 2\pi/\omega$.

The energy $e(t)$ of the harmonic oscillator is constant and is given by

$$e(t) = kx^2(t)/2 + mv^2(t)/2 = kx^2(0)/2 + mv^2(0)/2 = e(0) = e_0 \qquad (9.2.3)$$

The harmonic oscillator is computable by recursive functions from the discretization of the differential equations. The differential equations of the harmonic oscillator depend on the current time.

In the discrete form, there are the discrete current time t and the interval of time $\Delta t = h$.

The discrete time is defined as $t_k = t_0 + kh$, $k = 0, 1, 2, \ldots$,

where t_0 is the initial value of the time and k is the counter of the number of interval of time h.

The discrete position and velocity variables are defined as $x(k) = x(t_k)$ and $v(k) = v(t_k)$.

The discrete equations consists in computing firstly the first equation to obtain, $x(k + 1)$, and then compute the second equation in using the just computed, $x(k + 1)$, as follows

$$x(k + 1) = x(k) + hv(k)$$
$$v(k + 1) = v(k) - h\omega^2 x(k + 1) \qquad (9.2.4a, b)$$

In fact, the first equation used the forward derivative and the second equation used the backward derivative,

$$[x(k + 1) - x(k)]/h = v(k)$$
$$[v(k) - v(k - 1)]/h = -h\omega^2 x(k) \qquad (9.2.4c, d)$$

The position, $x(k + 1)$, and the velocity, $v(k)$, are not computed at the same time step.

I called such a system, an incursive system, for inclusive or implicit recursive system, e.g., [4].

A second possibility occurs if the second equation is firstly computed, and then the first equation is computed in using the just computed, $v(k + 1)$, as follows

$$v(k + 1) = v(k) - h\omega^2 x(k)$$
$$x(k + 1) = x(k) + hv(k + 1) \qquad (9.2.5a, b)$$

In fact, the first equation used the forward derivative and the second equation used the backward derivative,

$$[v(k+1) - v(k)]/h = -\omega^2 x(k)$$
$$[x(k) - x(k-1)]/h = v(k) \qquad\qquad (9.2.5c, d)$$

The position, $x(k)$, and the velocity, $v(k+1)$, are not computed at the same time step.

But in using the two incursive systems, we see that the position in the first incursion, $x(k+1)$, corresponds to the velocity in the second incursion, $v(k+1)$, at the same time step, $(k+1)$. And similarly, we see that the velocity in the first incursion, $v(k)$, corresponds to the position in the second incursion, $x(k)$, at the same time step k. So, both incursions give two successive positions and velocities at two successive time steps, k, and, $k+1$.

An important difference between the incursive and the recursive discrete systems is the fact that in the incursive system, the order in which the computations are made is important: this is a sequential computation of equations. In the recursive systems, the order in which the computations are made is without importance: this is a parallel computation of equations.

The two incursive harmonic oscillators are numerically stable, contrary to the classical recursive algorithms like the Euler and Runge–Kutta algorithms [7].

In the following paragraphs, it will be given a generalized equation that integrates both incursions to form a hyperincursive system.

In my paper [3], I defined a generalized forward-backward discrete derivative

$$D_w = wD_f + (1-w)D_b \qquad\qquad (9.2.6)$$

where w is a weight taking the values between 0 and 1, and where the discrete forward and backward derivatives on a function f are defined by

$$D_f(f) = \Delta^+ f/\Delta t = \left[f(k+1) - f(k)\right]/h$$
$$D_b(f) = \Delta^- f/\Delta t = \left[f(k) - f(k-1)\right]/h \qquad\qquad (9.2.7a, b)$$

The generalized incursive discrete harmonic oscillator is given by Dubois [3] and reprinted in the review paper [33]:

$$(1-w)x(k+1) + (2w-1)x(k) - wx(k-1) = hv(k)$$
$$wv(k+1) + (1-2w)v(k) + (w-1)v(k-1) = -h\omega^2 x(k) \qquad\qquad (9.2.8a, b)$$

When $w = 0$, $D_0 = D_b$, this gives the first incursive equations:

$$x(k+1) - x(k) = hv(k)$$
$$v(k) - v(k-1) = -h\omega^2 x(k) \qquad\qquad (9.2.9a, b)$$

When $w = 1$, $D_1 = D_f$, this gives the second incursive equations:

$$x(k) - x(k-1) = hv(k)$$
$$v(k+1) - v(k) = -h\omega^2 x(k) \qquad \text{(9.2.10a, b)}$$

When $w = 1/2$, $D_{1/2} = [D_f + D_b]/2$, this gives the hyperincursive equations:

$$x(k+1) - x(k-1) = +2hv(k)$$
$$v(k+1) - v(k-1) = -2h\omega^2 x(k) \qquad \text{(9.2.11a, b)}$$

where the discrete derivative is given by

$$D_s = D_{1/2} = [D_f + D_b]/2$$
$$D_s(f) = D_{1/2}(f) = [f(k+1) - f(k-1)]/2h \qquad \text{(9.2.7c)}$$

that defines a time-symmetric derivative noted, D_s.

NB: the time-symmetric derivative D_s in hyperincursive discrete equations

$$D_s(f) = [f(k+1) - f(k-1)]/2h$$

is not the same as the classical central derivative D_c given in classical difference equations theory

$$D_c(f) = [f(k+1/2) - f(k-1/2)]/h.$$

These Eqs. (9.2.11a, b) integrate the two incursive equations [4–6].

Let us remark that this first hyperincursive Eq. (9.2.11a) can be also obtained by adding the Eq. (9.2.9a) to the Eq. (9.2.10a), and the second hyperincursive Eq. (9.2.11b) by adding the Eq. (9.2.9b) to the Eq. (9.2.10b).

In putting the velocity, $v(k)$, of the first Eq. (9.2.11b)

$$v(k) = [x(k+1) - x(k-1)]/2h \qquad \text{(9.2.12a)}$$

to the second Eq. (9.2.11b),

$$x(k+2) - 2x(k) + x(k-2) = -4h^2\omega^2 x(k) \qquad \text{(9.2.12b)}$$

one obtains what I called "the second-order hyperincursive discrete harmonic oscillator", corresponding to the second-order differential equations of the harmonic oscillator given by Eq. (9.2.1a), with the velocity given by the Eq. (9.2.1b).

In this section, we have presented the second-order hyperincursive discrete harmonic oscillator given by the Eq. (9.2.12b) that is separable into 4 first-order incursive discrete equations of the harmonic oscillator. The next section will present the 4 dimensionless incursive discrete equations.

9.3 The 4 Dimensionless Incursive Discrete Equations of the Harmonic Oscillator

For the discrete harmonic oscillator, let us use the dimensionless variables, X and V, of Antippa and Dubois [16], for the variables, x and v, as follows:

$$X(k) = [k/2]^{1/2} x(k) \text{ and } V(k) = [m/2]^{1/2} v(k) \qquad (9.3.1\text{a, b})$$

with the dimensionless time

$$\tau = \omega t \qquad (9.3.2\text{a})$$

where the pulsation (9.2.1c) is given by

$$\omega = [k/m]^{1/2} \qquad (9.3.2\text{b})$$

and with the dimensionless interval of time given by

$$\Delta\tau = \omega\Delta t = \omega h = H \qquad (9.3.3)$$

So, the two incursive dimensionless harmonic oscillators are given by the following 4 first-order discrete equations: First Incursive Oscillator, from the dimensionless equations (9.2.4a, b):

$$X_1(k+1) = X_1(k) + HV_1(k)$$
$$V_1(k+1) = V_1(k) - HX_1(k+1) \qquad (9.3.4\text{a, b})$$

Second Incursive Oscillator, from the dimensionless equations (9.2.5a, b):

$$V_2(k+1) = V_2(k) - HX_2(k)$$
$$X_2(k+1) = X_2(k) + HV_2(k+1) \qquad (9.3.5\text{a, b})$$

These incursive discrete oscillators are non-recursive computing anticipatory systems.

Indeed, in Eq. (9.3.4b) of the first incursive oscillator, the velocity, $V_1(k+1)$, at the future next time step, $(k+1)$, is computed from the velocity, $V_1(k)$, at the current time step, k, and the position, $X_1(k+1)$, at the future next time step, $(k+1)$, which represents an anticipatory system represented by an anticipation of one time step, k. Similarly in Eq. (9.3.5b) of the second incursive oscillator, the position, $X_2(k+1)$, at the future next time step, $(k+1)$, is computed from the position, $X_2(k)$, at the current time step, k, and the velocity, $V_2(k+1)$, at the future next time step, $(k+1)$, which represents an anticipatory system represented by an anticipation of one time step, k.

These two incursive discrete harmonic oscillators define a discrete hyperincursive harmonic oscillator given by four incursive discrete equations.

A complete mathematical development of incursive and hyperincursive systems was presented in a series of papers by Adel F. Antippa and Daniel M. Dubois on the harmonic oscillator via the discrete path approach [9], on anticipation, orbital stability, and energy conservation in discrete harmonic oscillators [10], on the dual incursive system of the discrete harmonic oscillator [11], on the superposed hyperincursive system of the discrete harmonic oscillator [12], on the incursive discretization, system bifurcation, and energy conservation [13], on the hyperincursive discrete harmonic oscillator [14], on the synchronous discrete harmonic oscillator [15], on the discrete harmonic oscillator, a short compendium of formulas [16], on the time-symmetric discretization of the harmonic oscillator [17], and finally, on the discrete harmonic oscillator, evolution of notation and cumulative erratum [18].

The next section will present the constants of motion of the two incursive discrete harmonic oscillators.

9.4   The Constants of Motion of the Two Incursive Discrete Equations of the Harmonic Oscillator [33]

The constant of motion of the first incursive oscillator

$$X_1(k+1) = X_1(k) + HV_1(k)$$
$$V_1(k+1) = V_1(k) - HX_1(k+1) \tag{9.3.4a, b}$$

is given by

$$K_1(k) = X_1^2(k) + V_1^2(k) + HX_1(k)V_1(k) = K_1 = \text{constant} \tag{9.4.1a}$$

Theorem 9.1 [33] *The expression $K_1(k) = X_1^2(k) + V_1^2(k) + HX_1(k)V_1(k)$ is a constant of motion of the first incursive equations* (9.3.4a, b).

Proof Multiply the first Eq. (9.3.4a) by $X_1(k+1)$ at right and the second Eq. (9.3.4b) by $V_1(k+1)$ at left, then add the two equations, and one obtains successively

$$
\begin{aligned}
K_1(k+1) &= X_1(k+1)X_1(k+1) + V_1(k+1)V_1(k+1) + HX_1(k+1)V_1(k+1) \\
&= X_1(k+1)X_1(k) + HX_1(k+1)V_1(k) + V_1(k)V_1(k+1) \\
&= X_1(k)X_1(k) + HX_1(k)V_1(k) + HX_1(k+1)V_1(k) + V_1(k)V_1(k) - HV_1(k)X_1(k+1) \\
&= X_1(k)X_1(k) + HX_1(k)V_1(k) + V_1(k)V_1(k) = K_1(k) = K_1 = \text{constant}
\end{aligned}
$$

So the expression is constant because the expression is invariant in two successive temporal steps. ∎

In replacing the expression of the velocity $V_1(k)$ from Eq. (9.3.4a) to the H term in Eq. (9.4.1a), the term depending on H disappears, as follows

$$X_1(k)X_1(k) + V_1(k)V_1(k) + X_1(k)[X_1(k+1)-X_1(k)] = K_1$$

or

$$X_1(k)X_1(k+1) + V_1(k)V_1(k) = K_1 \tag{9.4.1b}$$

which looks like the conservation of the energy.

The constant of motion of the second incursive oscillator

$$V_2(k+1) = V_2(k) - HX_2(k)$$
$$X_2(k+1) = X_2(k) + HV_2(k+1) \tag{9.3.5a, b}$$

is given by

$$K_2(k) = X_2^2(k) + V_2^2(k) - HX_2(k)V_2(k) = K_2 = \text{constant} \tag{9.4.2a}$$

Theorem 9.2 [33] *The expression $K_2(k) = X_2^2(k) + V_2^2(k) - HX_2(k)V_2(k)$ is a constant of motion of the second incursive equations (9.3.5a, b).*

Proof Multiply the first Eq. (9.3.5a) by $V_2(k+1)$ at right and the second Eq. (9.3.5b) by $X_2(k+1)$ at left, then add the two equations, and one obtains successively

$$
\begin{aligned}
K_2(k+1) &= X_2(k+1)X_2(k+1) + V_2(k+1)V_2(k+1) - HX_2(k+1)V_2(k+1) \\
&= X_2(k+1)X_2(k) - HX_2(k)V_2(k+1) + V_2(k)V_2(k+1) \\
&= X_2(k)X_2(k) + HV_2(k+1)X_2(k) - HX_2(k)V_2(k+1) + V_2(k)V_2(k+1) - HV_2(k)X_2(k) \\
&= X_2(k)X_2(k) - HX_2(k)V_2(k) + V_2(k)V_2(k) = K_2(k) = K_2 = \text{constant}
\end{aligned}
$$

So the expression is constant because the expression is invariant in two successive temporal steps. ∎

In replacing the expression of the position $X_2(k)$ from Eq. (9.3.5a) to the H term in Eq. (9.4.2b), the term depending on H disappears

$$X_2(k)X_2(k) + V_2(k)V_2(k) - [V_2(k)-V_2(k+1))]V_2(k) = K_2$$

or

$$X_2(k)X_2(k) + V_2(k+1)V_2(k) = K_2 \tag{9.4.2b}$$

that also looks like the conservation of the energy.

These constants of motion (9.4.1a) and (9.4.2a) differ with the inversion of the sign of H, as follows

$$+H = +\omega h = +\omega \Delta t, \text{ and } -H = -\omega h = -\omega \Delta t \tag{9.4.3a, b}$$

because the inversion of the discrete time interval of the first incursion gives the second incursion.

NB: It is very important to notice that there is a fundamental difference between an inversion of the sign of the discrete time, Δt, in the discrete equations and an inversion of the sign of the continuous time, t, in the differential equations.

Let us now consider a simple example of the solution of the discrete position and the discrete velocity of the dimensionless discrete harmonic oscillator, given by the following analytical solution (synchronous solution)

$$X_1(k) = \cos(2k\pi/N) \text{ and } V_1(k) = -\sin((2k+1)\pi/N) \qquad (9.4.4a, b)$$

$$X_2(k) = \cos((2k+1)\pi/N) \text{ and } V_2(k) = -\sin(2k\pi/N) \qquad (9.4.5a, b)$$

where N is the number of iterations for a cycle of the oscillator,
   with the index of iterations $k = 0, 1, 2, 3, \ldots$
   The interval of discrete time H depends of N (for a synchronous solution):

$$H = 2\sin(\pi/N) \qquad (9.4.6)$$

For $N = 6$, for example,

$$H = 2\sin(\pi/6) = 1 \qquad (9.4.6a)$$

The two constants of motion, with the solutions (9.4.4a, b) and (9.4.5a, b) are given by

$$\cos^2(2k\pi/N) + \sin^2((2k+1)\pi/N) - H\cos(2k\pi/N)\sin((2k+1)\pi/N) = K_1$$
$$\cos^2((2k+1)\pi/N) + \sin^2(2k\pi/N) + H\cos((2k+1)\pi/N)\sin(2k\pi/N) = K_2$$

For $N = 6, H = 1, k = 0$, one obtains the same constant of motion for the two incursive oscillators:

$$\cos^2(0) + \sin^2(\pi/6) - \cos(0)\sin(\pi/6) = 1.0 + 0.25 - 0.5 = 0.75 = K_1$$
$$(9.4.7a)$$

$$\cos^2(\pi/6) + \sin^2(0) + \cos(\pi/6)\sin(0) = 0.75 + 0.0 + 0.0 = 0.75 = K_2$$
$$(9.4.7b)$$

And the averaged energy is a constant given by

$$[E_1(k) + E_2(k)]/2 = [X_1^2(k) + V_1^2(k) + X_2^2(k) + V_2^2(k)]/2$$
$$= \left[\cos^2((2k+1)\pi/N) + \sin^2(2k\pi/N) + \cos^2(2k\pi/N) + \sin^2((2k+1)\pi/N)\right]/2 = 1$$

A very interesting and important invariant, INV_{12}, is given by

$$INV_{12} = X_1(k)X_2(k) + V_2(k)V_1(k) = \text{constant} \tag{9.4.8}$$

With the values of the example, this gives a constant

$$INV_{12} = X_1(k)X_2(k) + V_2(k)V_1(k)$$
$$= \cos(2k/N)\cos((2k+1)/N) + \sin((2k+1)/N)\sin(2k/N) = \cos(\pi/N)$$

For N = 6,

$$INV_{12} \quad \cos(\pi/6) = 3^{1/2}/2 = 0.8660 \tag{9.4.8a}$$

For large value of N,

$$INV_{12} \approx 1 \tag{9.4.8b}$$

In the next section, we will give a numerical simulation of the two incursive discrete harmonic oscillators in view of comparing with the analytical solutions that we have presented in this section.

9.5 Numerical Simulations of the Two Incursive Discrete Harmonic Oscillators

This section gives the numerical simulations of the two incursive harmonic oscillators [33].

Firstly, the parameters for the simulation are given as follows.

The number of iterations is given by

$$N = 6 \tag{9.5.1}$$

The interval of discrete time is then given by

$$H = 2\sin(\pi/N) = 2\sin(\pi/6) = 1 \tag{9.5.2}$$

And the boundary conditions are given by

$$X_1(0) = \cos(0) = 1 \text{ and } V_1(0) = -\sin(\pi/6) = -0.5 \tag{9.5.3a, b}$$

Table 9.1a gives the simulation of the first incursive discrete equations (9.3.4a, b) of the harmonic oscillator.

In Table 9.1a, we give the energy

Table 9.1 (a) Simulation of the incursive discrete equations (9.3.4a, b). (b) Simulation of the discrete incursive equations (9.3.5a, b)

(a)

First incursive Discrete harmonic oscillator								Analytical solution	
N	H	k	$X_1(k)$	$V_1(k)$	$E_1(k)$	$E_{F1}(k)$	$K_1(k)$	$X_1(k) = \cos(2k\pi/N)$	$V_1(k) = -\sin((2k+1)\pi/N)$
6	1	0	1.000	−0.500	1.25	−0.50	0.75	$\cos(0\pi/6)$ $= 1$	$-\sin(1\pi/6)$ $-1/2$
		1	0.500	−1.000	1.25	−0.50	0.75	$\cos(2\pi/6)$ $= 1/2$	$-\sin(3\pi/6)$ $= -1$
		2	−0.500	−0.500	0.50	0.25	0.75	$\cos(4\pi/6)$ $= -1/2$	$-\sin(5\pi/6)$ $= -1/2$
		3	−1.000	0.500	1.25	−0.50	0.75	$\cos(6\pi/6)$ $= -1$	$-\sin(7\pi/6)$ $= 1/2$
		4	−0.500	1.000	1.25	−0.50	0.75	$\cos(8\pi/6)$ $= -1/2$	$-\sin(9\pi/6)$ $= 1$
		5	0.500	0.500	0.50	0.25	0.75	$\cos(10\pi/6)$ $= 1/2$	$-\sin(11\pi/6)$ $= 1/2$
		6	1.000	−0.500	1.25	−0.50	0.75	$\cos(12\pi/6)$ $= 1$	$-\sin(13\pi/6)$ $= -1/2$
		7	0.500	−1.000	1.25	−0.50	0.75	$\cos(14\pi/6)$ $= 1/2$	$-\sin(15\pi/6)$ $= -1$

(b)

Second incursive Discrete harmonic oscillator								Analytical solution	
N	H	k	$X_2(k)$	$V_2(k)$	$E_2(k)$	$E_{B2}(k)$	$K_2(k)$	$X_2(k) = \cos((2k+1)\pi/N)$	$V_2(k) = -\sin(2k\pi/N)$
6	1	0	0.866	0.000	0.75	0.00	0.75	$\cos(1\pi/6)$ $= \sqrt{3}/2$	$-\sin(0\pi/6)$ $= 0$
		1	0.000	−0.866	0.75	0.00	0.75	$\cos(3\pi/6)$ $= 0$	$-\sin(2\pi/6)$ $= -\sqrt{3}/2$
		2	−0.866	−0.866	1.50	−0.75	0.75	$\cos(5\pi/6) =$ $-\sqrt{3}/2$	$-\sin(4\pi/6)$ $= -\sqrt{3}/2$
		3	−0.866	0.000	0.75	0.00	0.75	$\cos(7\pi/6) =$ $-\sqrt{3}/2$	$-\sin(6\pi/6)$ $= 0$
		4	0.000	0.866	0.75	0.00	0.75	$\cos(9\pi/6)$ $= 0$	$-\sin(8\pi/6)$ $= \sqrt{3}/2$
		5	0.866	0.866	1.50	−0.75	0.75	$\cos(11\pi/6)$ $= \sqrt{3}/2$	$-\sin(10\pi/6)$ $= \sqrt{3}/2$
		6	0.866	0.000	0.75	0.00	0.75	$\cos(13\pi/6)$ $= \sqrt{3}/2$	$-\sin(12\pi/6)$ $= 0$

(continued)

Table 9.1 (continued)

(b)

Second incursive Discrete harmonic oscillator								Analytical solution	
N	H	k	$X_2(k)$	$V_2(k)$	$E_2(k)$	$E_{B2}(k)$	$K_2(k)$	$X_2(k) =$ $\cos((2k+1)\pi/N)$	$V_2(k) =$ $-\sin(2k\pi/N)$
		7	0.000	−0.866	0.75	0.00	0.75	$\cos(15\pi/6)$ $= 0$	$-\sin(14\pi/6)$ $= -\sqrt{3}/2$

$$E_1(k) = X_1^2(k) + V_1^2(k)$$

the forward energy

$$E_{F1}(k) = +HX_1(k)V_1(k)$$

and the constant of motion

$$K_1(k) = X_1^2(k) + V_1^2(k) + HX_1(k)V_1(k) = K_1 = \text{constant} \qquad (9.4.1a)$$

The numerical values correspond exactly to the analytical solutions

$$X_1(k) = \cos(2k\pi/N)$$
$$V_1(k) = -\sin((2k+1)\pi/N) \qquad (9.4.4a, b)$$

NB: see the correspondence of the variables with the hyperincursive harmonic oscillator at Table 9.3:

$$X_1(k) = X(2k), \; V_1(k) = V(2k+1) \qquad (9.5.4)$$

Secondly, the parameters for the simulation are given as follows. The number of iterations,

$$N = 6 \qquad (9.5.5)$$

The interval of discrete time is then given by

$$H = 2\sin(\pi/N) = 2\sin(\pi/6) = 1 \qquad (9.5.6)$$

The boundary conditions,

$$X_2(0) = \cos(\pi/6) = (3/4)^{1/2} = 0.8660$$
$$V_2(0) = -\sin(0) = 0 \qquad (9.5.7a, b)$$

Table 9.1b gives the simulation of the second incursive discrete equations (9.3.5a, b) of the harmonic oscillator.

In Table 9.1b, we give the energy

$$E_2(k) = X_2^2(k) + V_2^2(k)$$

the backward energy

$$E_{B2}(k) = -HX_2(k)V_2(k)$$

and the constant of motion

$$K_2(k) = X_2^2(k) + V_2^2(k) - HX_2(k)V_2(k) = K_2 = \text{constant} \qquad (9.4.2a)$$

The numerical values correspond exactly to the analytical solutions

$$X_2(k) = \cos((2k + 1)\pi/N)$$
$$V_2(k) = -\sin(2k\pi/N) \qquad (9.4.5a, b)$$

NB: see the correspondence of the variables with the hyperincursive harmonic oscillator at Table 9.3:

$$X_2(k) = X(2k + 1), \; V_2(k) = V(2k) \qquad (9.5.8)$$

In the next section, we demonstrate that the dimensionless hyperincursive discrete harmonic oscillator is separable into two incursive discrete harmonic oscillators.

9.6 The Dimensionless Hyperincursive Discrete Harmonic Oscillator Is Separable into Two Incursive Discrete Harmonic Oscillators

For the hyperincursive discrete harmonic oscillator, given by the Eqs. (9.2.11a, b), we use the dimensionless variables, X and V, for the variables, x and v, as follows:

$$X(k) = [k/2]^{1/2}x(k)$$
$$V(k) = [m/2]^{1/2}v(k) \qquad (9.3.1a, b)$$

with the dimensionless time

$$\tau = \omega t \qquad (9.3.2a)$$

where the pulsation (9.2.1c) is given by

$$\omega = [k/m]^{1/2} \tag{9.3.2b}$$

and with the dimensionless interval of time given by

$$\Delta\tau = \omega\Delta t = \omega h = H \tag{9.3.3}$$

So, the two Eqs. (9.2.11a, b) are then transformed to the following two dimensionless equations of the hyperincursive discrete harmonic oscillator

$$X(k+1) = X(k-1) + 2HV(k)$$
$$V(k+1) = V(k-1) - 2HX(k) \tag{9.6.1a, b}$$

for $k = 1, 2, 3, \ldots$,

with the 4 even and odd boundary conditions, $X(0)$, $V(1)$, $V(0)$, $X(1)$.

This hyperincursive discrete harmonic oscillator is a recursive computing system that is separable into two independent incursive discrete harmonic oscillators [33], as shown in Tables 9.2a, b.

Table 9.2a gives the first iterations of the hyperincursive discrete equations (9.6.1a, b).

It is well seen that there are two independent series of iterations defining two incursive discrete harmonic oscillators, as given in Table 9.2b.

As well seen in Table 9.2b, the first incursive harmonic oscillator, with the boundary conditions,

$X(0)$, $V(1)$, is given by

$$X(2k) = X(2k-2) + 2HV(2k-1)$$
$$V(2k+1) = V(2k-1) - 2HX(2k) \tag{9.6.2a, b}$$

and the second incursive harmonic oscillator, with the boundary conditions, $V(0)$, $X(1)$, is given by

$$V(2k) = V(2k-2) - 2HX(2k-1)$$
$$X(2k+1) = X(2k-1) + 2HV(2k) \tag{9.6.3a, b}$$

for $k = 1, 2, 3, \ldots$

Let us remark that the difference between the two incursive oscillators represented by the Eqs. (9.3.4a, b, 9.3.5a, b) and these Eqs. (9.6.2a, b, 9.6.3a, b), holds in the labeling of the successive time steps. In the incursive oscillators, (9.3.4a, b, 9.3.5a, b), the position and velocity are computed at the same time step while in the incursive oscillators, (9.6.2a, b, 9.6.3a, b), the position and the velocity are computed at successive time steps, but the numerical simulations of both give the same values. Each incursive oscillator is the discrete time inverse, $+\Delta t \to -\Delta t$, and, $-\Delta t \to +\Delta t$ of the other incursive oscillator, defined by time forward and time backward derivatives. So the two incursive oscillators are not reversible. But the superposition of the two

Table 9.2 (a) This table gives the first iterations of the hyperincursive discrete equations (9.6.1a, b). (b) This table shows the two independent incursive discrete harmonic oscillators

(a)

Hyperincursive discrete harmonic oscillator

	$X(k+1) = X(k-1) + 2HV(k)$	$V(k+1) = V(k-1) - 2HX(k)$
	Boundary conditions: $X(0) = C_1, V(1) = C_2, V(0) = C_3, X(1) = C_4$	
k	Iterations	
1	$X(2) = X(0) + 2HV(1)$	$V(2) = V(0) - 2HX(1)$
2	$X(3) = X(1) + 2HV(2)$	$V(3) = V(1) - 2HX(2)$
3	$X(4) = X(2) + 2HV(3)$	$V(4) = V(2) - 2HX(3)$
4	$X(5) = X(3) + 2HV(4)$	$V(5) = V(3) - 2HX(4)$
5	$X(6) = X(4) + 2HV(5)$	$V(6) = V(4) - 2HX(5)$
6	$X(7) = X(5) + 2HV(6)$	$V(7) = V(5) - 2HX(6)$
...	–	–

(b)

	First incursive discrete harmonic oscillator	Second incursive discrete harmonic oscillator
	Boundary conditions: $X(0) = C_1, V(1) = C_2$	Boundary conditions: $V(0) = C_3, X(1) = C_4$
k	Iterations	Iterations
1	$X(2) = X(0) + 2HV(1)$	$V(2) = V(0) - 2HX(1)$
2	$V(3) = V(1) - 2HX(2)$	$X(3) = X(1) + 2HV(2)$
3	$X(4) = X(2) + 2HV(3)$	$V(4) = V(2) - 2HX(3)$
4	$V(5) = V(3) - 2HX(4)$	$X(5) = X(3) + 2HV(4)$
5	$X(6) = X(4) + 2HV(5)$	$V(6) = V(4) - 2HX(5)$
6	$V(7) = V(5) - 2HX(6)$	$X(7) = X(5) + 2HV(6)$
...	–	–

incursive oscillators given by the hyperincursive discrete oscillator is reversible. In putting the expression of $V(k)$ from the Eq. (9.6.1b)

$$V(k) = [X(k+1) - X(k-1)]/2H \qquad (9.6.4)$$

to the Eq. (9.6.1a), one obtains the second-order hyperincursive discrete harmonic oscillator

$$X(k+2) - 2X(k) + X(k-2) = -4H^2X(k) \qquad (9.6.5)$$

With the dimensionless variables, the dimensionless energy is given by

$$E(k) = X^2(k) + V^2(k) \qquad (9.6.6)$$

The next section will give simulations of the hyperincursive discrete harmonic oscillator.

9.7 Numerical Simulations of the Hyperincursive Discrete Equations of the Harmonic Oscillator

This section gives the numerical simulations of the hyperincursive discrete harmonic oscillator.

Firstly, we will give explicitly the parameters for the simulation for the case corresponding to the simulations given at the preceding section for the two incursive discrete harmonic oscillators.

The number of iterations is given by,

$$N = 12 \tag{9.7.1}$$

The interval of discrete time is then given by

$$H = \sin(2\pi/N) = \sin(\pi/6) = 0.5 \tag{9.7.2}$$

NB: When N is large,

$$H = \sin(2\pi/N) \approx 2\pi/N = \omega\Delta t = 2\pi\,\Delta t/T \tag{9.7.3}$$

so the period T of the harmonic oscillator is

$$T = 2\pi/\omega = N\Delta t \tag{9.7.4}$$

The boundary conditions are given by

$$X(0) = C_1 = \cos(0) = 1 \text{ and } V(1) = C_2 = -\sin(\pi/6) = -0.5$$
$$V(0) = C_3 = -\sin(0) = 0 \text{ and } X(1) = C_4 = \cos(\pi/6) = (3)^{1/2}/2 = 0.8660$$
$$\text{(9.7.5a, b, c, d)}$$

Table 9.3 gives the simulation of the hyperincursive discrete equations (9.6.1a, b) of the harmonic oscillator.

NB: Let us remark that this hyperincursive discrete harmonic oscillator represents alternatively the values of the two incursive harmonic oscillators, given at Table 9.1a, b, with the following correspondence:

$$X_1(k) = X(2k),\ V_1(k) = V(2k+1),\ X_2(k) = X(2k+1),\ V_2(k) = V(2k) \tag{9.7.6}$$

Table 9.3 Numerical simulation of the Eqs. (9.6.1a, b)

N	H	k	X(k)	V(k)	E(k)	X(k) = cos(2kπ/N)	V(k) = −sin(2kπ/N)
						Hyperincursive Discrete harmonic oscillator	Analytical solution
12	0.5	0	1.0000	0.0000	1.0	cos(0) = 1	− sin(0) = 0
		1	0.8660	−0.5000	1.0	cos(2π/12) = √3/2	−sin(2π/12) = −1/2
		2	0.5000	−0.8660	1.0	cos(4π/12) = 1/2	−sin(4π/12) = −√3/2
		3	0.0000	−1.0000	1.0	cos(6π/12) = 0	− sin(6π/12) = −1
		4	−0.5000	−0.8660	1.0	cos(8π/12) = −1/2	− sin(8π/12) = −√3/2
		5	−0.8660	−0.5000	1.0	cos(10π/12) = −√3/2	− sin(10π/12) = −1/2
		6	−1.0000	0.0000	1.0	cos(12π/12) = −1	− sin(12π/12) = 0
		7	−0.8660	0.5000	1.0	cos(14π/12) = −√3/2	− sin(14π/12) = 1/2
		8	−0.5000	0.8660	1.0	cos(16π/12) = −1/2	− sin(16π/12) = √3/2
		9	0.0000	1.0000	1.0	cos(18π/12) = 0	− sin(18π/12) = 1
		10	0.5000	0.8660	1.0	cos(20π/12) = 1/2	− sin(20π/12) = √3/2
		11	0.8660	0.5000	1.0	cos(22π/12) = √3/2	− sin(22π/12) = 1/2
		12	1.0000	0.0000	1.0	cos(24π/12) = 1	− sin(24π/12) = 0
		13	0.8660	−0.5000	1.0	cos(26π/12) = √3/2	− sin(26π/12) = −1/2

Secondly, Figs. 9.1, 9.2, 9.3, 9.4, 9.5 and 9.6 give the simulations of the hyperincursive discrete harmonic oscillator from Eqs. 9.2.16 a, b, with $N = 3, 4, 6, 12, 24$ and 48 time steps.

The figures of the simulations of the hyperincursive discrete harmonic oscillator show the stability and the precision of the algorithm for values of time steps $N = 3, 4, 6, 12, 24$ and 48.

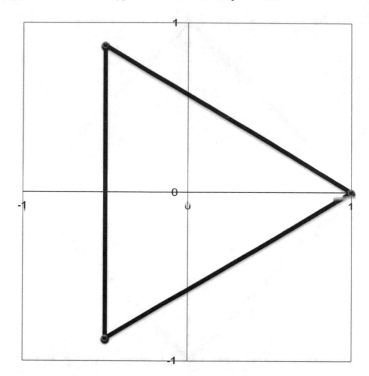

Fig. 9.1 Simulation of the Eqs. 9.6.1a, b of the hyperincursive discrete harmonic oscillator with N = 3 time steps. The horizontal axis gives the position $X(k)$ and the vertical axis gives the velocity $V(k)$ of the oscillator

The representation of the harmonic oscillator tends to a circle when the number of time steps increases.

In a recent paper [8], I introduced the concept of deterministic anticipation. The general case of the discrete harmonic oscillator is taken as a typical example of a discrete deterministic anticipation given by the hyperincursive discrete oscillator that is separable into two incursive discrete oscillators. The hyperincursive oscillator shows a conservation of energy. The incursive oscillators do not show such a conservation of energy but show a deterministic anticipation. It is proposed to add, to the energy equation, a forward energy depending on the positive discrete time, $+H$, for the first incursive oscillator, and a backward energy depending on the negative discrete time, $-H$. The figures of the simulations of the hyperincursive discrete harmonic oscillation show the stability of the oscillator and the high precision of the numerical computed values, even for very small values of time steps.

In the next section, we will present a new derivation of recursive discrete harmonic oscillator based on a rotation of the incursive discrete harmonic oscillator.

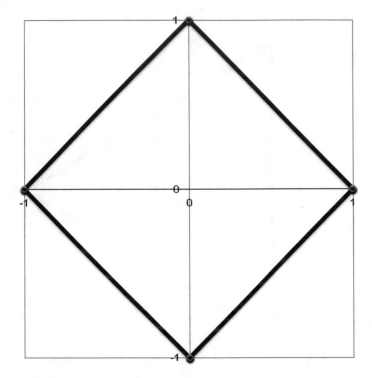

Fig. 9.2 Continuation of Fig. 9.1 with $N = 4$ time steps

9.8 Rotation of the Incursive Harmonic Oscillators to Recursive Discrete Harmonic Oscillators

In the expression of the constant of motion of the first incursive harmonic oscillator, a rotation on the position and velocity variables gives rise to a pure quadratic expression of the constant of motion, similarly to the constant of energy of the classical continuous harmonic oscillator [33, 34].

The constant of motion (9.4.1a)

$$X_1(k)X_1(k) + HX_1(k)V_1(k) + V_1(k)V_1(k) = K_1 \qquad (9.4.1a)$$

is an expression of a quadratic curve

$$Ax^2 + Bxy + Cy^2 + Dx + Ey + F = 0 \qquad (9.8.1)$$

with

$$A = 1, B = H, C = 1, D = 0, E = 0, F = -K_1$$

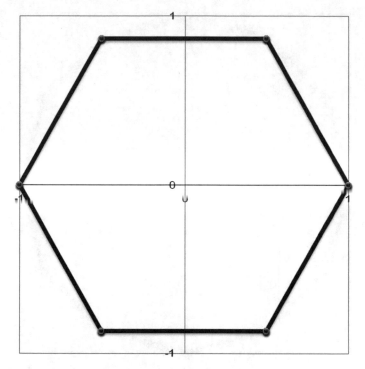

Fig. 9.3 Continuation of Fig. 9.2 with $N = 6$ time steps

$$x = X_1(k), y = V_1(k) \qquad (9.8.2)$$

The quantity

$$\Delta = B^2 - 4AC = INV \qquad (9.8.3)$$

is an invariant under rotations and is known as the discriminant of Eq. (9.5.1).
    The discriminant of the constant of motion is given by

$$\Delta = B^2 - 4AC = H^2 - 4 < 0 \qquad (9.8.4)$$

which defines an ellipse.
    NB: This inequality gives the maximum value of the discrete interval of time

$$H = \omega \Delta t < 2 \qquad (9.8.5)$$

and this is exactly the maximum value for the discrete harmonic oscillator:

$$H = 2\sin(\pi/N) \qquad (9.8.6)$$

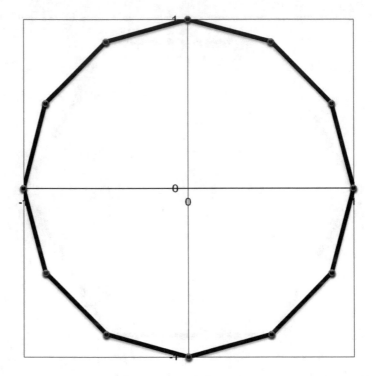

Fig. 9.4 Continuation of Fig. 9.3 with $N = 12$ time steps. This case corresponds to the numerical values given in Table 9.3

The equations for the rotation are given by

$$X_1(k) = \cos(\theta)u_1(k) - \sin(\theta)v_1$$
$$V_1(k) = \sin(\theta)u_1(k) + \cos(\theta)v_1 \qquad \text{(9.8.7a, b)}$$

With $A = C$, the angle θ is given by

$$\theta = \pi/4, \qquad \text{(9.8.8a)}$$

so

$$\cos(\pi/4) = 2^{-1/2} = \rho \qquad \text{(9.8.8b)}$$

and

$$\sin(\pi/4) = 2^{-1/2} = \rho \qquad \text{(9.8.8c)}$$

With the Eqs. (9.8.8b, 9.8.8c) the Eqs. (9.8.7a, b) of the rotation transformed to

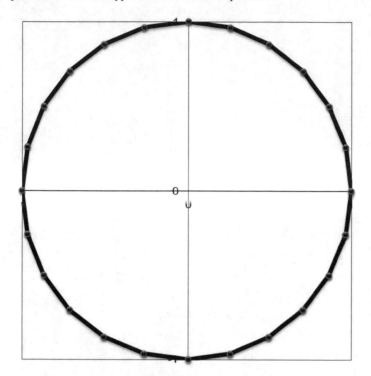

Fig. 9.5 Continuation of Fig. 9.4 with $N = 24$ time steps

$$X_1(k) = \rho(u_1(k) - v_1(k))$$
$$V_1(k) = \rho(u_1(k) + v_1(k)) \qquad (9.8.9a, b)$$

So the constant of motion becomes

$$(u_1(k) - v_1(k))^2 + H(u_1(k) - v_1(k))(u_1(k) + v_1(k)) + (u_1(k) + v_1(k))^2 = 2K_1$$
$$u_1^2(k) + v_1^2(k) - 2u_1(k)v_1(k) + Hu_1^2(k) - Hv_1^2(k) + u_1^2(k) + v_1^2(k) + 2u_1(k)v_1(k) = 2K_1$$
$$u_1^2(k) + v_1^2(k) + H[u_1^2(k) - v_1^2(k)]/2 = K_1(k) = K_1 \qquad (9.8.10a)$$

For the second incursion, the constant of motion is obtained by inversion the sign of H:

$$u_2^2(k) + v_2^2(k) - H[u_2^2(k) - v_2^2(k)]/2 = K_2(k) = K_2 \qquad (9.8.10b)$$

that is also a pure quadratic function.

Now let us give the discrete equations of the first oscillator

Let us make the rotation to the first incursive oscillator

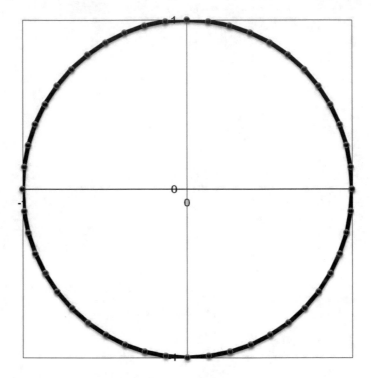

Fig. 9.6 Continuation of Fig. 9.5 with $N = 48$ time steps

$X_1(k+1) = X_1(k) + HV_1(k)$

$V_1(k+1) = V_1(k) - HX_1(k+1) = V_1(k) - HX_1(k) - H^2V_1(k)$

$\rho(u_1(k+1) - v_1(k+1)) = \rho(u_1(k) - v_1(k)) + H\rho(u_1(k) + v_1(k))$

$\rho(u_1(k+1) + v_1(k+1)) = \rho(u_1(k) + v_1(k)) - H\rho(u_1(k) - v_1(k)) - H^2\rho(u_1(k) + v_1(k))$

$$(9.3.4a, b)$$

Let us add the two equations

$$2\rho u_1(k+1) = 2\rho u_1(k) + 2H\rho v_1(k) - H^2\rho(u_1(k) + v_1(k))$$

and after division by 2ρ,

we obtain the first rotated equation of the first incursive oscillator:

$$u_1(k+1) = u_1(k) + Hv_1(k) - H^2(u_1(k) + v_1(k))/2 \qquad (9.8.11a)$$

Let us subtract the two equations

$$-2\rho v_1(k+1) = -2\rho v_1(k) + 2H\rho u_1(k) + H^2\rho(u_1(k) + v_1(k))$$

and after division by -2ρ,

we obtain the second rotated equation of the first incursive oscillator:

$$v_1(k+1) = v_1(k) - Hu_1(k) - H^2(u_1(k) + v_1(k))/2 \qquad (9.8.11b)$$

With a similar rotation, the two equations of the second incursive oscillator

$$V_2(k+1) = V_2(k) - HX_2(k)$$
$$X_2(k+1) = X_2(k) + HV_2(k+1) \qquad (9.3.5a, b)$$

are transformed to

$$v_2(k+1) = v_2(k) + Hu_2(k) - H^2(u_2(k) + v_2(k))/2 \qquad (9.8.12a)$$

$$u_2(k+1) = u_2(k) - Hv_2(k) - H^2(u_2(k) + v_2(k))/2 \qquad (9.8.12b)$$

These equations are the same as the equations of the first oscillator by inversion of the sign of H.

In conclusion, the 4 recursive equations of the discrete harmonic oscillator are given by [33]

$$u_1(k+1) = u_1(k) + Hv_1(k) - H^2(u_1(k) + v_1(k))/2 \qquad (9.8.13a)$$

$$v_1(k+1) = v_1(k) - Hu_1(k) - H^2(u_1(k) + v_1(k))/2 \qquad (9.8.13b)$$

$$u_2(k+1) = u_2(k) - Hv_2(k) - H^2(u_2(k) + v_2(k))/2 \qquad (9.8.14a)$$

$$v_2(k+1) = v_2(k) + Hu_2(k) - H^2(u_2(k) + v_2(k))/2 \qquad (9.8.14b)$$

with the corresponding constant of motion

$$u_1^2(k) + v_1^2(k) + H[u_1^2(k) - v_1^2(k)]/2 = K_1(k) = K_1 \qquad (9.8.15)$$

$$u_2^2(k) + v_2^2(k) - H[u_2^2(k) - v_2^2(k)]/2 = K_2(k) = K_2 \qquad (9.8.16)$$

This result is fundamental because it gives an explanation of the effect of the discretization of the time in discrete physics.

We have shown that the temporal discretization of the harmonic oscillator produces a rotation which gives rise to an anticipative effect with a reversible serial computation.

The information obtained from the discrete equations is richer than obtained by continuous physics.

9.9 The Space and Time-Symmetric Second-Order Hyperincursive Discrete Klein–Gordon Equation

In 1926, Klein [19] and Gordon [20] published independently their famous equation, called the Klein–Gordon equation.

The Klein–Gordon equation with the function $\varphi = \varphi(\mathbf{r}, t)$ in three spatial dimensions $\mathbf{r} = (x, y, z)$ and time t is given by

$$\hbar^2 \partial^2 \varphi(r, t)/\partial t^2 = -\hbar^2 c^2 \nabla^2 \varphi(r, t) + m^2 c^4 \varphi(r, t) \tag{9.9.1}$$

or, in the explicit form of the nabla operator ∇,

$$-\hbar^2 \partial^2 \varphi/\partial t^2 = -\hbar^2 c^2 \partial^2 \varphi/\partial x^2 - \hbar^2 c^2 \partial^2 \varphi/\partial y^2 - \hbar^2 c^2 \partial^2 \varphi/\partial z^2 + m^2 c^4 \varphi \tag{9.9.2}$$

where $\hbar$ is the constant of Plank, c is the speed of light, and m the mass.

As we will consider the discrete Klein–Gordon equation, we make the following usual change of variables

$$q(\mathbf{r}, t) = \varphi(\mathbf{r}, t) \tag{9.9.3}$$

$$a = \omega = mc^2/\hbar \tag{9.9.4}$$

where ω is a frequency, so the Klein–Gordon equation (9.9.2) becomes

$$\partial^2 q(\mathbf{r}, t)/\partial t^2 = +c^2 \partial^2 q(\mathbf{r}, t)/\partial x^2 + c^2 \partial^2 q(\mathbf{r}, t)/\partial y^2 + c^2 \partial^2 q(\mathbf{r}, t)/\partial z^2 - a^2 q(\mathbf{r}, t) \tag{9.9.5}$$

From the Klein–Gordon equation (9.5.5), the second-order hyperincursive discrete Klein–Gordon equation [31] is given by

$$q(x, y, z, t + 2\Delta t) - 2q(x, y, z, t) + q(x, y, z, t - 2\Delta t) =$$
$$+ B^2 [q(x + 2\Delta x, y, z, t) - 2q(x, y, z, t) + q(x - 2\Delta x, y, z, t)]$$
$$+ C^2 [q(x, y + 2\Delta y, z, t) - 2q(x, y, z, t) + q(x, y - 2\Delta y, z, t)]$$
$$+ D^2 [q(x, y, z + 2\Delta z, t) - 2q(x, y, z, t) + q(x, y, z - 2\Delta z, t)] - A^2 q(x, y, z, t) \tag{9.9.6}$$

where the following parameters A, B, C, and, D,

$$A = a(2\Delta t), B = c(2\Delta t)/(2\Delta x), C = c(2\Delta t)/(2\Delta y), D = c(2\Delta t)/(2\Delta z) \tag{9.9.7}$$

depend on the discrete interval of time Δt, and the discrete intervals of space, $\Delta x, \Delta y, \Delta z$, respectively. As usually made in computer science, let us now introduce the discrete time t_k, and the discrete spaces x_l, y_m, z_n, as follows

$$t_k = t_0 + k\Delta t, \ k = 0, 1, 2, \ldots, \quad (9.9.8)$$

where k is the integer time increment, and

$$x_l = x_0 + l\Delta x, l = 0, 1, 2, \ldots, y_m = y_0 + m\Delta y, m = 0, 1, 2, \ldots, z_n = z_0 + n\Delta z, n = 0, 1, 2, \ldots \quad (9.9.9)$$

where l, m, n, are the integer space increments. So, with these time and space increments, the second-order hyperincursive discrete Klein–Gordon equation (9.2.6) becomes

$$\begin{aligned}
q(l, m, n, k+2) &- 2q(l, m, n, k) + q(l, m, n, k-2) = \\
&+ B^2\big[q(l+2, m, n, k) - 2q(l, m, n, k) + q(l-2, m, n, k)\big] \\
&+ C^2\big[q(l, m+2, n, k) - 2q(l, m, n, k) + q(l, m-2, n, k)\big] \\
&+ D^2\big[q(l, m, n+2, k) - 2q(l, m, n, k) + q(l, m, n-2, k)\big] - A^2 q(l, m, n, k)
\end{aligned}$$

$$(9.9.10)$$

This equation without spatial components, corresponding to a particle at rest, is similar to the harmonic oscillator. For a particle at rest, the Klein–Gordon equation (9.9.5), with the function $q(t)$ depending only on the time variable, is given by

$$\partial^2 q(t)/\partial t^2 = -a^2 q(t) \quad (9.9.11)$$

with the frequency $a = \omega = mc^2/\hbar$, given by the Eq. (9.9.4). This Eq. (9.9.11) is formally similar to the equation of the harmonic oscillator for which $q(t)$ would represent the position $a = \omega = mc^2/\hbar$ and $\partial q(t)/\partial t$ would represent the velocity $v(t) = \partial x(t)/\partial t$, as shown in Sect. 9.2. So, with only the temporal component, the second-order hyperincursive discrete Klein–Gordon equation (9.9.10) becomes

$$q(k+2) - 2q(k) + q(k-2) = -A^2 q(k) \quad (9.9.12)$$

that is similar to the second-order hyperincursive equation of the harmonic oscillator.

This hyperincursive equation (9.9.12) is separable into a first discrete incursive oscillator depending on two functions defined by $q_1(k), q_2(k)$, and a second incursive oscillator depending on two other functions defined by $q_3(k), q_4(k)$, given by first-order discrete equations.

So the first incursive equations are given by:

$$q_1(2k) = q_1(2k-2) + Aq_2(2k-1)$$

$$q_2(2k + 1) = q_2(2k - 1) - Aq_1(2k) \qquad (9.9.13\text{a, b})$$

where $q_1(2k)$ is defined on the even steps of the time, and $q_2(2k + 1)$ is defined on the odd steps of the time. And the second incursive equations are given by:

$$q_3(2k) = q_3(2k - 2) - Aq_4(2k - 1)$$
$$q_4(2k + 1) = q_4(2k - 1) + Aq_3(2k) \qquad (9.9.14\text{a, b})$$

where $q_3(2k)$ is defined of the even steps of the time, and $q_4(2k + 1)$ is defined on the odd steps of the time. The second incursive system is the time reverse of the first incursive system in making the time inversion T

$$T : \Delta t \to -\Delta t \qquad (9.9.15)$$

this gives an oscillator and its anti-oscillator.

In the next sections, we will present the bifurcation of this Eq. (9.9.10) to the 4 hyperincursive discrete Majorana real equations which bifurcate to the 4 hyperincursive discrete Dirac equations.

9.10 The Hyperincursive Discrete Majorana Equations and Continuous Majorana Real 4-Spinors

We deduced the following 4 hyperincursive discrete Majorana equations, depending on the discrete Majorana functions $\tilde{q}_j = \tilde{q}_j(x, y, z, t) = \tilde{q}_j(l, m, n, k), j = 1, 2, 3, 4$, from the hyperincursive Klein–Gordon equation, [28–31],

$$\begin{aligned}
\tilde{q}_1(l, m, n, k + 1) &= \tilde{q}_1(l, m, n, k - 1) + \tilde{B}[\tilde{q}_4(l + 1, m, n, k) - \tilde{q}_4(l - 1, m, n, k)] \\
&\quad - \tilde{C}[\tilde{q}_1(l, m + 1, n, k) - \tilde{q}_1(l, m - 1, n, k)] + \tilde{D}[\tilde{q}_3(l, m, n + 1, k) - \tilde{q}_3(l, m, n - 1, k)] \\
&\quad - \tilde{A}\tilde{q}_4(l, m, n, k) \\
\tilde{q}_2(l, m, n, k + 1) &= \tilde{q}_2(l, m, n, k - 1) + \tilde{B}[\tilde{q}_3(l + 1, m, n, k) - \tilde{q}_3(l - 1, m, n, k)] \\
&\quad - \tilde{C}[\tilde{q}_2(l, m + 1, n, k) - \tilde{q}_2(l, m - 1, n, k)] - \tilde{D}[\tilde{q}_4(l, m, n + 1, k) - \tilde{q}_4(l, m, n - 1, k)] \\
&\quad + \tilde{A}\tilde{q}_3(l, m, n, k) \\
\tilde{q}_3(l, m, n, k + 1) &= \tilde{q}_3(l, m, n, k - 1) + \tilde{B}[\tilde{q}_2(l + 1, m, n, k) - \tilde{q}_2(l - 1, m, n, k)] \\
&\quad + \tilde{C}[\tilde{q}_3(l, m + 1, n, k) - \tilde{q}_3(l, m - 1, n, k)] + \tilde{D}[\tilde{q}_1(l, m, n + 1, k) - \tilde{q}_1(l, m, n - 1, k)] \\
&\quad - \tilde{A}\tilde{q}_2(l, m, n, k) \\
\tilde{q}_4(l, m, n, k + 1) &= \tilde{q}_4(l, m, n, k - 1) + \tilde{B}[\tilde{q}_1(l + 1, m, n, k) - \tilde{q}_1(l - 1, m, n, k)] \\
&\quad + \tilde{C}[\tilde{q}_4(l, m + 1, n, k) - \tilde{q}_4(l, m - 1, n, k)] - \tilde{D}[\tilde{q}_2(l, m, n + 1, k) - \tilde{q}_2(l, m, n - 1, k)] \\
&\quad + \tilde{A}\tilde{q}_1(l, m, n, k) \qquad (9.10.1\text{a, b, c, d})
\end{aligned}$$

with

$$\tilde{A} = A = a(2\Delta t), \tilde{B} = B = c\Delta t/\Delta x, \tilde{C} = C = c\Delta t/\Delta y, \tilde{D} = D = c\Delta t/\Delta z$$

$$(9.10.2a, b, c, d)$$

where Δt and Δx, Δy, Δz are the discrete intervals of time and space, respectively.

These 4 discrete equations (9.10.1a, b, c, d) can be transformed to partial differential equations.

Indeed, the discrete functions $\tilde{q}_j(x, y, z, t) = \tilde{q}_j(\mathbf{r}, t), j = 1, 2, 3, 4$ tend to the continuous functions

$\tilde{\Psi}_j(x, y, z, t) = \tilde{\Psi}_j(\mathbf{r}, t)$, when the discrete space and time intervals tend to zero. At the limit,

$$\tilde{\Psi}_j = \tilde{\Psi}_j(\mathbf{r}, t) = \lim_{\Delta \to 0, \Delta \to 0} \tilde{q}_j(\mathbf{r}, t), j = 1, 2, 3, 4 \qquad (9.10.3)$$

So, with the Majorana continuous functions

$$\tilde{\Psi}_j = \tilde{\Psi}_j(x, y, z, t), j = 1, 2, 3, 4, \qquad (9.10.4)$$

Equations (9.10.1a, b, c, d) are transformed to the following 4 first-order partial differential equations

$$+\partial\tilde{\Psi}_1/\partial t = +c\partial\tilde{\Psi}_4/\partial x - c\partial\tilde{\Psi}_1/\partial y + c\partial\tilde{\Psi}_3/\partial z - (mc^2/\hbar)\tilde{\Psi}_4 \qquad (9.10.5a)$$

$$+\partial\tilde{\Psi}_2/\partial t = +c\partial\tilde{\Psi}_3/\partial x - c\partial\tilde{\Psi}_2/\partial y - c\partial\tilde{\Psi}_4/\partial z + (mc^2/\hbar)\tilde{\Psi}_3 \qquad (9.10.5b)$$

$$+\partial\tilde{\Psi}_3/\partial t = +c\partial\tilde{\Psi}_2/\partial x + c\partial\tilde{\Psi}_3/\partial y + c\partial\tilde{\Psi}_1/\partial z - (mc^2/\hbar)\tilde{\Psi}_2 \qquad (9.10.5c)$$

$$+\partial\tilde{\Psi}_4/\partial t = +c\partial\tilde{\Psi}_1/\partial x + c\partial\tilde{\Psi}_4/\partial y - c\partial\tilde{\Psi}_2/\partial z + (mc^2/\hbar)\tilde{\Psi}_1 \qquad (9.10.5d)$$

which are identical to the original Majorana equations [25], e.g., Eqs. (4a, b, c, d) in Pessa [26].

In 1937, Ettore Majorana published this last paper, before his mysterious disappearance.

9.11 The Bifurcation of the Majorana Real 4-Spinors to the Dirac Real 8-Spinors

Recently, we demonstrated that the Majorana 4-spinors equations bifurcate simply to the Dirac real 8-spinors equations [29, 30, 32]. First, let us consider the inverse parity space, in inversing the sign of the space variables in the Majorana equations (9.6.3a, b, c, d),

$$+\partial\tilde{\Psi}_1/\partial t = -c\partial\tilde{\Psi}_4/\partial x + c\partial\tilde{\Psi}_1/\partial y - c\partial\tilde{\Psi}_3/\partial z - (mc^2/\hbar)\tilde{\Psi}_4$$

$$+\partial\tilde{\Psi}_2/\partial t = -c\partial\tilde{\Psi}_3/\partial x + c\partial\tilde{\Psi}_2/\partial y + c\partial\tilde{\Psi}_4/\partial z + (mc^2/\hbar)\tilde{\Psi}_3$$

$$+\partial\tilde{\Psi}_3/\partial t = -c\partial\tilde{\Psi}_2/\partial x - c\partial\tilde{\Psi}_3/\partial y - c\partial\tilde{\Psi}_1/\partial z - (mc^2/\hbar)\tilde{\Psi}_2$$

$$+\partial\tilde{\Psi}_4/\partial t = -c\partial\tilde{\Psi}_1/\partial x - c\partial\tilde{\Psi}_4/\partial y + c\partial\tilde{\Psi}_2/\partial z + (mc^2/\hbar)\tilde{\Psi}_1 \quad (9.11.1a, b, c, d)$$

In defining the 2-spinors real functions,

$$\varphi_a = \begin{pmatrix} \tilde{\Psi}_1 \\ \tilde{\Psi}_2 \end{pmatrix}, \ \varphi_b = \begin{pmatrix} \tilde{\Psi}_3 \\ \tilde{\Psi}_4 \end{pmatrix}, \quad (9.11.2a, b)$$

the two Eqs. (9.11.1a, b) and (9.11.1c, d) are transformed to the two 2-spinors real equations

$$+\partial\varphi_a/\partial t = -c\sigma_1\partial\varphi_b/\partial x + c\sigma_0\partial\varphi_a/\partial y - c\sigma_3\partial\varphi_b/\partial z + (mc^2/\hbar)\sigma_2\varphi_b \quad (9.11.3a)$$

$$+\partial\varphi_b/\partial t = -c\sigma_1\partial\varphi_a/\partial x - c\sigma_0\partial\varphi_b/\partial y - c\sigma_3\partial\varphi_a/\partial z + (mc^2/\hbar)\sigma_2\varphi_a \quad (9.11.3b)$$

where the real 2-spinors matrices $\sigma_1, \sigma_2, \sigma_3$, are defined by

$$\sigma_1 = \begin{pmatrix} 0 & 1 \\ 1 & 0 \end{pmatrix}, \sigma_2 = \begin{pmatrix} 0 & -1 \\ 1 & 0 \end{pmatrix}, \sigma_3 = \begin{pmatrix} 1 & 0 \\ 0 & -1 \end{pmatrix}, \quad (9.11.4a, b, c)$$

and 2-Identity

$$\sigma_0 = \begin{pmatrix} 1 & 0 \\ 0 & 1 \end{pmatrix} = I_2 \quad (9.11.4d)$$

With the inversion between σ_0 and σ_2, in introducing the tensor product by $-\sigma_2$, the functions $\tilde{\Psi}_j$

$$\tilde{\Psi}_j = \begin{pmatrix} \Psi_{j,1} \\ \Psi_{j,2} \end{pmatrix}, j = 1, 2, 3, 4, \quad (9.11.5)$$

bifurcate to two functions

$$-\sigma_2\Psi_j = -\sigma_2\begin{pmatrix} \Psi_{j,1} \\ \Psi_{j,2} \end{pmatrix} = -\begin{pmatrix} 0 & -1 \\ 1 & 0 \end{pmatrix}\begin{pmatrix} \Psi_{j,1} \\ \Psi_{j,2} \end{pmatrix} = \begin{pmatrix} +\Psi_{j,2} \\ -\Psi_{j,1} \end{pmatrix}, j = 1, 2, 3, 4$$

$$(9.11.6)$$

So the Majorana real 4-spinors equation bifurcates into the Dirac real 8-spinors equations

$$+\partial\Psi_{1,1}/\partial t = -c\partial\Psi_{4,1}/\partial x - c\partial\Psi_{4,2}/\partial y - c\partial\Psi_{3,1}/\partial z + (mc^2/\hbar)\Psi_{1,2} \quad (9.11.7a)$$

$$+\partial\Psi_{2,1}/\partial t = -c\partial\Psi_{3,1}/\partial x + c\partial\Psi_{3,2}/\partial y + c\partial\Psi_{4,1}/\partial z + (mc^2/\hbar)\Psi_{2,2} \quad (9.11.7b)$$

$$+\partial\Psi_{3,1}/\partial t = -c\partial\Psi_{2,1}/\partial x - c\partial\Psi_{2,2}/\partial y - c\partial\Psi_{1,1}/\partial z - (mc^2/\hbar)\Psi_{3,2} \quad (9.11.7c)$$

$$+\partial\Psi_{4,1}/\partial t = -c\partial\Psi_{1,1}/\partial x + c\partial\Psi_{1,2}/\partial y + c\partial\Psi_{2,1}/\partial z - (mc^2/\hbar)\Psi_{4,2} \quad (9.11.7d)$$

$$+\partial\Psi_{1,2}/\partial t = -c\partial\Psi_{4,2}/\partial x + c\partial\Psi_{4,1}/\partial y - c\partial\Psi_{3,2}/\partial z - (mc^2/\hbar)\Psi_{1,1} \quad (9.11.8a)$$

$$+\partial\Psi_{2,2}/\partial t = -c\partial\Psi_{3,2}/\partial x - c\partial\Psi_{3,1}/\partial y + c\partial\Psi_{4,2}/\partial z - (mc^2/\hbar)\Psi_{2,1} \quad (9.11.8b)$$

$$+\partial\Psi_{3,2}/\partial t = -c\partial\Psi_{2,2}/\partial x + c\partial\Psi_{2,1}/\partial y - c\partial\Psi_{1,2}/\partial z + (mc^2/\hbar)\Psi_{3,1} \quad (9.11.8c)$$

$$+\partial\Psi_{4,2}/\partial t = -c\partial\Psi_{1,2}/\partial x - c\partial\Psi_{1,1}/\partial y + c\partial\Psi_{2,2}/\partial z + (mc^2/\hbar)\Psi_{4,1} \quad (9.11.8d)$$

These 8 real first-order partial differential equations represent real 8-spinors equations that are similar to the original Dirac [21, 22] complex 4-spinors equations. In defining the wave function

$$\Psi_j(x, y, z, t) = \Psi_j = \Psi_{j,1} + i\Psi_{j,2}, j = 1, 2, 3, 4, \quad (9.11.9)$$

with the imaginary number i, we obtain the original Dirac equation as a complex 4-spinors equation

$$+\partial\Psi_1/\partial t = -c\partial\Psi_4/\partial x + ic\partial\Psi_4/\partial y - c\partial\Psi_3/\partial z - i(mc^2/\hbar)\Psi_1 \quad (9.11.10a)$$

$$+\partial\Psi_2/\partial t = -c\partial\Psi_3/\partial x - ic\partial\Psi_3/\partial y + c\partial\Psi_4/\partial z - i(mc^2/\hbar)\Psi_2 \quad (9.11.10b)$$

$$+\partial\Psi_3/\partial t = -c\partial\Psi_2/\partial x + ic\partial\Psi_2/\partial y - c\partial\Psi_1/\partial z + i(mc^2/\hbar)\Psi_3 \quad (9.11.10c)$$

$$+\partial\Psi_4/\partial t = -c\partial\Psi_1/\partial x - ic\partial\Psi_1/\partial y + c\partial\Psi_2/\partial z + i(mc^2/\hbar)\Psi_4 \quad (9.11.10d)$$

Following our recent papers [35, 36], in the non-relativistic limit $p \ll mc$, the particles are at rest, with a momentum $p \cong 0$. Let us consider the following Dirac 2-spinors

$$\widehat{\Psi}(t) = \begin{pmatrix} \Psi_1(t) \\ \Psi_4(t) \end{pmatrix}, \quad (9.11.11)$$

for which the temporal non-relativistic Dirac equation is given by

$$\partial_t \tilde{\Psi}(t) = -i\left(mc^2/\hbar\right)\sigma_z \tilde{\Psi}(t) \tag{9.11.12}$$

where $\partial_t = \partial/\partial t$, and $\sigma_z = \begin{pmatrix} 1 & 0 \\ 0 & -1 \end{pmatrix}$, is a Pauli matrix.

The analytical solution of the non-relativistic Dirac equation (9.11.12) is given by

$$\tilde{\Psi}(t) = \cos\left(mc^2 t/\hbar\right)\tilde{\Psi}(0) - i\sin\left(mc^2 t/\hbar\right)\sigma_z \tilde{\Psi}(0) \tag{9.11.13}$$

or in explicit form

$$\Psi_1(t) = \cos\left(mc^2 t/\hbar\right)\Psi_1(0) - i\sin\left(mc^2 t/\hbar\right)\Psi_1(0) \tag{9.11.14a}$$

$$\Psi_4(t) = \cos\left(mc^2 t/\hbar\right)\Psi_4(0) + i\sin\left(mc^2 t/\hbar\right)\Psi_4(0) \tag{9.11.14b}$$

We give, in the next section, the computing hyperincursive equations of the original Dirac complex 4-spinors equations [29].

9.12   The 4 Hyperincursive Discrete Dirac 4-Spinors Equations

Recently, we have presented the 4 hyperincursive discrete Dirac complex equations [29].

Let us define the discrete Dirac wave functions

$$Q_j(l, m, n, k) = Q_{j,1} + iQ_{j,2}, j = 1, 2, 3, 4, \tag{9.12.1a}$$

corresponding to the Dirac continuous wave functions (9.11.9), where i is the imaginary number.

The 4 hyperincursive discrete Dirac equations of the discrete wave functions are then given by

$$Q_1(l, m, n, k+1) = Q_1(l, m, n, k-1) - B[Q_4(l+1, m, n, k) - Q_4(l-1, m, n, k)]$$
$$+ iC[Q_4(l, m+1, n, k) - Q_4(l, m-1, n, k)] - D[Q_3(l, m, n+1, k) - Q_3(l, m, n-1, k)]$$
$$- iAQ_1(l, m, n, k) \tag{9.12.2a}$$

$$Q_2(l, m, n, k+1) = Q_2(l, m, n, k-1) - B[Q_3(l+1, m, n, k) - Q_3(l-1, m, n, k)]$$
$$- iC[Q_3(l, m+1, n, k) - Q_3(l, m-1, n, k)] + D[Q_4(l, m, n+1, k) - Q_4(l, m, n-1, k)]$$
$$- iAQ_2(l, m, n, k) \tag{9.12.2b}$$

$$Q_3(l, m, n, k + 1) = Q_3(l, m, n, k - 1) - B[Q_2(l + 1, m, n, k) - Q_2(l - 1, m, n, k)]$$
$$+ iC[Q_2(l, m + 1, n, k) - Q_2(l, m - 1, n, k)] - D[Q_1(l, m, n + 1, k) - Q_1(l, m, n - 1, k)]$$
$$+ iAQ_3(l, m, n, k) \tag{9.12.2c}$$

$$Q_4(l, m, n, k + 1) = Q_4(l, m, n, k - 1) - B[Q_1(l + 1, m, n, k) - Q_1(l - 1, m, n, k)]$$
$$- iC[Q_1(l, m + 1, n, k) - Q_1(l, m - 1, n, k)] + D[q_2(l, m, n + 1, k) - Q_2(l, m, n - 1, k)]$$
$$+ iAQ_4(l, m, n, k) \tag{9.12.2d}$$

with

$$A = 2\omega\Delta t, B = c\Delta t/\Delta x, C = c\Delta t/\Delta y, D = c\Delta t/\Delta z \tag{9.12.3}$$

where Δt and $\Delta x, \Delta y, \Delta z$ are the discrete intervals of time and space, respectively.

9.13   The Hyperincursive Discrete Klein–Gordon Equation Bifurcates to the 16 Proca Equations

Let us show that there are 16 complex functions associated with this second-order hyperincursive discrete Klein–Gordon equation [29].

For a particle at rest, the Klein–Gordon equation (9.9.5), with the function $q(t)$ depending only on the time variable, is given by

$$\partial^2 q(t)/\partial t^2 = -a^2 q(t) \tag{9.13.1}$$

with the frequency, given by the Eq. (9.9.4).

This Eq. (9.13.1) is formally similar to the equation of the harmonic oscillator for which $q(t)$ would represent the position $x(t)$, and $\partial q(t)/\partial t$ would represent the velocity $v(t) = \partial x(t)/\partial t$.

So, with only the temporal component, the second-order hyperincursive discrete Klein–Gordon equation (9.9.10) becomes

$$q(k + 2) - 2q(k) + q(k - 2) = -A^2 q(k) \tag{9.13.2}$$

that is similar to the second-order hyperincursive discrete equation of the harmonic oscillator [36].

This hyperincursive equation (9.13.2) is separable into a first discrete incursive oscillator depending on two functions defined by $q_1(k)$, $q_2(k)$, and a second incursive oscillator depending on two other functions defined by $q_3(k)$, $q_4(k)$, given by first-order discrete equations.

So the first incursive equations are given by:

$$q_1(2k) = q_1(2k - 2) + Aq_2(2k - 1)$$

$$q_2(2k + 1) = q_2(2k - 1) - Aq_1(2k) \qquad (9.13.3a, b)$$

where $q_1(2k)$ is defined of the even steps of the time, and $q_2(2k + 1)$ is defined on the odd steps of the time. And the second incursive equations are given by:

$$q_3(2k) = q_3(2k - 2) - Aq_4(2k - 1)$$
$$q_4(2k + 1) = q_4(2k - 1) + Aq_3(2k) \qquad (9.13.4a, b)$$

where $q_3(2k)$ is defined of the even steps of the time, and $q_4(2k + 1)$ is defined on the odd steps of the time. The second incursive system is the time reverse of the first incursive system in making the discrete time inversion T

$$T : \Delta t \rightarrow -\Delta t \qquad (9.13.5)$$

which gives an oscillator and its anti-oscillator.

In defining the following 2 complex functions, where i is the imaginary number,

$$q_{13}(2k) = q_1(2k) + iq_3(2k)$$
$$q_{24}(2k + 1) = q_2(2k + 1) - iq_4(2k + 1) \qquad (9.13.6a, b)$$

the 4 real incursive equations (9.13.3a, b) and (9.13.4a, b) are transformed to 2 complex incursive equations

$$q_{13}(2k) = q_{13}(2k - 2) + Aq_{24}(2k - 1)$$
$$q_{24}(2k + 1) = q_{24}(2k - 1) - Aq_{13}(2k) \qquad (9.13.7a, b)$$

So the hyperincursive equation for a particle at rest shows a temporal bifurcation into an oscillatory equation and an anti-oscillatory equation.

For a moving particle, the 3 discrete space-symmetric terms in Eq. (9.9.10)

$$q(l + 2, m, n, k) - 2q(l, m, n, k) + q(l - 2, m, n, k)$$
$$q(l, m + 2, n, k) - 2q(l, m, n, k) + q(l, m - 2, n, k)$$
$$q(l, m, n + 2, k) - 2q(l, m, n, k) + q(l, m, n - 2, k)$$

are similar to the discrete time-symmetric term (9.13.2)

$$q(l, m, n, k + 2) - 2q(l, m, n, k) + q(l, m, n, k - 2).$$

The two complex functions (9.13.6a, b) bifurcate for even and odd steps of space x, giving 4 complex functions depending on 4 discrete incursive equations. These 4 complex functions bifurcate for even and odd steps of space y, giving 8 complex functions depending on 8 discrete incursive equations. Finally, these 8 complex functions bifurcate for even and odd steps of space z, giving 16 complex functions depending on 16 incursive discrete equations.

But if we consider the space variable as a set of the 3 space variables

$$r = (x, y, z) \qquad (9.13.8)$$

the two complex functions bifurcate for even and odd steps of the space variable $r = (x, y, z)$, giving 4 complex functions depending on 4 discrete incursive equations, which correspond to a discrete parity inversion P

$$P : \Delta r \rightarrow -\Delta r \qquad (9.13.9)$$

So, with the discrete time inversion and the parity, we define a group of 4 incursive discrete equations with 4 functions. This is in agreement with the thesis of Proca. Indeed, as demonstrated by Proca [23, 24] in 1930 and 1932, the Klein–Gordon equation admits in the general case a total of 16 functions. Classically, for the well-known Dirac equation, there are 4 complex wave functions. Proca demonstrated that there are 4 fundamental equations of 4 wave functions for the Dirac equation

$$\varphi_{r,s} \text{ for } r = 1, 2, 3, 4, \text{ and } s = 1 \qquad (9.13.10)$$

and the other 3×4 other equations are similar to these 4 equations.

Proca classified the 16 equations in 4 groups of 4 functions:

I.   4 equations of the 4 functions $\varphi_{r,s}$ for $r = 1, 2, 3, 4$, and $s = 1$
II. 4 equations of the 4 functions $\varphi_{r,s}$ for $r = 1, 2, 3, 4$, and $s = 2$
III. 4 equations of the 4 functions $\varphi_{r,s}$ for $r = 1, 2, 3, 4$, and $s = 3$
IV. 4 equations of the 4 functions $\varphi_{r,s}$ for $r = 1, 2, 3, 4$, and $s = 4$

In each group, the 4 equations depend on 4 functions which are not separable except in particular cases.

In this chapter, we restricted our analysis to the first group of 4 functions in studying the case of the Majorana and Dirac equations.

9.14 Simulation of the Hyperincursive Discrete Quantum Majorana and Dirac Wave Equations

This last section deals with the numerical simulation of the hyperincursive discrete Majorana and Dirac wave equations depending on time and one spatial dimension (1D) and with a null mass.

The Majorana equations (9.10.5a, b, c, d) in one spatial dimension z and with a null mass $m = 0$ are given by the 2 following Majorana wave equations

$$+\partial \widetilde{\Psi}_1 / \partial t = +c \partial \widetilde{\Psi}_3 / \partial z \qquad (9.14.1a)$$

$$+\partial\widetilde{\Psi}_3/\partial t = +c\partial\widetilde{\Psi}_1/\partial z \qquad (9.14.1b)$$

and

$$+\partial\widetilde{\Psi}_2/\partial t = -c\partial\widetilde{\Psi}_4/\partial z \qquad (9.14.2a)$$

$$+\partial\widetilde{\Psi}_4/\partial t = -c\partial\widetilde{\Psi}_2/\partial z \qquad (9.14.2b)$$

The corresponding hyperincursive discrete Majorana equations (9.10.1a, b, c, d) are given by the 2 following hyperincursive discrete wave equations

$$\tilde{q}_1(n, k + 1) = \tilde{q}_1(n, k - 1) + D[\tilde{q}_3(n + 1, k) - \tilde{q}_3(n - 1, k)] \qquad (9.14.3a)$$

$$\tilde{q}_3(n, k + 1) = \tilde{q}_3(n, k - 1) + D[\tilde{q}_1(n + 1, k) - \tilde{q}_1(n - 1, k)] \qquad (9.14.3b)$$

and

$$\tilde{q}_2(n, k + 1) = \tilde{q}_2(n, k - 1) - D[\tilde{q}_4(n + 1, k) - \tilde{q}_4(n - 1, k)] \qquad (9.14.4a)$$

$$\tilde{q}_4(n, k + 1) = \tilde{q}_4(n, k - 1) - D[\tilde{q}_2(n + 1, k) - \tilde{q}_2(n - 1, k)] \qquad (9.14.4b)$$

with

$$D = c\Delta t/\Delta z \qquad (9.14.5)$$

where Δt and Δz are the discrete intervals of time and space, respectively.

The Dirac equations (9.11.10a, b, c, d) in one spatial dimension z with a null mass $m = 0$ are given by the 2 following Dirac wave equations

$$+\partial\Psi_1/\partial t = -c\partial\Psi_3/\partial z \qquad (9.14.6a)$$

$$+\partial\Psi_3/\partial t = -c\partial\Psi_1/\partial z \qquad (9.14.6b)$$

and

$$+\partial\Psi_2/\partial t = +c\partial\Psi_4/\partial z \qquad (9.14.7a)$$

$$+\partial\Psi_4/\partial t = +c\partial\Psi_2/\partial z \qquad (9.14.7b)$$

which are similar to the Majorana wave equations (9.14.1a, b) and (9.14.2a, b), where the space variable z is reversed to $-z$.

The corresponding hyperincursive discrete Dirac equations (9.12.2a, b, c, d) are given by the 2 following hyperincursive discrete wave equations

$$Q_1(n, k + 1) = Q_1(n, k - 1) - D[Q_3(n + 1, k) - Q_3(n - 1, k)] \qquad (9.14.8a)$$

$$Q_3(n, k + 1) = Q_3(n, k - 1) - D[Q_1(n + 1, k) - Q_1(n - 1, k)] \qquad (9.14.8b)$$

and

$$Q_2(n, k + 1) = Q_2(n, k - 1) + D[Q_4(n + 1, k) - Q_4(n - 1, k)] \qquad (9.14.9a)$$

$$Q_4(n, k + 1) = Q_4(n, k - 1) + D[Q_2(n + 1, k) - Q_2(n - 1, k)] \qquad (9.14.9b)$$

with

$$D = c\Delta t / \Delta z \qquad (9.14.10)$$

where Δt and Δz are the discrete intervals of time and space, respectively.

For the numerical simulations, it is sufficient to simulate the 2 wave Eqs. (9.14.9a, b), in talking the value of

$$D = c\Delta t / \Delta z$$

and its reversed sign value

$$D = -c\Delta t / \Delta z.$$

The numerical values of D is chosen as equal to

$$D = +1 \qquad (9.14.11a)$$

and

$$D = -1 \qquad (9.14.11b)$$

which correspond to the values of the interval of time given by

$$c\Delta t = \Delta z \qquad (9.14.11c)$$

For the simulations, we will consider the following generic names of the variables

$$Q(n, k) = Q_2(n, k) \qquad (9.14.12a)$$

$$P(n, k) = Q_4(n, k) \qquad (9.14.12b)$$

So the generic computing algorithms of the hyperincursive discrete wave equations are given by

$$Q(n, k + 1) = Q(n, k - 1) + D[P(n + 1, k) - P(n - 1, k)] \qquad (9.14.13a)$$

$$P(n, k + 1) = P(n, k - 1) + D\big[Q(n + 1, k) - Q(n - 1, k)\big] \qquad (9.14.13b)$$

with the 2 values of the parameter D

$$D = +1 \qquad (9.14.13c)$$

and

$$D = -1 \qquad (9.14.13d)$$

which represent the hyperincursive discrete Dirac relativistic quantum equations (9.14.9a, b) and (9.14.8a, b) and also the hyperincursive discrete relativistic quantum Majorana equations (9.14.3a, b) and (9.14.4a, b).

With those two values $D = 1$, the simulations are numerically stable and give discrete space and time periodic solutions similar to the continuous analytical solutions of the continuous wave equation.

Now, we will give a few examples of simulation of these hyperincursive quantum algorithms.

Table 9.4a, b gives the simulation of the hyperincursive discrete algorithms of the discrete quantum wave equations (9.14.13a, b) with the parameter $D = +1$ and $D = -1$ of two particles in a periodic spatial domain.

Table 9.6a, b deals with the simulation of the hyperincursive discrete algorithms of the discrete quantum wave equations (9.14.13a, b) with the parameter $D = +1$ of two particles in a box. The two particles reflect to the two opposite walls of the box.

Table 9.6 shows the simulation of the hyperincursive discrete algorithms of the discrete quantum wave equations (9.14.13a, b) with the parameter $D = +1$ of a packet of particles in a periodic spatial domain that separates to two opposite packets.

Table 9.5a, b deals with the simulation of the hyperincursive discrete algorithms of the discrete quantum wave equations (9.14.13a, b) with the parameter $D = +1$ of a packet of particles in a box. The two opposite packets reflect to the two opposite walls of the box.

In Table 9.4a, the columns represent alternatively the values of the two wave functions, $Q(n, k)$ and $P(n, k)$ of the Eqs. (9.14.13a, b), depending on the space parameter n and the time parameter k,

Vertically, the parameter $k = 0$–14 represents the time steps.

Horizontally, the parameter $n = 0$–11 represents the spatial intervals.

Table 9.4 (a) Simulation of the hyperincursive discrete algorithms of the discrete quantum wave equations (9.14.13a, b) with the parameter $D = +1$ of two particles. (b) Continuation of this table (a), with the parameter $D = -1$. Simulation of the hyperincursive discrete algorithms of the discrete quantum wave equations (9.14.13a, b) of two particles

(a)

Simulation of the hyperincursive algorithm
Of the discrete quantum wave equations with $D = +1$
Of two particles in a periodic space domain

n	0	0	1	1	2	2	3	3	4	4	5	5	6	6	7	7	8	8	9	9	10	10	11	11
k	Q	P	Q	P	Q	P	Q	P	Q	P	Q	P	Q	P	Q	P	Q	P	Q	P	Q	P	Q	P
0	0	0	0	0	2	0	0	0	0	0	0	0	0	0	0	0	0	0	0	0	0	0	0	0
1	0	0	1	1	0	0	1	-1	0	0	0	0	0	0	0	0	0	0	0	0	0	0	0	0
2	1	1	0	0	0	0	0	0	1	-1	0	0	0	0	0	0	0	0	0	0	0	0	0	0
3	0	0	0	0	0	0	0	0	0	0	1	-1	0	0	0	0	0	0	0	0	0	0	1	1
4	0	0	0	0	0	0	0	0	0	0	0	0	1	-1	0	0	0	0	0	0	1	1	0	0
5	0	0	0	0	0	0	0	0	0	0	0	0	0	0	1	-1	0	0	1	1	0	0	0	0
6	0	0	0	0	0	0	0	0	0	0	0	0	0	0	0	0	2	0	0	0	0	0	0	0
7	0	0	0	0	0	0	0	0	0	0	0	0	0	0	1	1	0	0	1	-1	0	0	0	0
8	0	0	0	0	0	0	0	0	0	0	0	0	1	1	0	0	0	0	0	0	1	-1	0	0
9	0	0	0	0	0	0	0	0	0	0	1	1	0	0	0	0	0	0	0	0	0	0	1	-1
10	1	-1	0	0	0	0	0	0	1	1	0	0	0	0	0	0	0	0	0	0	0	0	0	0

(continued)

Table 9.4 (continued)

(a)

Simulation of the hyperincursive algorithm
Of the discrete quantum wave equations with **D** = +1
Of two particles in a periodic space domain

n	0	0	1	1	2	2	3	3	4	4	5	5	6	6	7	7	8	8	9	9	10	10	11	11
k	Q	P	Q	P	Q	P	Q	P	Q	P	Q	P	Q	P	Q	P	Q	P	Q	P	Q	P	Q	P
11	0	0	0	-1	0	0	1	1	0	0	0	0	0	0	0	0	0	0	0	0	0	0	0	0
12	0	0	0	0	2	0	0	0	0	0	0	0	0	0	0	0	0	0	0	0	0	0	0	0
13	0	0	0	1	0	0	1	-1	0	0	0	0	0	0	0	0	0	0	0	0	0	0	0	0
14	1	1	0	0	0	0	0	0	1	-1	0	0	0	0	0	0	0	0	0	0	0	0	0	0

(b)

Simulation of the hyperincursive algorithm
Of the discrete quantum wave equations with **D** = -1
Of two particles in a periodic space domain

n	0	0	1	1	2	2	3	3	4	4	5	5	6	6	7	7	8	8	9	9	10	10	11	11
k	Q	P	Q	P	Q	P	Q	P	Q	P	Q	P	Q	P	Q	P	Q	P	Q	P	Q	P	Q	P
0	0	0	0	0	2	0	0	0	0	0	0	0	0	0	0	0	0	0	0	0	0	0	0	0
1	0	0	1	-1	0	0	1	1	1	0	0	0	0	0	0	0	0	0	0	0	0	0	0	0
2	1	-1	0	0	0	0	0	0	1	1	1	1	0	0	0	0	0	0	0	0	0	0	0	0
3	0	0	0	0	0	0	0	0	0	0	0	0	0	0	0	0	0	0	0	0	1	0	1	-1

(continued)

Table 9.4 (continued)

(b)

Simulation of the hyperincursive algorithm
Of the discrete quantum wave equations with $D = -1$
Of two particles in a periodic space domain

n	0	0	1	1	2	2	3	3	4	4	5	5	6	6	7	7	8	8	9	9	10	10	11	11
k	Q	P	Q	P	Q	P	Q	P	Q	P	Q	P	Q	P	Q	P	Q	P	Q	P	Q	P	Q	P
4	0	0	0	0	0	0	0	0	0	0	0	0	1	1	0	0	0	0	0	0	1	-1	0	0
5	0	0	0	0	0	0	0	0	0	0	0	0	0	0	1	1	0	0	1	-1	0	0	0	0
6	0	0	0	0	0	0	0	0	0	0	0	0	0	0	0	0	2	0	0	0	0	0	0	0
7	0	0	0	0	0	0	0	0	0	0	0	0	0	0	1	-1	0	0	1	1	0	0	0	0
8	0	0	0	0	0	0	0	0	0	0	0	0	1	-1	0	0	0	0	0	0	1	1	0	0
9	1	1	0	0	0	0	0	0	0	0	1	-1	0	0	0	0	0	0	0	0	0	0	1	1
10	1	1	0	0	0	0	0	0	1	-1	0	0	0	0	0	0	0	0	0	0	0	0	0	0
11	0	0	1	1	0	0	1	-1	0	0	0	0	0	0	0	0	0	0	0	0	0	0	0	0
12	0	0	0	0	2	**0**	0	0	0	0	0	0	0	0	0	0	0	0	0	0	0	0	0	0
13	0	0	1	-1	0	0	1	1	0	0	0	0	0	0	0	0	0	0	0	0	0	0	0	0
14	1	-1	0	0	0	0	0	0	1	1	0	0	0	0	0	0	0	0	0	0	0	0	0	0

Table 9.5 Simulation of the hyperincursive discrete algorithms of the discrete quantum wave equations (9.14.13a, b) with the parameter $D = +1$ of two particles in a box

(a)

Simulation of the hyperincursive algorithm
Of the discrete quantum wave equations with $D = +1$
Of two particles in a box

n	0		1		2		3		4		5		6		7		8		9		10		11	
	Q	P	Q	P	Q	P	Q	P	Q	P	Q	P	Q	P	Q	P	Q	P	Q	P	Q	P	Q	P
k																								
0	0	0	0	0	0	0	2	0	0	0	0	0	0	0	0	0	0	0	0	0	0	0	0	0
1	0	0	0	0	1	1	0	0	1	-1	0	0	0	0	0	0	0	0	0	0	0	0	0	0
2	0	1	1	1	0	0	0	0	0	0	-1	-1	0	0	0	0	0	0	0	0	0	0	0	0
3	1	1	0	0	0	0	0	0	0	0	0	0	0	0	0	0	0	0	0	0	0	0	0	0
4	0	0	0	0	0	0	0	0	1	1	0	0	0	0	0	0	0	0	0	0	0	0	0	0
5	1	1	0	0	0	0	0	0	0	0	-1	-1	0	0	0	0	0	0	0	0	0	0	0	0
6	0	0	-1	-1	0	0	0	0	0	0	0	0	1	1	0	0	0	0	1	-1	0	0	0	0
7	0	0	0	0	0	0	0	0	0	0	0	0	0	0	1	-1	0	0	0	0	1	-1	0	0
8	0	0	0	0	1	1	0	0	0	0	0	0	0	0	0	0	0	0	0	0	0	0	1	-1
9	0	0	0	0	0	0	0	0	1	1	-1	-1	0	0	0	0	0	0	0	0	0	0	0	0
10	0	0	0	0	0	0	0	0	0	0	0	0	-1	-1	0	0	0	0	-1	1	1	1	0	0
11	0	0	0	0	0	0	0	0	0	0	0	0	1	1	-1	-1	0	0	0	0	0	0	0	0
12	0	0	0	0	0	0	0	0	0	0	0	0	0	0	-1	-1	0	0	1	-1	0	0	0	0
13	0	0	0	0	0	0	0	0	0	0	0	0	0	0	0	0	0	2	0	0	0	0	0	0
14	0	0	0	0	0	0	0	0	0	0	0	0	0	0	1	-1	0	0	-1	-1	0	0	0	0

(continued)

Table 9.5 (continued)

(a)

Simulation of the hyperincursive algorithm
Of the discrete quantum wave equations with **D** = +1
Of two particles in a box

n	0	0	1	1	2	2	3	3	4	4	5	5	6	6	7	7	8	8	9	9	10	10	11	11
k	Q	P	Q	P	Q	P	Q	P	Q	P	Q	P	Q	P	Q	P	Q	P	Q	P	Q	P	Q	P
15	0	0	0	0	0	0	0	0	0	0	0	0	−1	1	0	0	0	0	0	0	1	1	0	0
16	0	0	0	0	0	0	0	0	0	0	1	−1	0	0	0	0	0	0	0	0	0	0	−1	−1
17	0	0	0	0	0	0	0	0	−1	1	0	0	0	0	0	0	0	0	0	0	0	0	0	0
18	0	0	0	0	0	0	1	−1	0	0	0	0	0	0	0	0	0	0	0	0	0	0	−1	−1
19	0	0	−1	0	−1	1	0	0	0	0	0	0	0	0	0	0	0	0	0	0	−1	−1	0	0
20	0	0	1	−1	0	0	0	0	0	0	0	0	0	0	0	0	0	0	−1	−1	0	0	0	0
21	−1	1	0	0	0	0	0	0	0	0	0	0	0	0	−1	−1	−1	−1	0	0	0	0	0	0
22	0	0	0	0	0	0	0	0	0	0	0	0	0	0	−1	−1	0	0	0	0	0	0	0	0
23	−1	1	0	0	0	0	0	0	0	0	0	0	−1	−1	0	0	0	0	0	0	0	0	0	0
24	0	0	−1	1	0	0	0	0	0	0	−1	−1	0	0	0	0	0	0	0	0	0	0	0	0
25	0	0	0	0	−1	1	0	0	−1	−1	0	0	0	0	0	0	0	−1	0	0	0	0	0	0

(continued)

Table 9.5 (continued)

(a)

Simulation of the hyperincursive algorithm
Of the discrete quantum wave equations with **D** = +1
Of two particles in a box

n	0	0	1	1	2	2	3	3	4	4	5	5	6	6	7	7	8	8	9	9	10	10	11	11
k	Q	P	Q	P	Q	P	Q	P	Q	P	Q	P	Q	P	Q	P	Q	P	Q	P	Q	P	Q	P
26	0	0	0	0	0	0	-2	0	0	0	0	0	0	0	0	0	0	0	0	0	0	0	0	0
27	0	0	0	0	-1	-1	0	0	-1	1	0	0	0	0	0	0	0	0	0	0	0	0	0	0

(b)

Continuation of the
Simulation of the hyperincursive algorithm
Of the discrete quantum wave equations
Of two particles in a box

n	0	0	1	1	2	2	3	3	4	4	5	5	6	6	7	7	8	8	9	9	10	10	11	11
k	Q	P	Q	P	Q	P	Q	P	Q	P	Q	P	Q	P	Q	P	Q	P	Q	P	Q	P	Q	P
26	0	0	0	0	0	0	-2	0	0	0	0	0	0	0	0	0	0	0	0	0	0	0	0	0
27	0	0	0	0	-1	-1	0	-1	1	0	0	0	0	0	0	0	0	0	0	0	0	0	0	0

(continued)

Table 9.5 (continued)

(b)

Continuation of the
Simulation of the hyperincursive algorithm
Of the discrete quantum wave equations
Of two particles in a box

n	0	0	1	1	2	2	3	3	4	4	5	5	6	6	7	7	8	8	9	9	10	10	11	11
28	0	0	-1	-1	0	0	0	0	0	0	-1	1	0	0	0	0	0	0	0	0	0	0	0	0
29	-1	-1	0	0	0	0	0	0	0	0	0	0	-1	1	0	0	0	0	0	0	0	0	0	0
30	0	0	0	0	0	0	0	0	0	0	0	0	0	0	-1	1	0	0	0	0	0	0	0	0
31	-1	-1	0	0	0	0	0	0	0	0	0	0	0	0	0	0	-1	1	0	0	0	0	0	0
32	0	0	1	1	0	0	0	0	0	0	0	0	0	0	0	0	0	0	-1	1	0	0	0	0
33	0	0	0	0	-1	-1	0	0	0	0	0	0	0	0	0	0	0	0	0	0	-1	1	0	0
34	0	0	0	0	0	0	1	1	-1	-1	0	0	0	0	0	0	0	0	0	0	0	0	-1	1
35	0	0	0	0	0	0	0	0	0	0	1	1	0	0	0	0	0	0	0	0	0	0	0	0
36	0	0	0	0	0	0	0	0	0	0	0	0	-1	-1	0	0	0	0	0	0	0	0	-1	1
37	0	0	0	0	0	0	0	0	0	0	0	0	0	0	0	0	0	0	-1	0	1	-1	0	0
38	0	0	0	0	0	0	0	0	0	0	0	0	0	0	1	1	0	0	0	1	0	0	0	0

(continued)

Table 9.5 (continued)

(b)

Continuation of the
Simulation of the hyperincursive algorithm
Of the discrete quantum wave equations
Of two particles in a box

n	0	0	1	1	2	2	3	3	4	4	5	5	6	6	7	7	8	8	9	9	10	10	11	11
39	0	0	0	0	0	0	0	0	0	0	0	0	0	0	0	0	-2	0	0	0	0	0	0	0
40	0	0	0	0	0	0	0	0	0	0	0	0	0	0	-1	1	0	0	1	1	0	0	0	0
41	0	0	0	0	0	0	0	0	0	0	0	0	1	-1	0	0	0	0	0	0	-1	-1	0	0
42	0	0	0	0	0	0	0	0	0	0	-1	1	0	0	0	0	0	0	0	0	0	0	1	1
43	0	0	0	0	0	0	0	0	1	-1	0	0	0	0	0	0	0	0	0	0	0	0	0	0
44	0	0	0	0	0	0	-1	1	0	0	0	0	0	0	0	0	0	0	0	0	0	0	1	1
45	0	0	0	0	1	-1	0	0	0	0	0	0	0	0	0	0	0	0	0	0	1	1	0	0
46	0	0	-1	1	0	0	0	0	0	0	0	0	0	0	0	0	0	0	1	1	0	0	0	0
47	1	-1	0	0	0	0	0	0	0	0	0	0	0	1	0	1	1	1	0	0	0	0	0	0
48	0	0	0	0	0	0	0	0	0	0	0	0	0	0	0	0	0	0	0	0	0	0	0	0
49	1	-1	0	0	0	0	0	0	0	0	0	0	1	1	0	0	0	0	0	0	0	0	0	0

(continued)

Table 9.5 (continued)

(b)

Continuation of the
Simulation of the hyperincursive algorithm
Of the discrete quantum wave equations
Of two particles in a box

n	0	0	1	1	2	2	3	3	4	4	5	5	6	6	7	7	8	8	9	9	10	10	11	11
50	0	0	1	-1	0	0	0	0	0	0	1	1	0	0	0	0	0	0	0	0	0	0	0	0
51	0	0	0	0	1	-1	0	0	1	1	0	0	0	0	0	0	0	0	0	0	0	0	0	0
52	0	0	0	0	0	2	0	2	0	0	0	0	0	0	0	0	0	0	0	0	0	0	0	0
53	0	0	0	0	1	1	0	0	1	-1	0	0	0	0	0	0	0	0	0	0	0	0	0	0

The initial conditions of $Q(n, k)$ and $P(n, k)$ are given by null values in all the space $n = 0$–11 for the time $k = 0$ and $k = 1$ except for the two particles
$Q(2, 0) = 2$ and $P(2, 0) = 0$
that represents two superposed particles, and
$Q(1, 1) = 1$ and $P(1, 1) = 1$
that represent the first particle moving to the left, and

$$Q(3, 1) = 1 \text{ and } P(3, 1) = -1$$

that represent the second particle moving to the right.

With the periodic boundary conditions of the space, the particles remain in the space domain.

The spatial domain is given by periodic boundary conditions: the first particle moving to the left moves from $n = 0$, $k = 2$ to $n = 11$, $k = 3$.

The second particle moving to the right moves from $n = 11$, $k = 9$ to $n = 0$, $k = 10$.

The two particles are superposed when they interfere, at $n = 2$, $k = 12$. The system is periodic in time, the values at times $k = 12$ and $k = 13$ are identical to the initial values at $k = 0$ and $k = 1$.

This Table 9.4b is the continuation of Table 9.4a, with the value of the parameter $D = -1$, for which the two particles move in the opposite directions.

The columns represent the two wave functions, $Q(n, k)$ and $P(n, k)$ of the Eqs. (9.14.13a, b).

The initial conditions of $Q(n, k)$ and $P(n, k)$ are given by null values in all the space $n = 0$–11 for the times $k = 0$ and $k = 1$ except for the two particles
$Q(2, 0) = 2$ and $P(2, 0) = 0$
that represent two superposed particles, and
$Q(3, 1) = 1$ and $P(3, 1) = 1$
that represent the first particle moving to the right, and

$$Q(1, 1) = 1 \text{ and } P(1, 1) = -1$$

that represent the second particle moving to the left.

With the periodic boundary conditions of the space, the particles remain in the space domain.

The spatial domain is given by periodic boundary conditions.

The first particle, of Table 9.4a, is now moving to the right and moves from $n = 11$, $k = 9$ to $n = 0$, $k = 10$. The second particle, of Table 9.4a, is now moving to the left and moves from $n = 0$, $k = 2$ to $n = 11$, $k = 3$.

The two particles are superposed when they interfere, at $n = 2$, $k = 12$.

The system is periodic in time, the values at times $k = 12$ and $k = 13$ are identical to the initial values at $k = 0$ and $k = 1$.

In this Table 9.5a, the boundary conditions of the two opposite walls of the 1D box are given by:

$$Q(-1, t) = 0, \quad P(-1, t) = 0, \quad Q(12, t) = 0, \quad P(12, t) = 0.$$

The two particles reflect on the two opposite walls of the box, and their values have reversed signs at $k = 26$ and $k = 27$. There is the continuation of this simulation at Table 9.5b.

In this Table 9.5b, the two particles reflect on the opposite walls of the box, and their values at $k = 52$ and $k = 53$ become identical to their initial values at $k = 0$ and $k = 1$ (see Table 9.5a).

In this Table 9.6, the spatial domain is given by periodic boundary conditions.

The initial packet of particles separates into two opposite packets of particles.

There is a stable propagation of the packets of particles.

Then the two packets of particles superpose and become the initial packet of particles.

The system is periodic in space and time, the values at times $k = 12$ and $k = 13$ are identical to the initial values at $k = 0$ and $k = 1$.

In this Table 9.7a, the boundary conditions of the two opposite walls of the 1D box are given by:

$$Q(-1, t) = 0, \quad P(-1, t) = 0, \quad Q(12, t) = 0, \quad P(12, t) = 0$$

The initial packet of particles separates into two opposite packets of particles.

The two packets of particles reflect on the two opposite walls of the box, and their values have reversed signs at $k = 26$ and $k = 27$. There is the continuation of the simulation at Table 9.7b.

In this Table 9.7b, the two packets of particles reflect on the opposite walls of the box. Then the two packets of particles become the initial packet of particles and their values at $k = 52$ and $k = 53$ become identical to the initial values at $k = 0$ and $k = 1$ (see Table 9.7a).

The simulations of the hyperincursive discrete algorithms of the quantum Majorana and Dirac wave equations presented in this last section demonstrate the power of these hyperincursive algorithms which are numerically stable.

Moreover these simulations are performed with discrete integer numbers.

9.15 Conclusion

This chapter presented algorithms for simulation of discrete space-time partial differential equations in classical physics and relativistic quantum mechanics.

We presented the second-order hyperincursive discrete harmonic oscillator that shows the conservation of energy. This recursive discrete harmonic oscillator is separable into two incursive discrete oscillators with the conservation of the constant of motion. The incursive discrete oscillators are related to forward and backward time derivatives and show anticipative properties. The incursive discrete oscillators are

Table 9.6 Simulation of the hyperincursive discrete algorithms of the discrete quantum wave equations (9.14.13a, b) with the parameter $D = +1$ of a packet of particles

Simulation of the hyperincursive algorithm
Of the discrete quantum wave equations with $D = +1$
Of a packet of particles in a periodic space domain

n	0		1		2		3		4		5		6		7		8		9		10		11	
k	Q	P	Q	P	Q	P	Q	P	Q	P	Q	P	Q	P	Q	P	Q	P	Q	P	Q	P	Q	P
0	0	0	0	0	0	0	2	0	4	0	8	0	4	0	2	0	0	0	0	0	0	0	0	0
1	0	0	0	0	1	1	2	2	5	3	4	0	5	-3	2	-2	1	-1	0	0	0	0	0	0
2	0	0	1	1	2	2	4	4	2	2	2	0	2	-2	4	-4	2	-2	1	-1	0	0	0	0
3	1	1	2	2	4	4	2	2	1	1	0	0	1	-1	2	-2	4	-4	2	-2	1	-1	0	0
4	2	2	4	4	2	2	1	1	0	0	0	0	0	0	1	-1	2	-2	4	-4	2	-2	2	0
5	5	3	2	2	1	1	0	0	0	0	0	0	0	0	0	0	1	-1	2	-2	5	-3	4	0
6	4	0	2	0	0	0	0	0	0	0	0	0	0	0	0	0	0	0	2	0	4	0	8	0
7	5	-3	2	-2	1	-1	0	0	0	0	0	0	0	0	0	0	1	1	2	2	5	3	4	0
8	2	-2	4	-4	2	-2	1	-1	0	0	0	0	0	0	1	1	2	2	4	4	2	2	2	0
9	1	-1	2	-2	4	-4	2	-2	1	-1	0	0	1	1	2	2	4	4	2	2	1	1	0	0
10	0	0	1	-1	2	-2	4	-4	2	-2	2	0	2	2	4	4	2	2	1	1	0	0	0	0
11	0	0	0	0	1	-1	2	-2	5	-3	4	0	5	3	2	2	1	1	0	0	0	0	0	0
12	0	0	0	0	0	0	2	0	4	0	8	0	4	0	2	0	0	0	0	0	0	0	0	0
13	0	0	0	0	1	1	2	2	5	3	4	0	5	-3	2	-2	1	-1	0	0	0	0	0	0
14	0	0	1	1	2	2	4	4	2	2	2	0	2	-2	4	-4	2	-2	1	-1	0	0	0	0
15	1	1	2	2	4	4	2	2	1	1	0	0	1	-1	2	-2	4	-4	2	-2	1	-1	0	0

Table 9.7 Simulation of the hyperincursive discrete algorithms of the discrete quantum wave equations (9.14.13a b) with the parameter $D = +1$ of a packet of particles in a box

(a)

Simulation of the hyperincursive algorithm
Of the discrete quantum wave equations with **D** = +1
Of a packet of particles in a box

n	0		1		2		3		4		5		6		7		8		9		10		11	
k	Q	P	Q	P	Q	P	Q	P	Q	P	Q	P	Q	P	Q	P	Q	P	Q	P	Q	P	Q	P
0	0	0	0	0	0	0	2	0	4	0	8	0	4	0	2	0	0	0	0	0	0	0	0	0
1	0	0	0	0	1	1	2	2	5	3	4	0	5	−3	2	−2	1	−1	0	0	0	0	0	0
2	0	0	1	1	2	2	4	4	2	2	2	0	2	−2	4	−4	2	−2	1	−1	0	0	0	0
3	1	1	2	2	4	4	2	2	1	1	0	0	1	−1	2	−2	4	−4	2	−2	1	−1	0	0
4	2	2	4	4	2	2	1	1	0	0	0	0	0	0	1	−1	2	−2	4	−4	2	−2	1	−1
5	5	5	2	2	1	1	0	0	0	0	0	0	0	0	0	0	1	−1	2	−2	4	−4	2	−2
6	4	4	0	0	0	0	0	0	0	0	0	0	0	0	0	0	0		1	−1	2	−2	5	−5
7	5	5	−2	−2	1	1	0	0	0	0	0	0	0	0	0	0	0		0	0	0	0	4	−4
8	2	2	−4	−4	2	2	−1	−1	0	0	0	0	0	0	0	0	0		1	−1	−2	2	5	−5
9	1	1	−2	−2	4	4	−2	−2	1	1	0	0	0	0	0	0	−1		2	−2	−4	4	2	−2
10	0	0	−1	−1	2	2	−4	−4	2	2	−1	−1	0	0	1	−1	−2		4	−4	−2	2	1	−1
11	0	0	0	0	1	1	−2	−2	4	4	−2	−2	0	2	2	−2	−4		2	−2	−1	1	0	0
12	0	0	0	0	0	0	−1	−1	2	2	−3	−5	0	4	3	−5	−2		1	−1	0	0	0	0
13	0	0	0	0	0	0	0	0	0	2	0	−4	0	8	0	−4	0		0	−1	0	0	0	0
14	0	0	0	0	0	0	1	−1	−2	2	3	−5	0	4	−3	−5	2		−1	−1	0	0	0	0

(continued)

Table 9.7 (continued)

(a)

Simulation of the hyperincursive algorithm
Of the discrete quantum wave equations with $D = +1$
Of a packet of particles in a box

n	0		1		2		3		4		5		6		7		8		9		10		11	
k	Q	P	Q	P	Q	P	Q	P	Q	P	Q	P	Q	P	Q	P	Q	P	Q	P	Q	P	Q	P
15	0	0	0	0	-1	1	2	-2	-4	4	2	-2	0	2	-2	-2	4	4	-2	-2	1	1	0	0
16	0	0	1	-1	-2	2	4	-4	-2	2	1	-1	0	0	-1	-1	2	2	-4	-4	2	2	-1	-1
17	-1	1	2	-2	-4	4	2	-2	-1	1	0	0	0	0	0	0	1	1	-2	-2	4	4	-2	-2
18	-2	2	4	-4	-2	2	1	-1	0	0	0	0	0	0	0	0	0	0	-1	-1	2	2	-5	-5
19	-5	5	2	-2	-1	1	0	0	0	0	0	0	0	0	0	0	0	0	0	0	0	0	-4	-4
20	-4	4	0	0	0	0	0	0	0	0	0	0	0	0	0	0	0	0	-1	-1	-2	-2	-5	-5
21	-5	5	-2	2	-1	1	0	0	0	0	0	0	0	0	0	0	-1	-1	-2	-2	-4	-4	-2	-2
22	-2	2	-4	4	-2	2	-1	1	0	0	0	0	0	0	-1	-1	-2	-2	-4	-4	-2	-2	-1	-1
23	-1	1	-2	2	-4	4	-2	2	-1	1	0	0	-1	-1	-2	-2	-4	-4	-2	-2	-1	-1	0	0
24	0	0	-1	1	-2	2	-4	4	-2	2	-2	0	-2	-2	-4	-4	-2	-2	-1	-1	0	0	0	0
25	0	0	0	0	-1	1	-2	2	-5	3	-4	0	-5	-3	-2	-2	-1	-1	0	0	0	0	0	0

(continued)

Table 9.7 (continued)

(a)

Simulation of the hyperincursive algorithm
Of the discrete quantum wave equations with $D = +1$
Of a packet of particles in a box

n	0		1		2		3		4		5		6		7		8		9		10		11	
	Q	P	Q	P	Q	P	Q	P	Q	P	Q	P	Q	P	Q	P	Q	P	Q	P	Q	P	Q	P
k																								
26	0	0	0	0	0	0	-2	0	-4	0	-8	0	-4	3	-2	2	0	0	0	0	0	0	0	0
27	0	0	0	-1	-1	-1	-2	-2	-5	-3	-4	0	-5	3	-2	2	-1	-1	0	0	0	0	0	0

(b)

Continuation of the simulation of the hyperincursive algorithm
Of the discrete quantum wave equations
Of a packet of particles in a box

n	0		1		2		3		4		5		6		7		8		9		10		11	
	Q	P	Q	P	Q	P	Q	P	Q	P	Q	P	Q	P	Q	P	Q	P	Q	P	Q	P	Q	P
k																								
26	0	0	0	0	0	0	-2	0	-4	0	-8	0	-4	3	-2	2	0	0	0	0	0	0	0	0
27	0	0	0	-1	-1	-1	-2	-2	-5	-3	-4	0	-5	3	-2	2	-1	-1	0	0	0	0	0	0

(continued)

Table 9.7 (continued)

(b)

Continuation of the simulation of the hyperincursive algorithm
Of the discrete quantum wave equations
Of a packet of particles in a box

n	0		1		2		3		4		5		6		7		8		9		10		11	
k	Q	P	Q	P	Q	P	Q	P	Q	P	Q	P	Q	P	Q	P	Q	P	Q	P	Q	P	Q	P
28	0	0	-1	-1	-2	-2	-4	-4	-2	-2	-2	0	-2	2	-4	4	-2	2	-1	1	0	0	0	0
29	-1	-1	-2	-2	-4	-4	-2	-2	-1	-1	0	0	-1	1	-2	2	-4	4	-2	2	-1	1	0	0
30	-2	-2	-4	-4	-2	-2	-1	-1	0	0	0	0	0	0	-1	1	-2	2	-4	4	-2	2	-1	1
31	-5	-5	-2	-2	-1	-1	0	0	0	0	0	0	0	0	0	0	-1	1	-2	2	-4	4	-2	2
32	-4	-4	0	0	0	0	0	0	0	0	0	0	0	0	0	0	0	0	-1	1	-2	2	-5	5
33	-5	-5	2	2	-1	-1	0	0	0	0	0	0	0	0	0	0	0	0	0	0	0	0	-4	4
34	-2	-2	4	4	-2	-2	1	1	0	0	0	0	0	0	0	0	0	0	-1	1	2	-2	-5	5
35	-1	-1	2	2	-4	-4	2	2	-1	-1	0	0	0	0	0	0	1	-1	-2	2	4	-4	-2	2
36	0	0	1	1	-2	-2	4	4	-2	-2	1	1	0	0	-1	1	2	-2	-4	4	2	-2	-1	1
37	0	0	0	0	-1	-1	2	2	-4	-4	2	2	0	-2	-2	2	4	-4	-2	2	1	-1	0	0
38	0	0	0	0	0	0	1	1	-2	-2	3	5	0	-4	-3	5	2	-2	-1	1	0	0	0	0

(continued)

Table 9.7 (continued)

(b)

Continuation of the simulation of the hyperincursive algorithm
Of the discrete quantum wave equations
Of a packet of particles in a box

n =	0	0	1	1	2	2	3	3	4	4	5	5	6	6	7	7	8	8	9	9	10	10	11	11
k	Q	P	Q	P	Q	P	Q	P	Q	P	Q	P	Q	P	Q	P	Q	P	Q	P	Q	P	Q	P
39	0	0	0	0	0	0	0	0	0	-2	0	4	0	-8	0	4	0	-2	0	0	0	0	0	0
40	0	0	0	0	0	0	-1	1	2	-2	-3	5	0	-4	3	5	-2	-2	1	1	0	0	0	0
41	0	0	0	0	1	-1	-2	2	4	-4	-2	2	0	-2	2	2	-4	-4	2	2	-1	-1	0	0
42	0	0	-1	1	2	-2	-4	4	2	-2	-1	1	0	0	1	1	-2	-2	4	4	-2	-2	1	1
43	1	-1	-2	2	4	-4	-2	2	1	-1	0	0	0	0	0	0	-1	-1	2	2	-4	-4	2	2
44	2	-2	-4	4	2	-2	-1	1	0	0	0	0	0	0	0	0	0	0	1	1	-2	-2	5	5
45	5	-5	-2	2	1	-1	0	0	0	0	0	0	0	0	0	0	0	0	0	0	0	0	4	4
46	4	-4	0	0	0	0	0	0	0	0	0	0	0	0	0	0	0	0	1	1	2	2	5	5
47	5	-5	2	-2	1	-1	0	0	0	0	0	0	0	0	0	0	1	1	2	2	4	4	2	2
48	2	-2	4	-4	2	-2	1	-1	0	0	0	0	0	0	1	1	2	2	4	4	2	2	1	1
49	1	-1	2	-2	4	-4	2	-2	1	-1	0	0	1	1	2	2	4	4	2	2	1	1	0	0

(continued)

Table 9.7 (continued)

(b)

Continuation of the simulation of the hyperincursive algorithm
Of the discrete quantum wave equations
Of a packet of particles in a box

n	0	0	1	1	2	2	3	3	4	4	5	5	6	6	7	7	8	8	9	9	10	10	11	11
	Q	P	Q	P	Q	P	Q	P	Q	P	Q	P	Q	P	Q	P	Q	P	Q	P	Q	P	Q	P
k																								
50	0	0	1	-1	2	-2	4	-4	2	-2	2	0	2	2	4	4	2	2	1	1	0	0	0	0
51	0	0	0	0	1	-1	2	-2	5	-3	4	0	5	3	2	2	1	1	0	0	0	0	0	0
52	0	0	0	0	0	0	2	0	4	0	8	0	4	0	2	0	0	0	0	0	0	0	0	0
53	0	0	0	0	1	1	2	2	5	3	4	0	5	-3	2	-2	1	-1	0	0	0	0	0	0

not recursive but time inverse of each other and are executed in series without the need of a work memory.

In simulation-based cyber-physical system studies, the main properties of the algorithms must meet the following constraints. The algorithms must be numerically stable and must be as compact as possible to be embedded in cyber-physical systems. Moreover the algorithms must be executed in real-time as quickly as possible without too much access to the memory.

The presented algorithms in this paper meet these conditions.

Then, we presented the second-order hyperincursive discrete Klein–Gordon equation given by space-time second-order partial differential equations for the simulation of the quantum Majorana real 4-spinors equations and of the relativistic quantum Dirac complex 4-spinors equations.

This chapter presented simulations of the hyperincursive discrete quantum Majorana and Dirac wave equations which are numerically stable.

One very important characteristic of these algorithms is the fact that they are space-time-symmetric, so the algorithms are fully invertible (reversible) in time and space.

The reversibility of the presented hyperincursive discrete algorithms is a fundamental condition to make quantum computing.

The development of simulation-based cyber-physical systems indeed evolves to quantum computing.

So the presented computing tools are well adapted to these future requirements.

References

1. Dubois DM (2016) Hyperincursive algorithms of classical harmonic oscillator applied to quantum harmonic oscillator separable into incursive oscillators. In: Amoroso RL, Kauffman LH et al (eds) Unified field mechanics, natural science beyond the Veil of spacetime: proceedings of the IXth symposium honoring noted French mathematical physicist Jean-Pierre Vigier, 16–19 november 2014, Baltimore, USA. World Scientific, Singapore, pp 55–65
2. Dubois DM (2018) Unified discrete mechanics: bifurcation of hyperincursive discrete harmonic oscillator, Schrödinger's quantum oscillator, Klein-Gordon's equation and Dirac's quantum relativist equations. In: Amoroso RL, Kauffman LH et al (eds) Unified field mechanics II, formulations and empirical tests: proceedings of the Xth symposium honoring noted French mathematical physicist Jean-Pierre Vigier, 25–28 July 2016, Porto Novo, Italy. World Scientific, Singapore, pp 158–177
3. Dubois DM (1995) Total incursive control of linear, non-linear and chaotic systems. In: Lasker GE (eds) Advances in computer cybernetics, vol II. The International Institute for Advanced Studies in Systems Research and Cybernetics, Canada, pp 167–171. ISBN 0921836236
4. Dubois DM (1998) Computing anticipatory systems with incursion and hyperincursion. In: Dubois DM (ed) Computing anticipatory systems: Conference on proceedings of CASYS— First international conference, vol 437, 11–15 August 1997, Liege Belgium. American Institute of Physics, Woodbury, New York, AIP CP, pp 3–29
5. Dubois DM (1999) Computational derivation of quantum and relativist systems with forward-backward space-time shifts. In: Dubois DM (ed) Computing anticipatory systems: conference on proceedings of CASYS'98–second international conference, vol 465, 10–14 August 1998, Liege Belgium. American Institute of Physics, Woodbury, New York, AIP CP, pp 435–456

6. Dubois DM (2000) Review of incursive, hyperincursive and anticipatory systems—foundation of anticipation in electromagnetism. In: Dubois DM (ed) Computing anticipatory systems: conference on proceedings of CASYS'99–third international conference, vol 517, 9–14 August 1999, Liege Belgium. American Institute of Physics, Melville, New York. AIP CP, pp 3–30

7. Dubois DM, Kalisz E (2004) Precision and stability analysis of Euler, Runge-Kutta and incursive algorithms for the harmonic oscillator. Int J Comput Anticipat Syst 14:21–36. ISSN 1373-5411. ISBN 2-930396-00-8

8. Dubois DM (2014) The new concept of deterministic anticipation in natural and artificial systems. Int J Comput Anticipat Syst 26:3–15. ISSN 1373-5411. ISBN 2-930396-15-6

9. Antippa AF, Dubois DM (2002) The harmonic oscillator via the discrete path approach. Int J Comput Anticipat Syst 11:141–153. ISSN 1373–5411

10. Antippa AF, Dubois DM (2004) Anticipation, orbital stability, and energy conservation in discrete harmonic oscillators. In: Dubois DM (ed) Computing anticipatory systems: conference on proceedings of CASYS'03–sixth international conference, vol 718, 11–16 August 2003, Liege Belgium. American Institute of Physics, Melville, New York. AIP CP, pp 3–44

11. Antippa AF, Dubois DM (2006) The dual incursive system of the discrete harmonic oscillator. In: Dubois DM (ed) Computing anticipatory systems: conference on proceedings of CASYS'05–seventh international conference, vol 839, 8–13 August 2005, Liege Belgium. American Institute of Physics, Melville, New York. AIP CP, pp 11–64

12. Antippa AF, Dubois DM (2006) The superposed hyperincursive system of the discrete harmonic oscillator. In: Dubois DM (ed) Computing anticipatory systems: conference on proceedings of CASYS'05–Seventh international conference, vol 839, 8–13 August 2005, Liege Belgium. American Institute of Physics, Melville, New York) AIP CP, pp 65–126

13. Antippa AF, Dubois DM (2007) Incursive discretization, system bifurcation, and energy conservation. J Mathemat Phys 48(1):012701

14. Antippa AF, Dubois DM (2008) Hyperincursive discrete harmonic oscillator. J Mathemat Phys 49(3):032701

15. Antippa AF, Dubois DM (2008) Synchronous discrete harmonic oscillator. In: Dubois DM (ed) Computing anticipatory systems: conference on proceedings of CASYS'07–eighth international conference, vol 1051, 6–11 August 2007, Liege Belgium. American Institute of Physics, Melville, New York. AIP CP, pp 82–99

16. Antippa AF, Dubois DM (2010) Discrete harmonic oscillator: a short compendium of formulas. In: Dubois DM (ed) Computing anticipatory systems: conference on proceedings of CASYS'09–ninth international conference, vol 1303, 3–8 August 2009, Liege Belgium. American Institute of Physics, Melville, New York. AIP CP, pp 111–120

17. Antippa AF, Dubois DM (2010) Time-symmetric discretization of the harmonic oscillator. In: Dubois DM (ed) Computing anticipatory systems: conference on proceedings of CASYS'09–ninth international conference, vol 1303, 3–8 August 2009, Liege Belgium. American Institute of Physics, Melville, New York. AIP CP, pp 121–125

18. Antippa AF, Dubois DM (2010) Discrete harmonic oscillator: evolution of notation and cumulative erratum. In: Dubois DM (ed) Computing anticipatory systems: conference on proceedings of CASYS'09–ninth international conference, vol 1303, 3–8 August 2009, Liege Belgium. American Institute of Physics, Melville, New York. AIP CP, pp 126–130

19. Klein O (1926) Quantentheorie und fünfdimensionale Relativitätstheorie. Zeitschrift für Physik 37:895

20. Gordon W (1926) Der Comptoneffekt nach Schrödingerschen Theorie. Zeitschrift für Physik 40:117

21. Dirac PAM (1928) The quantum theory of the electron. Proc R Soc A Mathemat Phys Eng Sci 117(778):610–624

22. Dirac PAM (1964) Lectures on quantum mechanics. Academic Press, New York

23. Proca A (1930) Sur l'équation de Dirac. J Phys Radium 1(7):235–248

24. Proca A (1932) Quelques observations concernant un article «sur l'équation de Dirac». J Phys Radium 3(4):172–184

25. Majorana E (1937) Teoria simmetrica dell'elettrone e del positrone. Il Nuovo Cimento 14:171

26. Pessa E (2006) The Majorana oscillator. Electron J Theo Phys EJTP 3(10):285–292
27. Messiah A (1965) *Mécanique Quantique* Tomes 1 & 2. Dunod, Paris
28. Dubois DM (2018) Deduction of the Majorana real 4-spinors generic dirac equation from the computable hyperincursive discrete Klein-Gordon equation. In: Dubois DM, Lasker GE (ed) Proceedings of the symposium on anticipative models in physics, relativistic quantum physics, biology and informatics: held as part of the 30th international conference on systems research, informatics and cybernetics, vol I, July 30–August 3 2018, Baden-Baden Germany. International Institute for Advanced Studies in Systems Research and Cybernetics, Canada, pp 33–38. ISBN 978-1-897546-80-2
29. Dubois DM (2018) The hyperincursive discrete Klein-Gordon equation for computing the Majorana real 4-spinors equation and the real 8-spinors Dirac equation. In: Dubois DM, Lasker GE (ed) Proceedings of the symposium on anticipative models in physics, relativistic quantum physics, biology and informatics: held as part of the 30th international conference on systems research, informatics and cybernetics, vol I, July 30–August 3 2018, Baden-Baden Germany. International Institute for Advanced Studies in Systems Research and Cybernetics, Canada, pp 73–78. ISBN 978-1-897546-80-2
30. Dubois DM (2018) "Bifurcation of the hyperincursive discrete Klein-Gordon equation to real 4-spinors Dirac equation related to the Majorana equation" *Acta Systemica*. Int J Int Inst Adv Stud Syst Res Cybernet (IIAS) XVIII(2):23–28. ISSN 1813-4769
31. Dubois DM (2019) Unified discrete mechanics II: the space and time symmetric hyperincursive discrete Klein-Gordon equation bifurcates to the 4 incursive discrete Majorana real 4-spinors equations. J Phys Conf Ser 1251:012001. Open access https://iopscience.iop.org/article/10.1088/1742–6596/1251/1/012001
32. Dubois DM (2019) Unified discrete mechanics III: the hyperincursive discrete Klein-Gordon equation bifurcates to the 4 incursive discrete Majorana and Dirac equations and to the 16 Proca equations. J Phys Conf Ser 1251:012002. Open access https://iopscience.iop.org/article/10.1088/1742-6596/1251/1/012002
33. Dubois DM (2019) Review of the time-symmetric hyperincursive discrete harmonic oscillator separable into two incursive harmonic oscillators with the conservation of the constant of motion. J Phys Conf Series 1251:012013. https://doi.org/10.1088/1742-6596/1251/1/012013. Article online open access—https://iopscience.iop.org/article/10.1088/1742-6596/1251/1/012013
34. Dubois DM (2019) Rotation of the two incursive discrete harmonic oscillators to recursive discrete harmonic oscillators with the Hadamard matrix. In: Dubois DM, Lasker GE (ed) Proceedings of the symposium on causal and anticipative systems in living science, biophysics, relativistic quantum mechanics, relativity: held as part of the 31st international conference on systems research, informatics and cybernetics, vol I, July 29–August 2, 2019, Baden-Baden Germany (IIAS), pp 7–12. ISBN 978-1-897546-41-3
35. Dubois DM (2019) Rotation of the relativistic quantum Majorana equation with the hadamard matrix and unitary matrix U. In: Dubois DM, Lasker GE (ed) Proceedings of the symposium on causal and anticipative systems in living science, biophysics, relativistic quantum mechanics, relativity: held as part of the 31st international conference on systems research, informatics and cybernetics, vol I, July 29–August 2, 2019, Baden-Baden Germany (IIAS), pp 13–18. ISBN 978-1-897546-41-3
36. Dubois DM (2019) Relations between the Majorana and Dirac quantum equations. In: Dubois DM, Lasker GE (ed) Proceedings of the symposium on causal and anticipative systems in living science, biophysics, relativistic quantum mechanics, relativity: held as part of the 31st international conference on systems research, informatics and cybernetics, vol I, July 29–August 2, 2019, Baden-Baden Germany) (IIAS), pp 19–24. ISBN 978-1-897546-41-3

Daniel M. Dubois received the Civil Engineer degree in Physics, from the University of Liège, Belgium, in 1970 and the Ph.D. degree in Applied Sciences from the University of Liège, Belgium, in 1975. He was Assistant then nominated Assistant Professor at the University of Liège. He was nominated Professor at the Business School HEC, at Liège, in 1980. His research interest includes the modeling and simulation of systems in biology and physics, the computing anticipatory systems, with the development of incursive, and hyperincursive algorithms in discrete classical and relativistic quantum mechanics. Since 1996, he is Founder and Director of the Centre for Hyperincursion and Anticipation in Ordered Systems (CHAOS), at the Institute of Mathematics of the University of Liège. He is the Founder and Editor of CASYS: the International Journal of Computing Anticipatory Systems. Dr Daniel M. Dubois' honors and awards include the Title of Doctor Honoris Causa of the University of Petrosani, Romania, the Insignia of Doctor Honoris Causa of the International Institute for Advanced Studies in Systems Research and Cybernetics, and the Adolphe Wetrems Award in Physical and Mathematical Sciences, presented by the Belgium Royal Academy, for his works in applied informatics.

Part III
Applications

Chapter 10
Offering Simulation Services Using a Hybrid Cloud/HPC Architecture

Thomas Bitterman

Abstract We outline the design and implementation of a system which implements the Simulation as a Service (SMaaS) model. SMaaS is based on the Software as a Service (SaaS) model, extending SaaS to include High-Performance Computing (HPC)-hosted applications. Simulations in an HPC context can be expensive, complex, and lengthy. The use of a cloud to provide and manage simulations as a service on an HPC cluster provides greater flexibility for users, in particular smaller businesses and educational institutions that might otherwise struggle to use simulation in their work. Adding a cloud to a standard HPC setup allows the HPC component to specialize in its strengths (e.g., performing calculations, storing Big Data), while the cloud can provide its own capabilities. We show how a cloud's ability to scale up/down and support heterogeneous environments provides support for all phases of simulation workflow—education, prototyping, and production. This chapter covers several different systems built at the Ohio Supercomputer Center. All systems have been deployed to production and used by paying customers. In addition to strictly technical concerns issues related to payment, licensing, and other business topics are covered.

10.1 Introduction

The trend over time has been toward using simulation to understand increasingly complex systems, including astrophysical phenomena [1], earthquakes [2], materials science [3], and many other fields. The term computational science has been coined to describe the multidisciplinary field that uses computer modeling and simulation to understand complex areas.

Simulating the real world has always been challenging. At any time, the computing power needed to simulate systems of interest has always outraced the ability of "standard" computer systems. To this end, computational science has relied on

T. Bitterman (✉)
Wittenberg University, Springfield, USA
e-mail: tom@bitterman.net

© Springer Nature Switzerland AG 2020

259

J. L. Risco Martín et al. (eds.), *Simulation for Cyber-Physical Systems Engineering*,
Simulation Foundations, Methods and Applications,
https://doi.org/10.1007/978-3-030-51909-4_10

supercomputers. Although originally composed of just faster than normal CPUs, supercomputers have evolved to utilize a high degree of parallelism.

In this chapter, we will use the terms "high-performance computing" or "HPC" to describe the combination of hardware and software that a user of a supercomputer typically has available. This includes programming languages (FORTRAN), job submission systems (batch-based systems), operating systems (primarily Unix/Linux-based), user interfaces (command line), and many other technologies. We do not wish to give the impression that the HPC toolkit is static. On the contrary, new tools are being adopted all the time, and old techniques are retired. This process is, however, slower than in many other fields. Web-based programming, for example, turns over infrastructure on what seems a weekly basis.

By contrast, HPC is a generally conservative world. Supercomputers are expensive to produce and run, so every cycle counts. HPC practices have, over time, evolved to minimize the "waste" of cycles on anything other than running the target simulation itself. This is true on both the front- and back end. On the front end Command-Line Interfaces (CLUIs) are preferred to cycle-wasting graphical user interfaces (GUIs). On the back end batch job submission mechanisms increase utilization compared to interactive job submission. Even programmer practices are constrained—FORTRAN is still in wide use as it is efficient when written as well as providing large numbers of optimized, bug-free libraries.

The majority of work is performed via HPC system access provided through SSH for text-based access and VNC for visualization access. A typical workflow when using a simulation that produces graphical output might resemble:

1. SSH into the HPC system using a CLUI client/command line
2. Use a text editor to create an initialization text file that contains the parameters for this run
3. Create a batch file that invokes the application on the initialization file along with a set of system commands (e.g., what directory to write log files to)
4. Submit the batch file to the batch scheduler
5. Wait for the job to be scheduled and run
6. Use VNC or transfer the resulting graphics files and examine the output

This workflow is efficient in terms of the use of computational resources and some variant of it is likely to be the standard for some time to come. However, other areas of applied computing have adopted more friendly interfaces, and use of the traditional SSH interface represents a barrier to entry for new users who must locate acceptable client software and learn command-line interfaces for file editing and job control. As web-based and mobile applications proliferate the gap between the interface that new researchers are used to and what HPC applications present will only grow.

An important goal of OnDemand was to bring capabilities present in other areas of computing to HPC applications and to do this with minimal overhead. The important areas for our discussion are web-based and cloud computing.

10.1.1   Web-Based Computing

The web has become the dominant access mechanism for remote compute services in every computing area except HPC. Web applications (gateways) have not truly proliferated in HPC for several reasons. Calegari et al. [4] provide an historical overview and touch on some HPC-specific issues, from which the following table is adapted (Table 10.1).

A couple of important projects that influenced OnDemand are not listed in the above table: XSEDE's web gateways, and HubZero. Both efforts provide a great deal of the functionality necessary for a successful HPC web portal, but in the end neither was judged to meet our needs. Explanations of both follow.

Our first instinct was to adopt XSEDE [5] and its web gateway technology. One of the first attempts to share HPC resources, XSEDE (or TeraGrid, as it was known at

Table 10.1   List of the major HPC portals available for internal on-premises usage

#	HPC portal name (with year of first release)	Comments
1	Agave ToGo with CyVerse Science APIs (2013)	Open source
2	Apache Airavata Django Portal (2003)	Open source
3	Compute Manager by Altair (2009)	Replaced by #14 in 2018
4	eCompute by Altair (2003)	Replaced by #3 in 2009
5	eBatch by Serviware/Bull (2004)	Replaced by XCS1 in 2011
6	EnginFrame by NICE/Amazon (1999)	
7	HPCDrive by Oxalya/OVH (2007)	Discontinued in 2015
8	HPC Gateway Appli. Desktop by Fujitsu (2015)	
9	HPC Pack Web Components by Microsoft (2008)	On Microsoft Windows only
10	JARVICE Portal by Nimbix (2012)	
11	MOAB Viewpoint by Adaptive Computing (2006)	
12	Open OnDemand (2017)	Open source
13	Orchestrate by RStor (2018)	
14	PBS Access by Altair (2018)	
15	Platform Application Center (PAC) by IBM (2009)	
16	ProActive Parallel Suite by ActiveEon (2014)	Open source
17	Sandstone HPC (2016)	Open source
18	ScaleX Pro by Rescale (2012)	
19	SynfiniWay by Fujitsu (2010)	Replaced by #8 in 2015
20	Sysfera-DS by Sysfera (2011)	Discontinued in 2015
21	UNICORE Portal (project:1997, v7:2014)	Open source since 2004
22	WebSubmit Portal by NIST (1998)	Last update in 1999
23	XCS1/XCS2 by Bull/Atos (2011/2014)	Replaced by #24 in 2018
24	XCS3 by Atos (2017)	

the time) employs a service-oriented architecture. Resources are provided as services with well-defined interfaces. Users at affiliated institutions can utilize a single-sign-on (SSO) system to access the services, with the XSEDE infrastructure handling authentication, authorization, and accounting. While the XSEDE framework is very powerful and connects many different systems, it is by no means a lightweight infrastructure. While there are many benefits to membership in XSEDE, the resources required to join and maintain membership in the XSEDE Federation go far beyond what is necessary to run an HPC web portal. For these reasons, XSEDE gateways were not adopted for OnDemand.

Another interesting option was provided by HubZero [6]. HubZero was developed at Purdue University and uses a number of open source packages (Linux, Apache, LDAP, PHP, MySQL, etc.) and middleware to provide a virtualized computing environment and connection to advanced computing systems. An individual site is known as a "hub". A hub can provide several "tools", each of which has its own GUI and functions as a more-or-less independent application. The GUI is exposed to the user through a VNC client embedded as a web browser, while the tool itself runs in an Open VZ virtual environment on the server. Tools running on the hub expect a typical X11 Window System environment. This graphical session is created by running, in each container, a special X server that also acts as a VNC server.

HubZero also provides middleware to control network operations. The middleware offers tools the ability to connect to non-local resources such as XSEDE or to monitor users for accounting purposes. Security is maintained by running each tool under the user's account rather than a shared account.

Between GUI support and this security model, already running Linux applications can be deployed on HubZero in a matter of hours. New hub developers are provided with a software project area which includes version control and support for development with Linux, Windows, Jupyter Notebooks, RStudio, and other web applications as publishing environments.

In the end, the OnDemand team felt that HubZero was still not lightweight enough. While it had some good ideas (especially with regards to security) the requirements for virtual machines, X11 interfaces, mandatory use of specified development interfaces, and other issues put too many restrictions on both the developer and the HPC center. While some of these restrictions have been loosened (for example, HubZero now supports more development environments than previously), OnDemand continues to provide superior flexibility with less overhead.

10.1.2 Cloud Computing

Cloud computing provides OnDemand access to a pool (a "cloud") of machines that are owned and managed by a third party. Widely known clouds include Amazon's Elastic Compute Cloud (EC2), Microsoft's Azure, and Google Cloud Platform. Cloud computing has important similarities to and differences from HPC systems and also

provides unique benefits of its own. We will look at three areas in which clouds and
HPC systems are similar—parallelism, storage, and services.

The most obvious similarity between a cloud computing platform (informally, a
cloud) and a supercomputer is the high degree of parallelism. Both a cloud and a
supercomputer are made of multiple copies of CPUs, storage units, network interfaces, and the like. In addition, both clouds and supercomputer systems can be rented
out by users, who can specify (and pay for) varying levels of performance. The
overlap in ability between clouds and supercomputers can be impressive—Descartes
Labs constructed the 136th fastest supercomputer from a standard Amazon EC2
instance and some custom software (as measured using LinPack and recognized by
the TOP500 site, top500.org). An important limitation to this overlap, however, is
inter-node latency. Supercomputers go to great lengths to improve the bandwidth and
reduce the communication time, between processing units. It is in the nature of HPC
to view the entire supercomputer as fundamentally a single unit which could potentially be entirely employed on a single calculation. A cloud has lower bandwidth
and higher inter-node communications latency as it is fundamentally designed as a
loosely coupled set of computing elements which will be rented to a great number
of different users at any time. These users will not, in general, wish their processes
to communicate with other users' and so are happy not to pay the overhead for a
supercomputer's high-performance internal connection network.

Both clouds and supercomputers provide access to large amounts of storage. The
Ohio Supercomputer Center (OSC), a medium-sized HPC installation, has over 5
PB of disk storage capacity. Each user has 900 TB available in their home directory.
The Blue Waters machine at the National Center for Supercomputing Applications
(NCSA) at the University of Illinois at Urbana-Champaign alone has more than
25 PB of disk. Numbers are harder to come by for commercial cloud vendors, but
Backblaze claims over 800 PB in storage. In a manner similar to the latency situation
in computing, a major difference between cloud and HPC storage is how tightly
the storage is coupled to the computing elements. An HPC system will typically
have (at least) 2 levels of disk. A larger level that stores data that is not needed
by an active process, and a set of smaller high-performance disks that are tightly
connected to the individual nodes that hold data required by the process running on
that node. Cloud machines tend to be commercial-grade hardware, with slower disks
and interconnects.

Access to infrastructure, known as Infrastructure as a Service (IaaS), means that
the provider buys and maintains computing infrastructure (generally, hardware). The
user can then rent the infrastructure rather than purchasing the hardware themselves.
Google Compute Engine is an example. A similar idea is Platform as a Service (PaaS).
A PaaS system supplies a platform for development—additional tools an application
might need. For example, an e-commerce platform might include a built-in shopping
card, credit card processing module, and the like. Google App Engine is an example.
Software as a Service (SaaS) means that the server provides an entire application,
ready-to-use. GoToMeeting is an example of SaaS.

Both cloud and HPC systems provide various levels of IaaS, PaaS, and SaaS.
However, they appeal to different audiences. Cloud computing appeals to the

computing public in general, with user-friendly interfaces and a pay-for-what-you-use model. The technical savvy and resource requirements of cloud users vary greatly. HPC systems appeal to serious researchers and are funded by grants aimed at well-defined problems. HPC users tend toward the high technical knowledge and large resource requirements. Both groups need services but vary greatly in their ability to use them.

10.2 Desired Functionality

OSC, where OnDemand was created, is a standard, medium-sized HPC center. It supports several supercomputers of various sizes and provides compute power, access to specialized software, and technical support. The primary customers are academic researchers from the Ohio State University (where the Center is located) and other universities in the state of Ohio. The infrastructure is standard for its size. On the hardware side OSC runs commodity processors with large amounts of attached RAM and high-bandwidth, low-latency interconnects. Many nodes have attached GPUs. The nodes run Linux and job submission is through a batch system. License servers provide access to specialized software such as ABAQUS, Turbomole, and MATLAB, in addition to a variety of free and open source software.

Before OnDemand started, OSC had experimented with small-scale web portals. A notable early success was the E-Weld Predictor [7] in partnership with EWI (ewi.org), which predicted temperature, microstructure, stress, and distortion for arc welding processes. The user interface enabled the user to enter information about the geometry and welding procedure parameters in both text-based and graphical modalities. The ability to graphically specify the location of a weld was an important feature that could not be easily duplicated in a command-line interface (Fig. 10.1).

A small group formed at OSC with the goal of creating more web portals as part of the National Digital Engineering and Manufacturing Consortium (NDEMC) program (now part of the Council on Competitiveness). This led to a series of portals with web front ends and HPC back ends, including portals that simulated the filling of bottles with fluid and the aerodynamic properties of truck add-ons.

While each portal provided useful experience in creating web-based applications that meshed well with the underlying supercomputer, there was no unifying body of knowledge that could be relied on by developers when they went to build the next portal. An effort was made to codify this knowledge in a Drupal-based infrastructure. While much knowledge was gained by building the infrastructure, potential clients found it too limiting to adopt. Our primary target adopters were in industry and unwilling to change their development practices to work within the framework. A fresh start was needed.

OnDemand began as part of an effort to provide supercomputer capabilities to traditionally underserved constituencies such as small- and medium-sized businesses. This led to a partnership with Nimbis Services, Inc. From the Nimbis website at

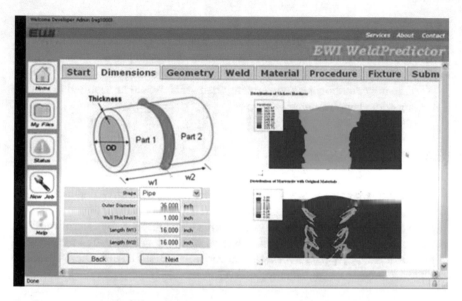

Fig. 10.1 The E-Weld user interface

https://nimbisservices.com/about/: "Nimbis was founded in 2008 by HPC industry veterans Robert Graybill and Brian Schott to act as the first nationwide brokerage clearinghouse for a broad spectrum of integrated cloud-based HPC platforms and applications." The division of labor was clear from the outset: Nimbis would act as a storefront allowing for the purchase and utilization of high-performance computing services for design, modeling, and simulation; while OSC would provide the resources that Nimbis would sell.

In the end, the following requirements were discovered to be necessary for the system to work:

1. Provide security and resource management (authentication, authorization, and accounting services)
2. Map identities across systems
3. Support a marketplace/user interface where app creators could sell apps, and users could buy (access to) them
4. Run web apps without any additional effort on the developer's part
5. Provide access to the user's storage
6. Allow users to access information on their resource usage/job status/system status/etc.
7. Provide interactive access to command line/GUI desktop/3rd-party apps (MATLAB/Abaqus/etc.)
8. Do this all seamlessly from a systems point of view. In particular:

   a. It must run on unmodified HPC hardware
   b. No new user IDs or login required above the standard system login

    c. No change to authentication, authorization, or accounting (AAA) mechanisms from the standard system mechanisms

    d. Would use the software licensing systems already in place

    e. A minimum of software required to run OnDemand itself

    f. The only additional software being that which would be required to run the individual applications.

10.3   The HPC Solution

In this section, we will describe the solution that OnDemand arrived at and explain how that solution met the requirements laid out in the previous section.

The solution was deceptively simple: when a user logs in at the central web site, they are authenticated using the HPC center's standard authentication mechanism. After authentication, a web server is started under that user ID. All further interaction with the user occurs through that web server.

10.3.1   Security and Resource Management

A developer wishing to make an HPC app accessible via the web must do so securely. The data and algorithms in HPC work can be very confidential for the organizations involved or subject to security restrictions in the case of projects related to national defense. Standard PC applications rely on the Operating System (OS) to provide authentication, authorization, and accounting (collectively, security) services. This ensures that only authorized users can access an application, that each user who uses the app can access only their own data, and that billing for the use of specialized software (e.g. sophisticated fluid dynamics solvers) is done correctly. This greatly simplifies the work of the application developer, who can concentrate on the functionality proper to the application and allow the operating system to handle security. In addition, the security functionality in the OS was written by experts and has been tested by many users over the years, so it is higher quality than anything the developer would be able to create.

A web app, on the other hand, handles security differently. The only process running on the system is the web server itself. The capability of the web server to launch processes, access resources, and act on the system in general is limited to the privileges the OS grants the user process that is running the web server. On most systems, the web server runs under a special user account that is dedicated to running the server. This account has privileges to access all folders under a particular directory (e.g. WEBROOT) and can run only those applications in a particular directory (e.g. cgi-bin). To the extent that a web app recognizes different users, this is entirely handled at the application (or web infrastructure) level, not the OS level. This way of doing things provides an extra level of security on its own—no matter what level of access a web site user might have to the web site itself, it can never have more access to the underlying system than the web server does.

To summarize, when a user logs onto a web site, that user is logging onto the software that runs the site, not the computer that is running the web server. All security functions are being handled by the software being run by the web server, not the OS.

This is a problem for a web application on an HPC system. In order to launch an application or access a resource, the HPC system depends on the OS to have cleared the user through security. This is exactly what a standard web app does not do. This limits the developer to creating two types of applications.

First, applications for which security is unimportant. For example, an app that displayed up time for a center's machines could be created that would run under a web server's account and display a page with that information when requested. The center would not care who accessed such a page, so there would be no need for security.

Second, applications for which the identity of the user that runs the app is unimportant. For example, an application that calculates stresses on an object where the user uploads a CAD file and enters the forces upon the object. A company might run such a web site under a special account as a way of demonstrating a new product, for example.

Both of these types of application are useful, but fall short of the functionality required to take full advantage of an HPC center. OnDemand, in particular, needed to keep track of individual users for accounting purposes. Running the web server under that user's ID provides exactly that ability. Whatever the user does (requesting a web page, starting an application, accessing storage, etc.) they do by issuing a request to a web server running under their user ID. Whatever action the web server takes to fulfill the request are performed under the user's ID. This means that all standard operating system-level security mechanisms work automatically with OnDemand. There is no need for the application or the infrastructure to insert code to interface with a custom security system. Even if the user were to somehow break through the web server and start executing code at the command line, the command line would simply be running under the user's ID. The hacker would gain no privileges in such an attack.

10.3.2   Identity Mapping

An application market requires the vendor to be able to control access to applications and charge for their use—in short, to run their own AAA. The vendor must allow users to sign up and register to use applications and be able to charge the user for that use.

The issue is that Nimbis users and OSC users overlapped but were not identical. Nimbis and OSC had their own user databases, permissions, and the like. In order for OSC to track resource usage in a way that Nimbis could understand, OSC had to report the usage in terms of Nimbis's users, not OSC's.

The solution was for Nimbis to keep a database that mapped Nimbis usernames to OSC usernames. When Nimbis signed up a new user they simply sent an out-of-band request for a new OSC username and stored the mapping in the database. Requests for the use of OSC resources required Nimbis to consult that mapping so the request could be sent under the OSC username. This had the advantage that multiple resource providers were easy to support—just add another username to the mapping.

10.3.3 Market Support

Decoupling the vendor's AA from the resource providers is crucial to allowing the vendor to run a market. For example, if Nimbis were to sell a customer the rights to use MATLAB at OSC, it would be Nimbis's responsibility to ensure that the user was presented with MATLAB as an option on their interface (while simultaneously not being presented with products that they had not purchased). Nimbis users could buy and cancel products at will, without involving OSC in any way. Should the user try to run MATLAB, an out-of-band message would have to pass between Nimbis and OSC notifying OSC that the user was authorized to run MATLAB and that OSC should start up a web server under that user's ID and redirect it to the MATLAB page.

In this model, the vendor is free to add or remove customers, set prices at will, issue credits and refunds, and generally provide any marketplace functionality desired without regard to the resource provider. The need to communicate with the resource provider, outside of requesting an application be launched for a user, is limited to just a few cases: when usage statistics are required, when a new user is created, and when the status of an application has been changed on the resource provider (so the market can display a new application, or delete one that is no longer available).

10.3.4 Developer Effort

No marketplace will succeed without something to sell, and the more the better. In the application world, this means making it as easy as possible for developers to develop for your platform. Early experiments with e-Weld and other portals taught the OnDemand team a lot about developing web portals that worked well in an HPC setting. The first lesson was that batch job submission worked differently than the "normal" job submission process that almost all web portals use. Similar lessons were learned about the use of shared resources, licensing issues, and the general mismatch between web and HPC environments. A large infrastructure—similar in spirit to Ruby on Rails—was built to make navigating these issues easier on web developers. It really made developing web applications on our supercomputer much easier, as we discovered by using it to build several portals. Each was built more quickly and had more features than the last. It was a good tool.

No one wanted to use it. When we started work with outside developers we discovered that they already had their preferred building procedures—a preferred language, framework, set of libraries, etc.—and they were not going to change a thing to work on a supercomputer. This was true even for companies that did work both on the web and supercomputers. The work was done for separate projects and no cross-fertilization occurred. Our framework was not so good, nor OSC so powerful, that we could force them to adapt.

The only solution to the impasse was to make developing a web application on an HPC system as much like developing it on any other sort of system as was possible. To this end, starting a dedicated web server for every user was no close as we could come to replicating a standard setup. Some differences were unchangeable—for example, OSC was not going to abandon its batch job submission model. Most differences were subtler and had less obvious positive and negative impacts on development.

Possibly the most important initial difference between running on a web server dedicated to a single-user (single-user) and the usual practice of all web users sharing a single system account (shared account) was that resources were slightly more complicated to share. An important distinction underlying this is the difference between a "web user ID" and a "system user ID". A web user ID is the ID that someone who logs into a web site has. A system user ID is the ID that someone who logs onto the server itself has. A web user gets a web user ID and has their AAA provided by the web server, but not the system. Conversely, a system user gets a system user ID and has their AA provided by the system. A shared-account system depends on web user IDs to deal with portal-based resources while using the system ID of the web server to interact with the underlying server, a single-user system depends solely on system user IDs.

A shared-account system, for example, can use a single database to hold all user records. Since every request to the database will come from the single account that runs the web server, all that is left is to simply grant that user the requisite permissions on the database and its tables. It is the responsibility of the web application to keep the various users' data secure. In a single-user system, every user of the system would be attempting to log into the database from a different account. Granting permissions every time a new user tried to use an application would be an administrative nightmare. Allowing the application to log in as a special user to the database is its own security problem. The solution in this case is to store each user's information in a storage mechanism owned by that user: a database owned by that user or a file in that user's directory structure. If anything, this is an even more secure setup than standard, as there is no central repository that contains everyone's data.

It is not only the case that a single-user system made things harder. Some interesting things were made easier, even possible. For example, software licensing systems often depend on the user ID of the requestor in order to determine if a request should be allowed. This is done because the software licensing system trusts the host's operating system to perform authentication. In a shared-user system, one would have to grant all (or none) of the users access to a software license, as all requests would come from the same system-level user ID. A single-user system allows the licensing system to operate the way it was intended. The request comes from a process (the web server) running under the appropriate user's system-level ID. Other examples of important functionality that become possible, or just easier, are included in the following sections.

10.3.5 User Storage

A lot of simulation software requires input in the form of configuration files. This is a standard HPC practice: all the inputs to the simulation, plus any information about running the simulation itself, are placed into files. When the application starts, the files are read in and the simulation can run. A similar workflow is implemented for output, where the results of the simulation are written to files that can be examined after the application is finished. This fits well with the batch nature of HPC—the user is not expected to be around for the start of the application nor the end so inputs and outputs must be placed somewhere persistent rather than being entered into a form.

Web portals were not meant to entirely replace this workflow, so users still wanted to be able to access their files through the portal so they could edit inputs and examine outputs. In a shared-account system this is difficult. One ends up with some sort of shared storage space, a giant directory or series of sub-directories, where the web server is responsible for determining what files belong to who. A user who logged into their system account could see none of their portal files, nor could the portal see any user files. This was unacceptable for many users who had established workflows and wanted to simply augment them with the portal.

A single-user system makes file access easy. Since the user ID being presented upon the request to access a directory/file is a system user ID the operating system itself can do its usual job of authorizing and accounting for the access. The user is presented with the same files and directories whether they access them through the portal or an SSH connection because they are the same files either way. The same is true for any other storage mechanism, such as a database.

10.3.6 Resource Usage

An important use case for supercomputers is when users want to access information on their resource usage, job status, system status, or other general information relating to the system itself and the user's use of its resources. It is system's responsibility to collect and provide this information, and a single-user system is able to directly access them because it can use the user's system ID to request them. A shared-user system would need elevated privileges to access the data of its users.

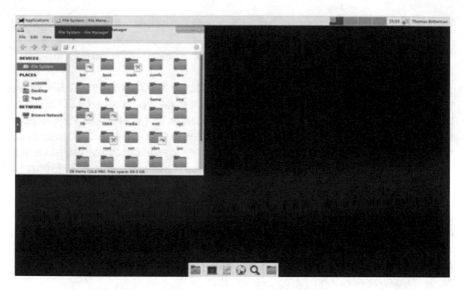

Fig. 10.2   A sample desktop

10.3.7   Access to the Command Line and Desktop

We did not want to limit prospective users to interacting with the system only through portals. The use of a single-user web server meant that they could be provided with direct access to a command line, a GUI desktop, and all the applications that could be used on a standard computer (that just happened to be hooked up to a supercomputer). These all share the property that they are processes started by the operating system for a system user, so the single-user solution starts them without issue. The most difficult part was relaying the GUI back to the user's web page. VNC worked well for this and is still used (Fig. 10.2).

10.3.8   Seamless System Integration

The Ohio Supercomputer Center is not large as HPC centers go, and had neither the inclination nor the budget to change its infrastructure—there was no time or money to adapt already existing systems to support OnDemand, and no prospect of a new, dedicated system. In addition to minimizing hardware change, there were two groups that wanted minimal disruption from OnDemand: system administrators and users. The system administrators had their hands full operating the center using a standard HPC setup and did not need any additional work. Many OnDemand users also used an SSH client to log into their accounts. An important goal was to make logging in via the web and via SSH as close to indistinguishable as possible. In order to please both groups, an OnDemand user had to be treated identically to a non-OnDemand user to the extent that was possible.

To conserve hardware, OnDemand had to reuse as much of the current infrastructure as possible.

At a base level, this meant running on the current OSC machines. This was the easiest integration requirement to meet. OnDemand does not require any hardware more specialized than being able to run a web server. In the end, OnDemand used some publicly available nodes and some machines in OSC's DMZ. OnDemand did not even require full use of the machines, so the impact was limited to increased load. Of course, should usage increase, it would be possible to dedicate hardware to OnDemand.

As discussed above, access to resources must be controlled. The standard setup did this by utilizing mechanisms provided by the operating system, which depend on the user being logged in. OnDemand uses the standard system login server and attaches the resulting system user ID to everything the user does from that point on. This meant no change to log in procedures, password change policies, and other login-related issues.

Because the OnDemand user is attached to a system user ID, all the AAA mechanisms currently in place could be applied to everything a user did. System administrators could handle adding/removing privileges, creating usage reports, creating/deleting accounts, and all the usual user maintenance activities for OnDemand users identically to non-OnDemand users. This typically meant using operating system utilities, although there were important exceptions, such as third-party licensing software.

One important authorization capability was software licensing. This was often provided by an off-site server which OSC would connect to, send some information about the user, and receive a token that would allow the user to activate the software in question. At minimum, requesting a software license would require that the user's system user ID be added to a list of permitted users. Often, depending on the server, must more elaborate steps needed to be taken. Ensuring that OSC could interoperate with all of the necessary licensing servers took a non-trivial amount of the administrators' time. Having OnDemand use actual system IDs at least did not make this problem any worse.

In addition to running on standard hardware, OnDemand runs on standard software. On the server side, a basic installation could be up and running with just a web server and a VNC server. The initial version of OnDemand was written in Java and Python, while the current version is Ruby and node.js. On the client side any modern browser should work, with the ability to run NoVNC a plus. There are a few minor parts of OnDemand (in particular, system configuration files and scripts) that consist of text files and shell scripts (currently, bash scripts). These may need to be rewritten on a site-by-site basis. These parts tend to touch on more sensitive security matters, and system administrators would want to inspect and approve this code in any case.

In addition to the software required to run OnDemand, there will be software that OnDemand provides access to. This software is treated no differently (for the most part) than any other software that is to be installed on the system. All of the usual installation procedures must be followed: approval by administrators, formalized installation and update procedures, authorization of users, possible licensing server issues, and so forth, are handled in exactly the same way as any standard application. In addition, some OnDemand-specific configuration might be required: registration with the OnDemand server, any required setup/shutdown scripts required for use over the web, and so on. Most of this work devolves onto the application writer rather than OSC staff.

10.4   OnDemand in a Cloud

Our initial experience working with Nimbis showed that OnDemand worked well with a cloud. Nimbis was interested in running a storefront that would enable customers to buy access to resources for design, modeling, simulation, and analytics.

The sort of hardware that is required to perform high-end simulation is expensive. Most small-to-medium-sized companies (SMC) cannot afford such an expense, especially for a capability that would be used infrequently. In addition to the cost of the hardware, expertise in high-end simulation software and techniques is rare and expensive. These issues kept a lot of companies that could have benefitted from using simulation tools from being able to use them for product design and testing.

On the other hand, centers that house hardware capable of performing advanced simulations (HPC centers and some cloud providers) have traditionally been attached to large consumers,[1] academia (OSC and others) or government.[2] As hardware became cheaper and new technologies (including the cloud) matured, these centers started to look outside the confines of their usual customer base in an attempt to bring the benefits of simulation to a larger audience. A wider audience also meant greater demand for the specialized knowledge contained at these centers.

The opportunity Nimbis saw was that SMCs did not know what the benefits of simulation were, nor how to work with a simulation center to take advantage of them. On the other side, the simulation centers were not skilled in dealing with commercial entities and their particular needs. Nimbis could build a market for simulation-based computing if they could bridge this gap. The end goal was to provide users with a marketplace from which they could choose various high-end design, modeling, simulation, and analytical tools (e.g., Abaqus or Mathematica) and access experts who could help them use these tools. The users would be freed from the expense of maintaining the hardware and hiring the experts, while the resource providers would have access to a new pool of customers.

---

[1] See [8] for an example of industry using HPC for simulation.

[2] The Department of Energy maintains a large stable of supercomputing resources including Argonne National Laboratory, Oak Ridge National Laboratory, Los Alamos National Laboratory, and many others. See https://press3.mcs.anl.gov/sc13-internal/department-of-energy-high-performance-com puting/ for a listing.

The marketplace itself would consist of a set of resource providers, each of which would offer some suite of services (software or consulting). Customers would be able to pick and choose which they wanted based on price, availability, and other considerations. It would be possible to, for example, create a model using Abaqus at OSC, run the simulation at Argonne, and analyze the results using Mathematica on AWS.

In order for this to work the users had to be insulated from the particulars of working with the individual resource providers. Nimbis needed to be able to perform authentication, authorization, and accounting across all the different providers in a manner transparent to the customer. While Sects. 10.3.1–10.3.3 discussed this in terms of Nimbis's connection to OSC, it should be emphasized that this same situation was replicated a number of times, one for each vendor Nimbis wanted to work with.

The solution was to agree on a common protocol that each vendor would implement to pass out-of-band information. The protocol took the form of XML templates that were sent over a secure connection. Each template corresponded to a request or response between the systems. For example, when a user paid Nimbis for access to a piece of software that was provided at OSC, Nimbis would send a message (following the XML template) that would inform OSC which user it was and what software they had bought access to. This would allow OSC to authorize access to that software for that user in the future.

As this example shows, speaking the protocol required more than simply following the syntactic conventions. Each message specified that an action be taken or some information be provided. The prerequisites and effects of each action, and the meaning of each piece of information, were spelled out in the protocol definition document. For example, when reporting back on how much compute time a user has consumed during a session it was not enough to simply following the message syntax. Also important were details such as single-core vs. multi-core use, CPU type (big-memory vs. standard), use of specialized resources (e.g., GPUs), and so forth. These details tended to complicate the semantics of the protocol, but, in the end, were not insurmountable.

10.4.1 HPC as a Cloud

An unexpected insight provided by the protocol was that, to some extent, it made the OSC supercomputer look like a cloud to the customer. An important feature of a cloud is its flexibility—if the user wants more of some resource (compute power, memory) they can just request it. The cloud will provision the resource (and charge for it) dynamically, so the user/application can respond to changing conditions.

Using OSC's resources through OnDemand made OSC somewhat like a cloud. If the user had a scalable application, they could simply request more resources so that their job would finish faster. Users could request specialized resources (GPUs, big-memory nodes) and so shape the execution profile of their jobs—using these resources could result in the job running more quickly, while potentially costing more.

The biggest practical difference was that requests for resources at OSC were static—there was no way for a user/application to, while running, request additional resources as they could in a cloud. Regardless of this difference, the degree to which OnDemand could make an HPC system look like a cloud was striking.

10.4.2 Using a Cloud as a Resource Provider

The use of this protocol had a more far-reaching effect than just uncovering a previously underappreciated similarity between HPC systems and clouds. While the initial goal of this protocol was just to smooth over the differences between different HPC centers, each of which had its own different ways of doing things, it was quickly realized that the protocol allowed a nearly complete decoupling of the front and back ends.

Defining the back end as "anything that supported the protocol" meant that the service provider no longer had to be an HPC center—it could be anything that could run the software that users wanted. In particular, as clouds became more powerful, it became feasible to include clouds as possible back ends. As long as the instance provided a publicly accessible web server that spoke the protocol and provided some service that resource could join the set of service providers.

This freed up Nimbis to include clouds as resource providers. That this worked was demonstrated in an internal demo created at OSC. An Amazon Web Services account was created and the protocol software was installed, along with a web server and simple application. This AWS instance was added to an internal version of OnDemand and it was able to launch an AWS instance and provide the user with access to an application installed on the cloud. The effort to make this happen was minimal—between 40 and 80 h of development time by developers new to the AWS framework.

10.5 Evolving OnDemand

OSC fits into Nimbis's business model as a resource provider. Independent of Nimbis, but in a similar fashion, OSC was looking to expand its customer base. The first step was to create a new front end, independent of Nimbis, that would run at OSC. This would in turn allow OSC to serve two new customer bases: commercial users, and "regular" OSC users.

To describe these new systems, we will need some new terminology. The term "AweSim" will be used to denote a system that OSC created based on OnDemand technology to partner with its commercial clients. "OSC OnDemand" will be used to denote a system that OSC created for use by its traditional users.

10.5.1 Building a Front End at OSC

The first step was to replace Nimbis's front end with one that ran at OSC. There were two important features that made this possible:

- The decoupling of the front and back ends made possible by the protocol
- The use of the system ID to identify users

The decoupling of the front and back end meant that OSC could set up a new back end and a new front end. The front end would be completely new, while the back end would be a copy of the already existing back end configured to listen to the new front end rather than Nimbis.

Setting up a new back end took little effort. Every user session, whether it was initiated through Nimbis or OSC, required its own port for communication between the application and the user. Before such communication began, the back end already checked whether the port was already in use. This was necessary even with a single front end back end combination due to the shared nature of OSC's system. While it was rare, there was nothing preventing an OSC user (or other OSC-based project) from starting up a web server on a publicly accessible node. It was part of the requirements for OnDemand that it has minimal impact on the current system, so we decided against setting aside a range of port numbers for OnDemand use as that might impact users. As such, OnDemand had to check whether a port was clear before starting each session. This paid dividends when running multiple back ends, as they would automatically avoid interfering with each other.

The only required port configuration involved which port the initial connection used. Understanding the idea of an initial connection requires a brief discussion of the steps taken when starting an OnDemand application. These steps are as follows:

1. The user logs into the front end
2. The user is presented with a list of applications
3. One is chosen by the user
4. The front end sends a message to the back end requesting that the chosen application be started for the user in question
5. The back end fires up a new web server on a free port
6. The back end sends the URL back to the front end
7. The front end redirects the user's browser to this URL

Step 4 requires that there be an agreed-upon machine + port that the front end can communicate to the back end on. This is what is meant by the "initial connection"—it is the connection over which the message to initialize a user's connection is made. It is important that this connection be surrounded by appropriate security measures so that the back end can trust that requests to start an application on a user's behalf comes from a party authorized to make that request, and the requisite setup is non-trivial. The same process needs to occur every time a new front end is set up (e.g., if Nimbis moved to an architecture in which multiple machines could make requests, each machine would have to go through this security configuration).

10.5.2   Working with AweSim

The first group of users that were interested in working with OSC were commercial entities with specific needs. For example, a local fluid dynamics company[3] wanted to build an app that would allow their customers to experiment with the fuel efficiency effects of various truck add-ons (skirts, tails, etc.). While the calculations required a supercomputer, it was felt that the target audience would be more able to more productively engage with the application through a web interface. The company was very familiar with the use of fluid dynamics tools on OSC's system and had done some web development, so they were a perfect trial project for AweSim.

As Nimbis was out of the picture, the OSC front end needed a different way to do authentication and authorization (accounting continued to be done by the usual operating system-level mechanisms). The primary change to authentication was that the Nimbis user ID no longer needed to be mapped to an OSC user ID. When new customers signed up to use the application, they were given standard OSC user IDs for authentication. A new Unix group was created for the application to which only user IDs that had paid for the application were added.

When a user wanted to run the app they would point their browser at the main AweSim page for login. Upon successful login, the system would check to see if the user ID was in the appropriate group. If it were, the user would be presented with an icon that would allow them to launch the app. If not, no icon would be displayed. Using a Unix group meant that authorization to launch the application took place at the operating system level, not in the AweSim portal. This helped secure the application. For example, this form of authorization meant than unauthorized users would not be able to bypass the front end and attempt to launch the application from the command line and connect via browser.

This first portal was successful enough that several more followed.[4] Each followed the same basic template as the first, using OSC IDs and Unix groups to keep everything straight. The process was not without issues. In particular, it became cumbersome for the commercial partner and OSC to coordinate on issuing (and canceling) user IDs every time some started (or stopped) using a portal.

An interesting side effect was that, as users became more used to the portal, they started to want more power over what was going on. In effect, they moved from regular users to power users. Giving users more power meant giving them more tools. For traditional users, this would mean learning to use the command line. The portal users, however, were not interested in doing this. Their experience was primarily with web-based and GUI systems, such as Windows. These users expected a similar experience when accessing files and other resources at OSC.

To some extent, this was possible by providing an alternate web-based interface. For example, see the following image for an example of OSC OnDemand presenting a user's files as a web page (Fig. 10.3):

---

[3]TotalSim, at https://www.totalsim.us/.

[4]The current lineup can be found at https://www.awesim.org/products.

Fig. 10.3 The file interface

With this interface, a user can navigate through their files, and both view and edit them at will using a built-in viewer and editor.

As users grew more confident with the intricacies of creating and submitting batch jobs rather than depending on the portal the ability to work with batch jobs was added (Fig. 10.4).

Fig. 10.4 Creating and submitting a job

10.5.3   OSC OnDemand

As the AweSim users requested more functionality, it became clear that the line
between a commercial user and a traditional user was largely imaginary. Commercial
users were increasingly comfortable around standard command-line tools, while
traditional users were starting to clamor for more graphical, interactive tools. The
result was OSC OnDemand.

Is a sense, OSC OnDemand is just AweSim for traditional (educational and
academic) HPC users. It grew out of the realization that the software that OSC
supported for its userbase (Abaqus, MATLAB, Stata, etc.) was not, at heart, different
from the applications that commercial entities were creating for their customers.
In the OnDemand framework there is little difference between starting a copy of
BolometerSIM for a customer and starting a copy of R Studio for a student in a class.

OSC OnDemand grew past that, however. Eventually it was realized that OSC
OnDemand was not just a way to offer software to traditional users, it was an entirely
new way to access OSC, equal in importance to the tradition SSH command-line
interface. This entailed allowing browser users to employ GUI desktops, graphical
applications, specialized servers, and even command lines (Fig. 10.5).

There were some changes from AweSim to OSC OnDemand. In particular,
accounting was simplified. Traditional users were well-understood before the new
web interface began—the same mechanisms that worked previously worked without
change when they moved from SSH to web-based access. Preliminary experience
has shown that users new to the HPC paradigm have been able to use OSC OnDemand as an "on-ramp" to the full command-line experience. This has been especially

Fig. 10.5   OSC OnDemand options

helpful for classwork as a professor can create a class using OSC OnDemand and be confident that the students, familiar with GUIs, can spend more of their time on the problem at hand, and less time on figuring out the command-line interface.

10.5.4   The Future of OnDemand

While high-performance computing is a conservative field, other areas of computing continue to progress. While not all advances are suitable for adoption by HPC—cycles are too precious to allow for users to waste time with GUIs on compute nodes—some new technologies hold promise. OSC OnDemand started as an attempt to use one of those technologies: cloud computing.

The primary promise of cloud computing was its use as a platform for simulation and modeling. Once the province of supercomputers, cloud systems have started to make inroads into simulations [9]. Cloud computing architectures have numerous advantages over standard HPC systems, including: scalable availability of resources, lower barrier to entry due to greater user familiarity with cloud-style interfaces as opposed to HPC interfaces, and the backing of such industry giants as Google, Amazon, and Microsoft.

Cloud computing does not obsolete the standard HPC paradigm, however. There are plenty of applications which run better on a standard supercomputer architecture, particularly those in which large amounts of data need to be shared between computing elements.

In the end, there is no need for the cloud and HPC to be enemies. There is more than enough work out there for both. What users need is a hybrid architecture that would allow them to utilize the strong points of both cloud and HPC systems.

The OnDemand framework provides a way to seamlessly blend cloud and HPC systems into a single whole in such a way that a user can access resources without worrying about what sort of system is providing them, cloud or HPC. It does this through two important policies: separation of front end, back end, and application; and extensive use of operating system-level security mechanisms.

Separation of the front end and back end is accomplished by requiring that the front end and back end run on separate web servers owned by different users. All communication between the two is accomplished through message sent between them. The portals themselves are standard web applications that are, in turn, run on separate web server owned by the user that requested the application. The use of separate web servers and messaging means that the front and back ends can run on either an HPC or cloud machine. As long as the machine follows the protocol, the other partner (and the applications) does not know what sort of architecture it is communicating with. This allows a mix of a cloud front end and HPC back end, for example, in addition to any other combination desired. In addition, the same front end could potentially talk to multiple back ends, allowing simultaneous access to both (possibly multiple) cloud and HPC systems.

Operating system-level mechanisms are used throughout for authentication, authorization, and accounting and any other necessary security. While this might seem to limit the cross-platform nature of OnDemand, it turned out to free it from platform dependency. Any process run on a computer will at some point need to be approved by the operating system. Any framework that attempts to provide a comprehensive AA solution will eventually come to face with this fact and require some sort of adapter for each different type of system it will run its applications on. OnDemand reduced this problem to simple ID mapping as follows:

1. The front end requires the user to log into the front end
2. When the user launches an application, the front end looks up the user's ID on the requested host
3. The front end sends that ID (which is the user's ID on that system) along with the request to start the application
4. The back end starts the application under the user's system ID

The application then runs as a standard application on that system. All AAA functionality is handled as usual. If the front end wants any AAA information (e.g., compute time used), it is easy enough to write a script for that system to pull such information from standard system files that are being kept in any case. When dealing with architectures as different as clouds and HPC any opportunity to avoid cross-platform work should be taken.

In the process of creating OnDemand it was surprising to find out how closely, at an appropriate level of abstraction, supercomputers and clouds resemble each other. In the end, they are both ways of providing resources (processors, memory, storage, networking, applications, etc.) to users that such users would otherwise be economically incapable of accessing. Both allow the user to specify what resources they would like, how much they would like, and for how long. Both provide security for accounts and data.

OnDemand's biggest breakthrough may have been the way it allows users to concentrate on the things clouds and supercomputers have in common, while abstracting away the differences. This has allowed users to get on with the real problem: understanding the world through computation. The author and the OnDemand team hope that continued adoption and development of the OnDemand framework will continue to support breakthroughs in modeling and simulation regardless of the underlying computing architecture: supercomputer, cloud, or whatever comes next.

References

1. Liska M, Tchekhovskoy A, Ingram A, van der Klis M (2019) Bardeen-Petterson alignment, jets, and magnetic truncation in GRMHD simulations of tilted thin accretion discs. Mon Not R Astron Soc 487(1):550–561. https://doi.org/10.1093/mnras/stz834
2. Aagaard BT, Graves RW, Schwartz DP, Ponce DA, Graymer RW (2010) Ground-motion modeling of Hayward Fault scenario earthquakes, Part I: Construction of the suite of scenarios. Bull Seismol Soc Am 100(6):2927–2944. https://doi.org/10.1785/0120090324

3. Schneider J, Hamaekers J, Chill ST, Smidstrup S, Bulin J, Thesen R, Blom A, Stokbro K (2017) ATK-ForceField: a new generation molecular dynamics software package. Model Simul Mater Sci Eng. 25 085007. https://doi.org/10.1088/1361-651X/aa8ff0
4. Calegari P, Levrier M, Balczyński P (2019) Web portals for high-performance computing: a survey. ACM Trans Web 13(1), Article 5:36 pp. https://doi.org/10.1145/3197385
5. Towns J, Cockerill T, Dahan M, Foster I, Gaither K, Grimshaw A, Hazlewood V, Lathrop S, Lifka D, Peterson GD, Roskies R, Ray Scott J, Wilkins-Diehr N (2014) XSEDE: accelerating scientific discovery. Comput Sci Eng 16(5):62–74. https://doi.org/10.1109/mcse.2014.80
6. McLennan M, Kennell R (2010) HUBzero: a platform for dissemination and collaboration in computational science and engineering. Comput Sci Eng 12(2):48–52
7. Zhang W, Yang Y-P (2009) Development and application of on-line weld modelling tool. Weld World 53(1–2):67–75. https://doi.org/10.1007/bf03266693
8. Rachakonda SK, Wang Y, Grover RO, Moulai M, Baldwin E, Zhang G, Schmidt DP (2018) A computational approach to predict external spray characteristics for flashing and cavitating nozzles. Int J Multiph Flow 106:21–33. https://doi.org/10.1016/j.ijmultiphaseflow.2018.04.012
9. Cayirci E (2013) Modeling and simulation as a cloud service: a survey. In: Proceedings of the 2013 Winter simulation conference: simulation: making decisions in a complex world (WSC '13). IEEE Press, Piscataway, NJ, USA, pp 389–400

Thomas Bitterman received the B.S. degree in Mathematics from Kent State University, Kent, Ohio, in 1988 and a Ph.D. in Computer Science from Louisiana State University, Baton Rouge, Louisiana, in 1992. From 2008 to 2014 he was an architect and lead developer at the Ohio Supercomputer Center. He is currently on faculty at Wittenberg University. His research interests include genetic algorithms and software engineering.

Chapter 11
Cyber-Physical Systems Design Flow to Manage Multi-channel Acquisition System for Real-Time Migraine Monitoring and Prediction

Kevin Henares, José L. Risco Martín, Josué Pagán, Carlos González, José L. Ayala, and Román Hermida

Abstract Chronic diseases represent the major health problems of the twenty-first century. These diseases kill 41 million people each year, equivalent to 71% of all deaths globally. The major chronic diseases listed by the World Health Organization are cardiovascular diseases, cancer, chronic respiratory diseases, diabetes mellitus, and neurodegenerative disorders. Monitoring and maintaining normal values for key health metrics play a primary role in reducing chronic disease risk. Powerful mechanisms based on prevention to combat the chronic disease crises are currently present and continue to evolve. A new healthcare delivery model is needed to implement these mechanisms effectively. This model implies the utilization of wearable devices connected to the Cloud, allowing continuous monitoring and prevention of chronic disease crises. Standard Modeling and Simulation (M&S) methodologies created to design Cyber-Physical Systems (CPS) and deploy them into the Cloud can help design and implement complex scenarios. In this chapter, we show the automatic

K. Henares · J. L. Risco Martín (✉) · C. González · J. L. Ayala · R. Hermida
Department of Computer Architecture and Automation, Complutense University of Madrid,
C/Prof. José García Santesmases 9, 28040 Madrid, Spain
e-mail: jlrisco@ucm.es

K. Henares
e-mail: khenares@ucm.es

C. González
e-mail: carlosgo@ucm.es

J. L. Ayala
e-mail: jayala@ucm.es

R. Hermida
e-mail: rhermida@ucm.es

J. Pagán
Department of Electronic Engineering, Technical University of Madrid,
Avda. Complutense 30, 28040 Madrid, Spain
e-mail: j.pagan@upm.es

© Springer Nature Switzerland AG 2020
J. L. Risco Martín et al. (eds.), *Simulation for Cyber-Physical Systems Engineering*,
Simulation Foundations, Methods and Applications,
https://doi.org/10.1007/978-3-030-51909-4_11

CPS implementation process of a robust migraine prediction system that allows the generation of alarms before the appearance of new pain episodes. This method is used to implement the device in an FPGA and to study the scalability of the proposed infrastructure, the integration of the designed device into an Internet of Things (IoT) ecosystem is demonstrated.

11.1   Introduction

Migraine causes recurrent headaches and it is one of the most disabling neurological diseases. It affects around 10% of population worldwide [10] and 15% in Europe [21]. Besides, in Europe it has a cost of 1,222 € per patient per year [10], taking into account direct and indirect costs. They include both medical costs and the ones caused by the decline in the productivity at work or school.

Migraines are characterized by recurrent headaches of medium and high intensity. Nevertheless, the pain is not the only symptom of a migraine. Some of them are prodromes, auras, and postdromes. Premonitory or prodromic symptoms may occur from three days to hours before the pain starts [3]. They are subjective, varied, and include changes in mood, appetite, sleep, etc. Auras occur in one-third of the cases [5] and appear within 30 min before the onset of pain. It consists in a short period of visual disturbance. Postdromes are symptoms that occur after the headache. Some of the most common are tiredness, head pain, or cognitive difficulties. They are present in 68% of the patients and they have an average duration of 25.2 h [9].

The medicines used to neutralize the migraine-related pain do not have an immediate response but rather its constituents are slowly absorbed by the organism. Also, most migraine patients tend to wait until the onset of pain to take their medications (since the prodromic symptoms do not allow them to know for sure when will the pain start and the auras have too short durations). This delayed intake reduces the effectiveness of the treatment. In this way, the prediction of the beginning of the pain episode can help the patients to reduce or even cancel their pain episodes.

Migraine prediction has been modeled in our previous works [13, 14]. These predictive models based their pain levels estimations in a set of hemodynamic variables. These variables are controlled by the Autonomous Nervous System (ANS) and are altered when a migraine occurs. Specifically, four variables were analyzed: skin temperature (TEMP), electrodermal activity (EDA), oxygen saturation (SpO2) and heart rate (HR). In [12], the Discrete Event System Specification (DEVS) formalism was used to describe the prediction and define fault tolerance mechanisms for the system. We also proved in [6] that the DEVS model could be easily translated into a hardware specification language like VHSIC[1] Hardware Description Language (VHDL).

This chapter describes the prediction system developed so far in the literature and shows the details about the final implementation of a predictive device in a Field-Programmable Gate Array (FPGA). The FPGA technology facilitates the translation

---

[1] Very High Speed Integrated Circuit.

of the conceptual model into a hardware platform. An FPGA is an integrated circuit that can be programmed or reprogrammed to the required functionality or application after manufacturing. Physical sensors and an additional control board complement the FPGA design. Finally, a model of this device is incorporated into a novel Modeling and Simulation (M&S) framework, designed to analyze the deployment of the final device within an IoT environment, where a huge set of monitoring devices are continuously used for predicting migraine crises. When one of the devices accumulates an error that is not acceptable, the predictive model is re-trained at the Cloud that stores the patient's entire dataset. After that, the updated model is sent back to the physical device. Although the prediction methodology used in this chapter is applied to the migraine disease, it can be applied to other diseases with symptomatic crisis [15], developing equivalent prediction systems. Some examples would be strokes, epileptic seizures, or Parkinson. Since the whole design of the cyber-physical systems is conducted through the DEVS M&S formalism [24]; this chapter defines a reproducible design flow for the design and deployment of these monitoring devices.

This chapter is organized as follows: the set of technologies involved in the design flow is presented in Sect. 11.2. The hardware implementation of the migraine prediction system is shown in Sect. 11.3. The framework implemented to incrementally deploy the resultant IoT system is described in Sect. 11.4. Finally, the chapter is concluded in Sect. 11.5.

11.2 Technologies Involved in the Design Flow

In this Section, we explain the main concepts behind the methods and hardware components used to implement the monitoring device, as well as those methods used to simulate and analyze the impact on scalability. Firstly, we provide an introduction of FPGAs, why we have selected this development platform and its utility as a prototyping architecture in the healthcare domain. Next, we explain the DEVS M&S formalism, used to conduct the whole design of the hardware device on the one hand and the software platform to analyze scalability on the other. Finally, we also briefly introduce some basic concepts used to build and train our predictive models that will be later recorded in the physical device.

11.2.1 FPGAs and Healthcare Monitoring Systems

There is a wide range of applications for Healthcare Monitoring System (HMS) in supporting medical and healthcare services. By attaching portable devices to patients, vital healthcare data can be automatically collected, which is then forwarded to a nurse center for patient state monitoring. The benefit of this scenario is that it can reduce the working load of physicians and result in increased efficiency in patient management. For the patient, the most important qualities of HMS must be small size, easy-to-use, lightweight, and portable.

For the development of a successful HMS, the cost-effectiveness of the solution is one of the main driving factors. The development of a system that only has the necessary functions will help to reduce expenses on the design. Another aspect that can reduce costs is using easily accessible, widely used, and fully configurable components. Using programmable components removes the likelihood that an inconvenient, non-cost-effective device be chosen for the system. The best solution is to build an HMS based on FPGA. It is an integrated circuit that is programmable by developers or customers after production. This is why it is called field-programmable. FPGA is designed to be programmable by changing the functional logic of the principal circuit using a hardware description language. An HMS using this technology would contain a low cost, Analog-to-Digital Converter (ADC), which is used to transform an analog signal into a digital one. Digitization allows users to connect the FPGA to the entire system. The main advantage of the FPGA is the ability to reconfigure it after it has been manufactured. This helps fix bugs easier and more quickly. Moreover, the array takes less time to go from the drawing board to the market. FPGA also has lower non-recurring engineering costs. This means that manufactures only pay for research, design, building, and testing once.

Another major challenge in these applications is power consumption. All these systems require a high level of integration due to the necessary portability. High power efficiency is needed for systems that are truly portable and thus battery operated. Here, lower power consumption will increase the operation time of the device without recharging or replacing batteries. FPGAs implement optimized data paths for each computational task. Using optimized data paths not only improves the performance but also reduces the power and energy consumption when compared with General Purpose Processors (GPPs). Executing a given task in GPPs involves adjusting the required computations to the GPP instruction set and carrying out an expensive instruction decoding process, which leads to important power and energy overheads. In contrast, the power and energy efficiency of FPGAs has significantly improved during the last decade. FPGA vendors have achieved this goal by improving the FPGA architectures, including optimized hardware modules, and taking advantage of the most recent silicon technology. For instance, Xilinx reports a 50% reduction in the power consumption when moving from their previous Xilinx 6 FPGAs (implemented using 40-nm technology) to their most recent Xilinx 7 FPGAs (a new architecture implemented using 28-nm technology).

For these reasons, and those discussed below, the trend is to implement these HMS in FPGAs instead of microprocessors, as traditionally done [1, 2, 17]. The first systems were simpler and the number of input modules provided by the microcontrollers was sufficient [23]. Unfortunately, simply switching to a bigger microcontroller only helps to a certain extent as the focus is usually on more memory and general-purpose pins rather than I/O modules. Adding more microcontrollers complicates the overall system design due to the communication overhead. FPGAs, in contrast, do not have these limitations. Further, the microprocessors are capable of executing only one instruction per cycle while in FPGAs, due to a large number of available resources, it is also possible to execute several tasks in parallel to achieve further speedups.

11.2.2 Discrete Event System Specification (DEVS)

DEVS is a general formalism for discrete event system modeling based on Set theory [24]. The DEVS formalism provides the framework for information modeling which gives several advantages to analyze and design complex systems: completeness, verifiability, extensibility, and maintainability. Once a system is described in terms of the DEVS theory, it can be easily implemented using an existing computational library.

The parallel DEVS formulation (PDEVS), which is the one used in our approach, enables the representation of a system by three sets and five functions: input set (X), output set (Y), state set (S), external transition function (δ_{ext}), internal transition function (δ_{int}), confluent function (δ_{con}), output function (λ), and time advance function (ta).

DEVS models are of two types: atomic and coupled. Atomic models are directly expressed in the DEVS formalism specified above. Atomic DEVS processes input events based on their model's current state and condition, generates output events and transition to the next state. The coupled model is the aggregation/composition of two or more atomic and coupled models connected by explicit couplings. Particularly, an atomic model is defined by the following equation:

$$A = \langle X, Y, S, \delta_{ext}, \delta_{int}, \delta_{con}, \lambda, ta \rangle \tag{11.1}$$

where:

- X is the set of inputs described in terms of pairs port-value: $\{p \in IPorts, v \in X_p\}$.
- Y is the set of outputs, also described in terms of pairs port-value: $\{p \in OPorts, v \in Y_p\}$.
- S is the set of states.
- $\delta_{ext} : Q \times X^b \to S$ is the external transition function. It is automatically executed when an external event arrives to one of the input ports, changing the current state, if needed.

  - $Q = (s, e)s \in S, 0 \le e \le ta(s)$ is the total state set, where e is the time elapsed since the last transition.
  - X^b is the set of bags over elements in X.

- $\delta_{int} : S \to S$ is the internal transition function. It is executed right after the output (λ) function and is used to change the state S.
- $\delta_{con} : Q \times X^b \to S$ is the confluent function, subject to $\delta_{con}(s, ta(s), \emptyset) = \delta_{int}(s)$. This transition decides the next state in cases of collision between external and

internal events, i.e., an external event is received and elapsed time equals time-advance. Typically, $\delta_{con}(s, ta(s), x) = \delta_{ext}(\delta_{int}(s), 0, x)$.

- $\lambda : S \to Y^b$ is the output function. Y^b is the set of bags over elements in Y. When the time elapsed since the last output function is equal to $ta(s)$, then λ is automatically executed.
- $ta(s) : S \to \Re_0^+ \cup \infty$ is the time advance function.

The formal definition of a coupled model is described as

$$M = \langle X, Y, C, EIC, EOC, IC \rangle \tag{11.2}$$

where

- X is the set of inputs described in terms of pairs port-value: $\{p \in IPorts, v \in X_p\}$.
- Y is the set of outputs, also described in terms of pairs port-value: $\{p \in OPorts, v \in Y_p\}$.
- C is a set of DEVS component models (atomic or coupled). Note that C makes this definition recursive.
- EIC is the external input coupling relation, from external inputs of M to component inputs of C.
- EOC is the external output coupling relation, from component outputs of C to external outputs of M.
- IC is the internal coupling relation, from component outputs of $c_i \in C$ to component outputs of $c_j \in C$, provided that $i \neq j$.

Given the recursive definition of M, a coupled model can itself be a part of a component in a larger coupled model system giving rise to a hierarchical DEVS model construction.

11.2.3 Predictive Models

11.2.3.1 Migraine Predictive Modeling

The migraine predictive system presented by the authors in [12, 16] included the possibility of using different types of predictive mathematical models such as Grammatical Evolutionary algorithms and state-space models, respectively. In this work we have considered the later, a Subspace State-Space System Identification (N4SID) model to generate the prediction of the characteristics of the acute migraine attack, including time of onset, time to peak, duration, intensity, quality, etc.

N4SID is a state-space based algorithm [22]. It describes immeasurable states and specifies differential equations that relate future outputs with current and past inputs. It is formally described in (11.3) and (11.4).

$$x_{k+1} = Ax_k + Bu_k + w_k \tag{11.3}$$

$$y_k = Cx_k + Du_k + v_k \tag{11.4}$$

where

- u_k are our $U = 4$ hemodynamic inputs—body temperature, sweating, heart rate, oxygen saturation—at time k.
- y_k is the output at time k. In this project it will be the predicted pain level.
- A, B, C, and D are the state-space matrices.
- v_k and w_k represent white immeasurable noises. More details can be found in [13].

11.2.3.2   History-Based Signal Repair

If a sensor breaks, prediction worsens. Being able to temporarily repair a signal keeps the predictive system active until the error is fixed. A DEVS-based *Sensor Status Detector* with a subsystem for signal repairing based on Gaussian Process Machine Learning (GPML) has already been reported [13, 18]. Due to the complexity of a procedure like GPML, an FPGA would consume numerous resources to replicate it. In this chapter, we have explored simpler ways to optimize this task. As a result, an Auto-Regressive model with eXogenous inputs (ARX) model is used to repair the signal (in case of sensor failure).

Auto-Regressive (AR) models assume that current values of a variable depend on a polynomial combination of its own past values. In addition, ARX models consider the influence of past data of exogenous (external) variables as well. As our hemodynamic variables relate each other through the ANS, we contemplate these polynomial models as Eq. 11.5 shows

$$y_k + a_1 \cdot y_{k-1} + \cdots + a_{n_a} \cdot y_{k-n_a} = b_1 \cdot u_{k-n} + \cdots + b_{n_b} \cdot u_{t-n_b-n+1} + e_k \tag{11.5}$$

where

- y_k is the output—one hemodynamic variable—at time k.
- n_a and n_b are the number of poles and, zeros plus one, of the polynomial.
- n is the number of input samples that occur before the input affects the output.
- $y_{t-1}, \ldots, y_{t-n_a}$ are the previous outputs on which the current output depends.
- $u_{k-n}, \ldots, u_{k-n-n_b+1}$ are the previous exogenous inputs on which the current output depends—the remaining non-damaged hemodynamic variables.
- e_k is a white-noise disturbance value.

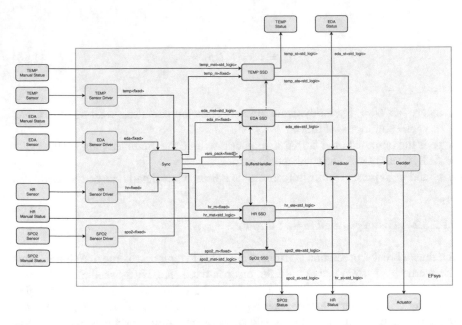

Fig. 11.1 Root component of the migraine prediction system implemented in an FPGA with VHDL

11.3 System I: Migraine Predictive Device

In this section, we show the details about the implementation of the VHDL system in the FPGA and the sensor management board used to group the input. Moreover, it shows the options proposed for decimal numbers representation and a study of the precision loss resulting from the use of fixed-point decimal numbers.

11.3.1 FPGA Implementation

The earlier developed DEVS model [12] validated the design and behavior of the system. As both share a hierarchical nature, the transformation from DEVS to VHDL was a straightforward process.

After validating the system in the DEVS environment, an HW implementation using VHDL was implemented in an FPGA. A general view of this system can be seen in Fig. 11.1 (coupled models are TEMP SSD, EDA SSD, HR SSD, SpO2 SSD, and Predictor). It has the following main components:

- Drivers: Two of these modules were specified. One of them collects oxygen saturation and heart rate data. It receives that data from a NONIN OEM III module[2]

---

[2]http://www.nonin.com/OEM-III-Module.

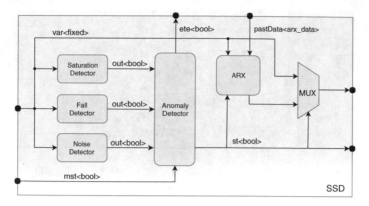

Fig. 11.2 SSD component. It detects errors in the signal

using serial communication. The other one receives the data relative to electrodermal activity and body temperature measures. As the sensors used for this purpose generate an analog output, an Analog-to-Digital Converter (ADC) module has been used. Specifically, a Pmod AD2 has been selected.[3] It counts with 4 channels, 12 bits of precision and communication through the Inter-Integrated Circuit (I2C) protocol.

- Synchronizer (Sync): Packs the input values of the different sensors into a unified data structure. As the sensors can have different sampling frequency, the synchronizer has to decide how to process the situations where there is sparse data. In this way, the values received by the synchronizer are grouped into accumulators. Once per minute, a pulse is raised and a new data packet is created. It contains the average of all the values received in each one of the variables. When that happens, the accumulators are reset to store the data of the next packet.
- Sensor Status Detectors (SSDs, Fig. 11.2): They check if different error types are present in the input signal (saturation, fall, or noise). Where one of them is detected, the status signal related to the controlled variable is raised, so that the patient can restore the sensor. In the meantime, a module to regenerate the signal is activated (ARX). It generates input estimations using previous samples of both the controlled variable and the exogenous ones. When too much time since the error detection passes, an Elapsed Time Exceeded (ETE) signal is raised. That signal points out that the variable implied can not be used reliably to generate predictions, so the Predictor module will discard it.
- BuffersHandler: This was added to the implementation because all the SSDs use the information of all the biomedical variables of the system (in their ARX modules). Therefore, it is convenient to centralize the management of these data.
- Predictor (Fig. 11.3): Generates a numeric indicator that predicts the probability of occurrence of a new pain episode. For that, it contains five sets of three N4SID models each, 12 models in total. Each set is trained using a subset of the

---

[3]https://reference.digilentinc.com/reference/pmod/pmodad2/start.

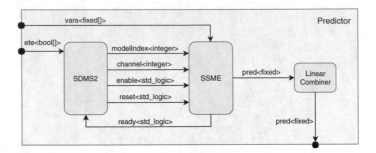

Fig. 11.3 `Predictor` component. It generates predictions about the probability of a migraine occurring

four biomedical variables. There is one set of models for the situation in which all the variables are operational and four more sets for the cases where one of the variables fails. When there are failures in two or more variables the prediction is not considered representative and is not supported by the system. The three models generated for each combination of variables are trained taking into account different prediction horizons. There are models specialized in generating short-term, medium-term and long-term predictions. By combining these predictions more accurate predictions are generated. In this way, when the group of variables is used to predict change, the corresponding set of models is loaded. This happens when the number of damaged variables changes (an additional error is found or one of the damaged variables is regenerated).

- `Decider`: Activates the alarm when the output generated by the `Predictor` module exceeds a certain threshold (previously trained with several hours of data).

Among these components, the SSD and the `Predictor` deserve a detailed explanation. Both contain several sub-components, as shown in Figs. 11.2 and 11.3.

The SSD, as discussed previously, aims to detect the possible errors present in the signal and to temporarily fix them. For that, it is composed of the following components:

- `Noise Detector`: Detect noise present in the input signal following the mean squared error.
- `Fall Detector`: Establish the existence of a fall when the average of the values of a sensor is lower than a certain threshold.
- `Saturation Detector`: Establish the existence of a saturation when the average of the values of a sensor exceeds a threshold.
- `Anomaly Detector`: Raises a status signal when one of the three errors previously explained is detected. In this way, the patient can check it and replace it, if necessary. From that moment the GPML module will be activated to estimate the signal temporally. Moreover, when this situation continues for a certain time, the Estimation Time Exceeded (ETE) signal is raised, reporting that the input signal can no longer be estimated.

- ARX: Estimates the signal based on previous samples. This estimation is generated using an Auto-Regressive model with eXogenous inputs (ARX) model.

The `Predictor` estimates the probability of a migraine attack and has the following components:

- Sensor-Dependant Model Selection System (`SDSM2`): The SDMS2 module is responsible for selecting the correct models and requesting the generation of the three predictions, which are carried out sequentially. This has been done to reduce the resources consumed by the system.
- Sensor-Dependant Model Selection System (`SSME`): This generates the predictions based on state-space models. Since this process is sequential, it has three channels to select which one of the predictions is being calculated. In this way, each one of the channels is associated with one state. When the set of variables are used to predict changes, the SDMS2 detects it and informs the SSME through a reset pulse. Then, the SSME module loads the three models related to the new set of operational variables and resets the states that were being used to generate predictions in the previous conditions.
- `Linear Combiner`: This performs the average of each one of the groups of three predictions generated by the SSME module.

System synchronization is controlled by several pulses and clock signals. The used FPGA has a base frequency of 125MHz. It is used to manage the communication protocols and to generate the operational clock: a clock is used to control the operation of the components of the system. Since the system does not have high time requirements it has been set to 100KHz. This allows reducing system resource consumption. Using the operation clock, several pulses are generated to inform the modules of the existence of newly available data. Extra details about how synchronization works are presented in [6].

11.3.2   HW Setup

In this work, both the N4SID and the ARX algorithms have been computed using the System Identification Toolbox of the MATLAB software.[4]

For the implementation of the DEVS simulator, the framework xDEVS has been used [20].

To deal with decimal numbers in VHDL the FLOAT32 data type was first used. It is included in the FLOAT_PKG package of the IEEE_PROPOSAL library. However, when the system is synthesized it needs many resources to handle the operations with this data type (especially, if the used FPGA does not have dedicated circuits to deal with floating-point numbers). For this reason, a fixed 20-bit precision data type was used instead.

---

[4]MATLAB 2015. version 8.5.0.197613 (R2015a). Natick, Massachusetts, The MathWorks Inc.

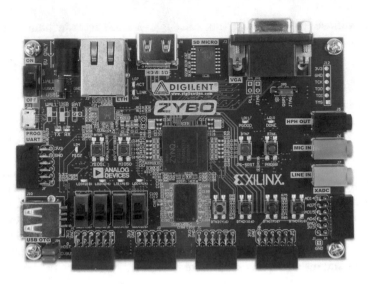

Fig. 11.4 Figure 7: Zybo Zynq 7000: Board used to test the migraine prediction system

To develop the VHDL implementation of the migraine prediction system the design software Xilinx ISE 14.7 has been used. Firstly, it was loaded in a Xula2-LX9, but due to memory limitations was ported to a Zybo Zynq 7000 development board (Fig. 11.4). Therefore, a Zynq 7000 has been established as target FPGA (XC7Z010 device, CLG400 package). This FPGA has 240 KB of Block RAM, 28000 logic Cells, and 80 DSP slices.

Finally, for the integration of the sensors with the FPGA an additional board was designed as an interface. It has a DB9 connector (for the oxygen saturation input), a 3.5 mm Jack connector (for the temperature sensor), and two Snap connectors (for the entry of electrodermal activity data). To obtain electrodermal activity values, the conductance between the two electrodes must be measured. To do this, a Wheatstone bridge is used. It allows us to read a potential difference dependent on the impedance variation of a sensor.

11.3.3 Validation

To check the correct functioning of the system two testing phases were implemented. On the one hand, a software emulation was carried out using the ISE simulator (ISim), to check the operation of all system modules. On the other hand, after the synthesis of the system, data was generated directly in a Zybo Zynq 7000 FPGA and compared with those obtained in the DEVS simulation. To perform these tests real data was used. The data was obtained monitoring the activity of migraine patients in ambulatory conditions with a wireless body sensor network (WBSN) [13].

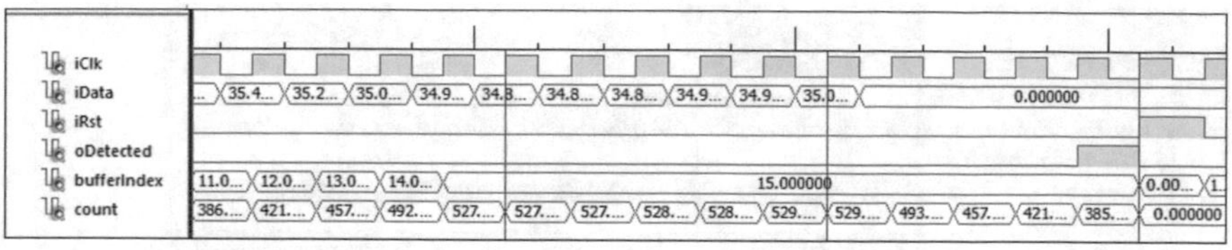

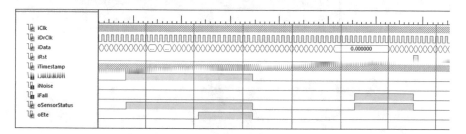

Fig. 11.5 Fall detector emulation. In it we can see how a fall is detected and how it restarts the operation when the reset button is pressed

Fig. 11.6 Anomaly detector emulation. It groups the errors of the three detector modules

Figure 11.5 corresponds to a `FallDetector` module. It can be seen how the error detection works. The data present in the input on each rising edge are stored in a buffer. When the buffer is full, the error detection goes into operation, calculating the mean of the last values and checking if it is lower than a threshold. When an error is detected the `oDetected` signal is raised. It will remain in this state until either data in the buffer is considered valid or the associated reset button is pressed. When that button is pressed the operation is restarted and the buffer is emptied.

`SaturationDetector` and `NoiseDetector` modules work similarly, raising their output in the situations explained above.

Those three detectors are connected with the `AnomalyDetector` module. That module manages two output signals. `oSensorStatus` is activated when at least one of the detectors reports an error. `oETE` is activated when `oSensorStatus` is in high state during a certain time. Figure 11.6 shows the aforementioned situation.

Finally, Fig. 11.7 shows the emulation of the root component of the prediction system. It shows the four input variables corresponding to a real episode, that was previously stored, and the prediction generated by the system. Processing that information the system generates predictions and raises the `oAlarm` signal when it exceeds a threshold (32 in the shown figure). This signal will return to low level when the predictions are back below the limit.

Once all the components of the system were emulated, patients' data were injected directly on the VHDL system (implemented in an FPGA). Figure 11.8 depicts the comparison between the pain levels relative to a real migraine episode. The blue curve corresponds to the one generated with the actual patient, with its subjective data. The red curve represents the predictions generated by the DEVS simulator

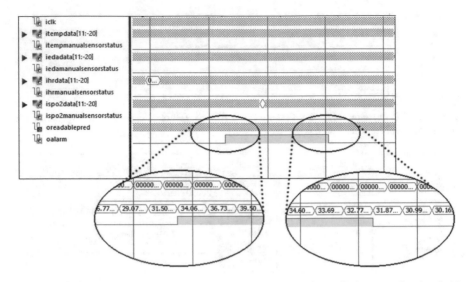

Fig. 11.7 Root emulation. We can see how the alarm raises when the prediction exceeds a threshold

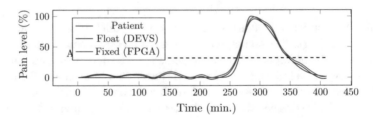

Fig. 11.8 Comparisons between the patient curve (generated with its own subjective pain level information) and the predictions generated by both the DEVS simulator (RMSE: 3,6%) and the FPGA implementation (RMSE: 4,94%). The alarm raises when the prediction raises the 32 threshold

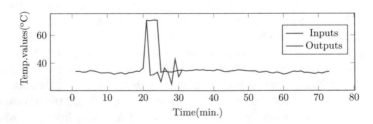

Fig. 11.9 ARX response after the appearance of a saturation error in a temperature input

using the IEEE754 standard. Finally, the brown one represents the output of the Predictor module in the FPGA (using fixed-point data types).

In Fig. 11.9, the appearance of saturation in the input of an SSD that manages temperature values (red line) can be seen. The saturation is detected by the

Table 11.1 Summary of Resource Utilization for the FPGA-Based Implementation of the migraine prediction system on a Zynq XC7Z010-CLG400 FPGA

Component	SSD module	Predictor module	Complete system
# DSP48E1s	28	24	80
# slice registers	1530	1743	9618
# slice LUTs	1549	2705	13056
# LUT flip flop pairs	2648	4172	20059
Percentage of total	21	24	71
Maximum frequency (MHz)	41	64	40

`SaturationDetector` module and the `AnomalyDetector` module activates the `ARX` to start recovering the signal. From this moment, the `ARX` module generates temporal estimations adapted to the previous trend of the input (blue line). To this end, it uses the previously registered values of that variable.

Table 11.1 shows the percentage of hardware utilization for the main modules and the complete system. The percentage of total hardware utilization is 71%, so we have to still leave room in the FPGA for additional algorithms.

11.4   System II: DEVS-based Framework to Deploy Cyber-Physical Systems Over IoT Environments

11.4.1   Framework Design

Once the nodes have been developed, it is time to think about how to manage the generation of models and where to perform the simulations and analyses, i.e., the M&S deployment infrastructure. In some situations, models are generated statically with datasets and configurations good enough to cover the needs of the system to develop. As a consequence, they are trained offline once and do not have to be re-trained. However, it is also common to evolve the models periodically with new data so that they can learn progressively how to generate the best outcomes. This happens in a large variety of scenarios. One example is when medical events or features of some disease are determined based on the data generated through the use of monitoring devices, especially in the ones that present different symptoms by each patient. This is the situation in the migraine scenario presented before. At first, grouped data of other patients with the same disease can be used to generate a general model. After that, to improve the outcomes in individual patients, their specific events are taken into account to improve the models gradually.

All these operations can be performed in different layers, depending on the power consumption constraints and the selected maximum latency. These layers are usually

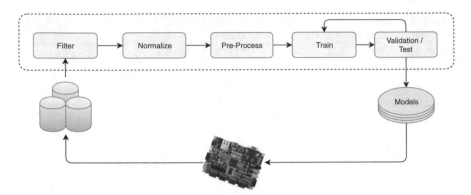

Fig. 11.10 Centralized representation of the framework. This framework is able to (re-)train predictive models based on data gathered from monitoring devices

the edge, the fog, and the cloud. Edge devices usually have limited processing capabilities and battery constraints. On the other hand, the use of data centers implies the acceptance of higher delays. The use of the fog layer to perform the training, as long as they do not require large computing loads, usually offer a good trade-off between these two layers.

Before training and evaluating, data needs to be filtered, Formatted, and information recovery operations must be performed. The use of Model-Based Systems Engineering (MBSE) allows better management of all these operations. In this way, they can be easily reused and combined and brings greater scalability to the system. Also, as these operations usually imply similar tasks, they can be encapsulated so that only a parameterization is needed. In this way, a model training workflow could be specified by describing and parameterizing individual predefined modules and the connections among them. After that, these modules can be instantiated and tuned to deploy a functional system based on this description. This methodology reduces development times and facilitates the generation of predictive models. Also, a formalism like DEVS can be used as a background to validate the operation of the system while controlling its operation. In DEVS, modeling and simulation layers are independent. The modular and hierarchical structure makes it a great option to support this methodology.

The resulting parameterized modules can easily communicate with each other. Moreover, this is still possible even if the modules are configured and instantiated in different devices across the edge, fog, and cloud layers. An easy way to achieve this is using a DEVS library supporting distribution. In this way, couplings between local and remote modules can be specified, sending the intermediate values properly codified over the network.

We have developed a DEVS-based M&S framework to perform the automatic creation of predictive models [8]. Figure 11.10 shows the common phases followed in this process. First, data from all the monitoring devices are collected and stored in the database (this process can be performed online or offline). In the Filter module, these data are filtered following several criteria, to assure quality discarding wrong or incomplete records. Next, normalization operations are applied in the `Normalize` module, unifying units of the variables, or performing categorization tasks and adjustments in the implied values or distributions. In the `Pre-Process` module, some transformations are applied to adapt the data format to the use case. For example, single samples coming from a patient monitoring device can be too granular to be directly related to the event to predict. In this situation, it is common to split clinical data into time-windows of a predefined size so that a certain number of consecutive samples matches the class to predict. Finally, the `Pre-Process` module separates the input data in several datasets: train datasets that are used to fit the model, and validation/test datasets that are used to adjust the models and select the best ones. Once the separation is done, different sets of models are generated with the training dataset using the suitable machine learning algorithms in the `Train` module. The set of algorithms to use depends as much on the problem characteristics as on the data types. Once trained, models are verified and compared using the previously separated validation/test datasets in the `Validation/Test` module. The selected models will be subsequently evaluated to generate valuable information about the area to be modeled. New samples can be used both to evaluate the models and to feed the data sources. This allows us to generate new models periodically so that its accuracy can be increased over time.

This framework has been implemented using xDEVS [20], a DEVS M&S library. Particularly, the Python branch of xDEVS allows us to deploy a DEVS execution over the network. Hence, both FPGAs and their corresponding DEVS models can be integrated into the distributed co-simulation framework and to analyze the impact of (i) scalability as the number of monitoring devices is increased and (ii) performance optimization, distributing the computation power of the framework between the fog and the cloud layers.

We can use this framework with actual devices, simulated devices, or both. Co-Simulation of physical devices and software models can be easily performed [19]. Figure 11.11 shows an example of a deployment that uses the three layers aforementioned. In the edge layer, we may find the monitoring devices. Data collected are sent to a database located in the fog layer. This database represents the reference hospital of each patient. Data are filtered, normalized and pre-processed at the fog layer. Resulting sets are sent to the cloud, where more computational power is needed to train (or re-train) the existing models. This DEVS framework, currently a preliminary version developed in Python, has been tested both in a migraine and stroke predictive IoT system [7, 8].

11.4.2   From Sensors to the Cloud: Scalability Issues

IoT frameworks like the one presented in this chapter (Fig. 11.11) combine three computing layers, i.e., edge, fog and cloud computing. Large monitoring networks using this paradigm are a reality nowadays. In the sports area, for instance, the

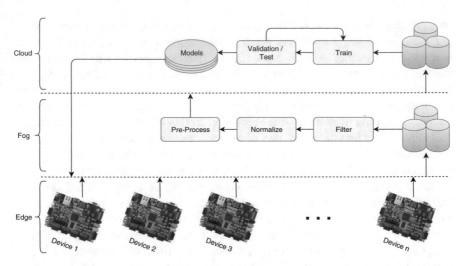

Fig. 11.11 Example of a combined deployment with VHDL and DEVS subsystems. The same two DEVS subsystems are instantiated in the fog and cloud layers

monitoring is being used to generate predictions in team sports [4]. In the medical area, applications like remote diagnosis, disease alarms generation, or prediction of atrial fibrillation are also following this trend [11].

This IoT architectures typically consist of three main parts: the sensing nodes, the coordinators, and the data centers. Each one of the three network elements can operate in various modes. Sensing nodes can collect, transmit, and process data. The controller can receive, transmit, and process data, and perform predictions, if necessary. The data center can process data and perform predictions. The energy efficiency policies take these possibilities into account to minimize the power consumption of the whole system and to maximize the accuracy of the prediction.

Hence, sensor devices can collect and transmit data, coordinators can receive and transmit data, and data centers can process data and perform predictions. Another possible scenario would be sensors to collect and transmit data, coordinators to receive, process data, perform predictions and transit data, and data centers would not be necessary here, with a possible loss of accuracy. Further details about these features and the tasks to be performed are available in [7].

As can be seen, these IoT frameworks pose important challenges because of the large volumes of data that must be gathered and analyzed. Among all the possible IoT applications, population monitoring in e-health includes important constraints. These applications also demand intelligent strategies to develop massive healthcare solutions in massive population scenarios, to minimize energy consumption while maximizing models' accuracy, ensuring reasonable levels of scalability.

11.5   Conclusions

This chapter describes a design flow to conceive, design, and implement cyber-physical systems oriented to monitoring neurological diseases and predict possible attacks or crises.

The first step consists of defining a DEVS model of the whole device. The use of the DEVS formalism allows us to quickly debug our design and make decisions to optimize performance and accuracy. Once the model has been verified and validated, the second step consists of automatically translating the DEVS model into a VHDL description, which is a straightforward operation. This VHDL description can be synthesized into an FPGA to test the real-world device in the field. We have discussed the multiple benefits of using FPGA as Health Monitoring Systems.

We have tested this design flow in a real neurological disorder, the migraine disease. Unfortunately, the floating-point representation has occupied too much space inside the FPGA, so we have implemented an ad-hoc fixed-point model. We compared both the DEVS floating-point approach against the fixed-point approach, increasing the number of bits used in the decimal part until obtaining a low error. Future work includes the pre-computation of the design size so that this issue can be fixed from the DEVS model.

Finally, the DEVS device model has been integrated into a novel M&S framework. This framework has been implemented to analyze the scalability of the monitoring device when it is used in an IoT environment. It is used to simulate the coexistence of a huge set of health monitoring systems, analyzing the effects of re-training the models saved inside the devices when the prediction error is increased.

Future work includes the automation of the whole design flow since currently many design steps must be performed manually. We are also considering the use in our design flow of other monitoring devices currently present in the market, like smart wristband or watches.

Acknowledgements This work has been partially supported by the Education and Research Council of the Community of Madrid (Spain), under research grant P2018/TCS-4423.

References

1. Arshak K, Jafer E, Ibala CS (2007) FPGA based system design suitable for wireless health monitoring employing intelligent RF module. In: 2007 IEEE Sensors, pp 276–279
2. Chou CC, Fang WC, Huang HC (2012) A novel wireless biomedical monitoring system with dedicated FPGA-based ECG processor. In: 2012 IEEE 16th International Symposium on Consumer Electronics, pp 1–4
3. Giffin NJ, Ruggiero L, Lipton RB, Silberstein SD, Tvedskov JF, Olesen J, Altman J, Goadsby PJ, Macrae A (2003) Premonitory symptoms in migraine an electronic diary study. Neurology 60(6):935–940
4. Groh BH, Reinfelder SJ, Streicher MN, Taraben A, Eskofier BM (2014) Movement prediction in rowing using a dynamic time warping based stroke detection. In: 2014 IEEE Ninth International

Conference on Intelligent Sensors, Sensor Networks and Information Processing (ISSNIP). IEEE, pp 1–6

5. Headache Classification Subcommittee of the International Headache Society et al (2004) The international classification of headache disorders. Cephalalgia 24(1):1–160

6. Henares K, Pagán J, Ayala JL, Risco-Martín JL (2018) Advanced migraine prediction hardware system. In: Proceedings of the 50th Summer Computer Simulation Conference, SummerSim '18. Society for Computer Simulation International, San Diego, CA, USA, pp 7:1–7:12

7. Henares K, Pagán J, Ayala JL, Zapater M, Risco-Martín JL (2019) Complexity challenges in cyber physical systems: using Modeling and Simulation (M&S) to support intelligence, adaptation and autonomy, chapter cyber-physical systems design methodology for the prediction of symptomatic events in chronic diseases

8. Henares K, Risco-Martín JL, Hermida R, Reig-Roselló G (2019) Modular framework to model critical events in stroke patients. In: Proceedings of the 2019 Summer Simulation Conference (SummerSim'19)

9. Kelman L (2006) The postdrome of the acute migraine attack. Cephalalgia 26(2):214–220

10. Linde M, Gustavsson A, Stovner LJ, Steiner TJ, Barré J, Katsarava Z, Lainez JM, Lampl C, Lantéri-Minet M, Rastenyte D et al (2012) The cost of headache disorders in europe: the eurolight project. Eur J Neurol 19(5):703–711

11. Milosevic J, Dittrich A, Ferrante A, Malek M, Quiros CR, Braojos R, Ansaloni G, Atienza D (2014) Risk assessment of atrial fibrillation: a failure prediction approach. In: Computing in Cardiology Conference (CinC), 2014. IEEE, pp 801–804

12. Pagán J, Moya JM, Risco-Martín JL, Ayala JL (2017) Advanced migraine prediction simulation system. In: Summer Computer Simulation Conference (SCSC)

13. Pagán J, De Orbe MI, Gago A, Sobrado M, Risco-Martín JL, Mora JV, Moya JM, Ayala JL (2015) Robust and accurate modeling approaches for migraine per-patient prediction from ambulatory data. Sensors 15(7):15419–15442

14. Pagán J, Risco-Martín JL, Moya JM, Ayala JL (2016) Grammatical evolutionary techniques for prompt migraine prediction. In: Genetic and Evolutionary Computation Conference, 2016. ACM

15. Pagán J, Risco-Martín JL, Moya JM, Ayala JL (2016) Modeling methodology for the accurate and prompt prediction of symptomatic events in chronic diseases. J Biomed Inform 62:136–147

16. Pagán J, Risco-Martín JL, Moya JM, Ayala JL (2016) A real-time framework for a DEVS-based migraine prediction simulator system. In: MAEB 2016

17. Ragit C, Shirgaonkar SS, Badjate S (2015) Reconfigurable FPGA chip design for wearable healthcare system. Int J Comput Sci Netw 4(2):208–212

18. Rasmussen CE, Williams CK (2005) Gaussian processes for machine learning (adaptive computation and machine learning). The MIT Press

19. Risco-Martín JL, Mittal S, Fabero JC, Malagón P, Ayala JL (2016) Real-time hardware/software co-design using DEVS-based transparent M&S framework. In *Proceedings of the 2016 Summer Simulation Multi-conference (SummerSim 2016)*

20. Risco-Martín JL, Mittal S, Fabero JC, Zapater M, Hermida R (2017) Reconsidering the performance of DEVS modeling and simulation environments using the DEVStone benchmark. Transactions of the SCS 93(6):459–476

21. Stovner LJ, Andree C (2010) Prevalence of headache in europe: a review for the eurolight project. J Headache Paint 11(4):289–299

22. Van Overschee P, De Moor B (1994) N4SID: subspace algorithms for the identification of combined deterministic-stochastic systems. Automatica 30(1):75–93

23. Warbhe S Karmore S (2015) Wearable healthcare monitoring system: a survey. In: 2015 2nd International Conference on Electronics and Communication Systems (ICECS), pp 1302–1305

24. Zeigler BP, Muzy A, Kofman E (2000) Theory of modeling and simulation. Integrating discrete event and continuous complex dynamic systems, 2nd ed. Academic Press

Kevin Henares is a Ph.D. candidate at the Complutense University of Madrid. He received the B.S. degree in computer engineering from Vigo University, Orense, Spain, in 2016, and the M.S. degree in computer engineering from Complutense University, Madrid, Spain, in 2018. He is currently pursuing the Ph.D. degree in computer engineering, also at Complutense University. He was a Research Assistant at the Federal Institute of Technology in Lausanne (EPFL), Lausanne, Switzerland in 2019, and at Carleton University, Ottawa, Canada in 2020. His research interest includes the development of validation tools for simulation environments and the modeling and prediction of events in medical scenarios.

José L. Risco Martín is an Associate Professor of the Computer Science Faculty at Universidad Complutense de Madrid, Spain, and head of the Department of Computer Architecture and Automation at the same University. He received his M.Sc. and Ph.D. degrees in Physics from Universidad Complutense de Madrid, Spain in 1998 and 2004, respectively. His research interests focus on Computer Simulation and Optimization, with emphasis on Discrete Event Modeling and Simulation, Parallel and Distributed Simulation, Artificial Intelligence in Modeling and Optimization and Feature Engineering. In these fields, he has co-authored more than 150 publications in prestigious journals and conferences, several book chapters, and three Spanish patents. He has received the SCS Outstanding Service Award in 2017, and the HiPEAC Technology Transfer Award in 2018. He is associate editor of SIMULATION: Trans. of Soc. Mod. and Sim. Int., and has organized several Modeling and Simulation conferences like DS-RT, ANSS, SCSC, Summer-Sim or SpringSim. He is ACM Member and SCS Senior Member. Prof. José L. Risco Martín has participated in more than 15 research projects and more than 10 contracts with industry. He has elaborated simulation and optimization models for companies like Airbus, Repsol or ENAGAS. He has been CTO of DUNIP Technologies LLC, and co-founder of BrainGuard SL.

Josué Pagán is a Teaching Assistant at the Technical University of Madrid. He received his doctorate from the Complutense University of Madrid in 2018 with honors. His work focuses on the development of robust methodologies for information acquisition, modeling, simulation and optimization in biophysical and critical scenarios. He has worked on the development of models for early crisis prediction and classification of neurological and oncological diseases. In 2016 he was a visiting researcher at Washington State University, and in 2015 at Friedrich Alexander Universität. He obtained his M.S. in Telecommunications Engineering from the Technical University of Madrid in 2013, and his B.S. from the Public University of Navarra in 2010. He is proactive in the field of technology transfer, and founder of a start-up. His teaching is focused on HW-SW systems for IoT and AmI. He is a member of the Spanish Association of Artificial Intelligence since 2016. He is a member of the Quality Committee of the Health Research Institute of Madrid Hospitals and has participated as a member of the program committee of international congresses such as DS-RT and SummerSim.

Carlos González received the M.S. and Ph.D. degrees in computer engineering from the Complutense University of Madrid, Madrid, Spain, in 2008 and 2011, respectively. He is currently a Associate Profesor with the Department of Computer Architecture and Automation, Complutense University of Madrid. As a research member of GHADIR group, he mainly focuses on applying run-time reconfiguration in aerospace applications. His research interests include remotely sensed hyperspectral imaging, signal and image processing, and efficient implementation of large-scale scientific problems on reconfigurable hardware. He is also interested in the acceleration of artificial intelligence algorithms applied to games. Dr. González received the Design Competition of the 2009 and 2010 IEEE International Conferences on Field Programmable Technology and obtain the second prize in 2012. He received the Best Paper Award of an Engineer under 35 years old at the 2011 International Conference on Space Technology.

José L. Ayala received the M.Sc. and Ph.D. degrees in telecommunication engineering from the Technical University of Madrid (UPM), Spain, in 2001 and 2005, respectively. He is currently an Associate Professor in the Department of Computer Architecture and Automation, Universidad Complutense de Madrid, Madrid, Spain. He is also an Associate Member of the Center for Computational Simulation, UPM. He has served organizing committee of many international conferences, including DATE, GLSVLSI, VLSI-SoC, ISC2, ISLPED. He has led a large number of international research projects and bilateral projects with industry, in the fields of power and energy optimization of embedded systems, and noninvasive health monitoring. His research interests include automatic diagnostic tools, predictive modeling in the biomedical field, and wearable non-invasive monitoring.

Román Hermida received a Ph.D. degree in Physics in 1984 from the Complutense University (CU) of Madrid. He has been a professor in the Department of Computer Architecture and System Engineering of the Complutense University since 1994. He served as vice-dean of the School of Computer Science of CU during the period 1995–97, and head of the departmental section of Computer Architecture of the same university from 1997 to 1999. During the period 1999–01 he served as the coordinator of the degree in Computer Science at the Felipe II College of Aranjuez. From July 2001 to June 2003 he served as Vice-Chancellor for New Technologies of the Complutense University. In the period 2006–2010 he was the Dean of the School of Computer Science.

Chapter 12
Significance of Virtual Battery in Modern Industrial Cyber-Physical Systems

Mayank Singh and Hemant Gupta

Abstract Many companies are facing challenges in today's competitive environment while handling big data issues for rapid decision-making and productivity. These companies are not ready due to lack of smart analysis tools. Cyber-physical system's information generation algorithm should be able to sense and address all visible and invisible issues of industry like component wear, machine degradation, and power issue (battery). One of the newly introduced concepts of CPS is the virtual battery. Due to the rise in demand for the electric and hybrid vehicle and the storage of power in power plant, batteries have become more and more critical. Batteries power-level are impacted by many environmental factors like, temperature, humidity, charging level, discharge rate. We need a battery model which evaluate the battery health and failure prediction, through simulation under different conditions. A virtual battery helps to collect health, reliability, operational readiness information in real time in visualization form possible, which helps the designer to address the flaws and design issues and improve life expectancy. However, there are many challenges with the CPS system as well. To make the correct decision, it is essential to send the correct data at the right time for the right reason, data security, system security, and quality. In this chapter, we will study the history of the battery, how virtual battery works, its application in a different field, and about its security.

Keywords CPS · Industry · Virtual battery

M. Singh (✉)
University of KwaZulu-Natal, Durban, South Africa
e-mail: dr.mayank.singh@ieee.org

H. Gupta
School of Computer Science, Carleton University, Ottawa, Canada
e-mail: hemantgupta@ccsl.carleton.ca

© Springer Nature Switzerland AG 2020
J. L. Risco Martín et al. (eds.), *Simulation for Cyber-Physical Systems Engineering*,
Simulation Foundations, Methods and Applications,
https://doi.org/10.1007/978-3-030-51909-4_12

305

12.1 Introduction

Many companies are facing challenges in today's competitive environment while handling big data issues for rapid decision-making and productivity. These companies are not ready due to lack of smart analysis tools. Cyber-physical system's information generation algorithm should be able to sense and address all visible and invisible issues of an industry like component wear, machine degradation, and power issue (battery).

There are many challenges we are facing today with respect to the resources due to the increasing demand and usage of embedded systems. Advancements in the fields of biomedical sensors, Internet of Things (IoT), and other similar technologies require a powering mechanism to obtain, process, or transmit information. In recent years, the significance of batteries has become more and now we desire high performance, quality, and reliability from these batteries to ensure the smooth operation of these devices.

The changes in electric power systems that are undergoing today have brought renewed interest in this idea. Advances in power electronics help us to monitor the precise control over how much power load consumes. This shows the need for some smart power source termed as "Virtual Battery". Due to the rise in demand of the electric and hybrid vehicle and for the storage of power in power plant, batteries have become more and more important. Batteries' power-level is impacted by many environmental factors like, temperature, humidity, charging level, and discharge rate. We need a battery model which evaluates the battery health and failure prediction, through simulation under different conditions. The output of this model is fed back to the manufacturers to enhance the manufacturing process on battery performance. Many smart appliances like thermostat and electric car, can, collectively act as a massive battery. There is always a trade-off between the battery's capacity and the rates it can be charged or discharged. Most of the battery models focus on single battery cell which is not sufficient. In case of multiple cell battery, all cells and environmental factors play a significant role in battery performance which require further study. A virtual battery helps to collect health, reliability, operational readiness information in real time in visualization form possible which helps the designer to address the flaws and design issues and improve life expectancy.

But there are many challenges with CPS system as well. To make correct decision, it is important to send the correct data at the right time for the right reason, data security, system security, and quality. The goal of CPS system is to reach zero downtime and to enhance efficiency of machine and reduce cost. In the case of attack from outside, machine would fail and cause financial loss or sometimes loss of human life.

12.2 Related Work

Several researchers and companies are working in the field to develop and improve the virtual battery.

Ju et al. [1] suggest a virtual battery model framework which is used to simulate performance of the battery during its usage for electric vehicles. Hughes [2] proposed a technique based on stress testing to create the battery type models for residential HVAC system. Authors suggested that asymmetric regulation markets are extremely helpful in allowing the full usage of the virtual battery.

Hewlett-Packard [3] presented the designing parameters and measured the parameters from the practical circuit of the virtual battery for the RF tags. Hentunen [4] proposed the simulation tool which can be used in conjunction with a battery cycler to emulate a battery in full-scale hardware-in-the-loop testing. Wu [5] proposed the design proposal of the virtual battery for the U.S, department of energy.

Boukhal et al. [6] implement the battery management system which is used to estimate the state of charge of battery cells which helps to protect battery from early degradation and damages. Similarly, Brandl et al. [7] also proposed an architecture for the battery management system for the electric vehicles which includes charge estimation and charge balancing.

Abdelraheem et al. [8] suggested that for low-power circuit it is better to reduce battery size and improve user comfort. Authors presented the design process of a wearable energy harvester which seems a solution for replacing conventional battery with a virtual battery.

12.3 History of Cell and Batteries

Batteries are considered to be the main source of power before the development of generators and grids. Even today we use batteries or cells in many of the IoT and embedded devices. Continuous growth in battery technologies helped in major electrical advances, from early scientific studies to the rise of telephones, eventually leading to mobile phones, electric cars, and many other electrical devices.

In 1749, Benjamin Franklin, first used the term "battery" to explain a set of linked capacitors he used for his experiments with electricity. Scientists and engineers developed several commercially important types of battery. In 1800, Volta invented the first working battery, which is known as the voltaic pile. It consisted of pairs of copper and zinc discs piled on top of each other, separated by a layer of cloth or cardboard soaked in brine, the electrolyte (Fig. 12.1).

In 1836, John Frederic Daniell invented the Daniell cell, which consists of a unglazed earthenware container which is filled with sulfuric acid, and a zinc electrode is immersed into the pot filled with copper sulfate solution. It has an operating voltage of approximately 1.1 V. Another cell is invented by John Dancer in 1839 named porous pot cell. It consists of a central zinc anode dipped into a porous earthenware

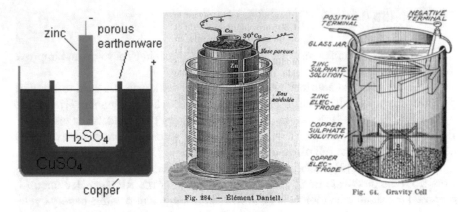

Fig. 284. — Élément Daniell.

Fig. 64. Gravity Cell

Fig. 12.1 Schematic representation of Daniell's cell and porous pot cell and gravity cell

pot containing a zinc sulfate solution and this immersed in a copper sulfate solution contained in a copper can, which acts as the cell's cathode.

Many other cells were designed by getting motivation from the design. Gravity Cell, another variant of Daniell's cell was invented in 1950s by Frenchman named Callaud. Poggendorff Cell was invented by German scientist Johann Christian Poggendorff in 1842. This cell solves the problem by separating the electrolyte and the depolariser using a porous earthenware pot. This cell provides 1.9 V. Grove cell and Dunn Cell was also invented in the nineteenth century.

Several researchers and companies are working in the field to develop and improve the virtual battery.

Ju et al. [1] suggest a virtual battery model framework which is used to simulate performance of the battery during its usage for electric vehicles.

MANET is well studied for the last 10 years. The growth of the internet of all ideas in recent years significantly improves the usage and usefulness of MANET. IoT-MANET is a popular topic as devices are wirelessly linked and many of them are power-constrained and networks are self-organized. Hence, the safety and reliability criteria for these networks must be re-examined. In an ad hoc networking situation, multiple trust-based routing protocols were developed and tested. The state-of-the-art study was summarized in this section. The central point of attraction is trust management schemes for MANET. We are also talking about secure routing and propagation of trust.

Most reputation-based trust management systems are structured by identification of nodes that are either greedy or harmful for safe collaborative routing. Researchers presumed a priori trust connections between mobile nodes.

(a) **Rechargeable batteries and dry cells**: All existing batteries discussed so far would be permanently drained when all their chemical reactions were spent. Gaston Planté invented the first lead–acid battery in 1859 which could be recharged by passing a reverse current through it (Fig. 12.2).

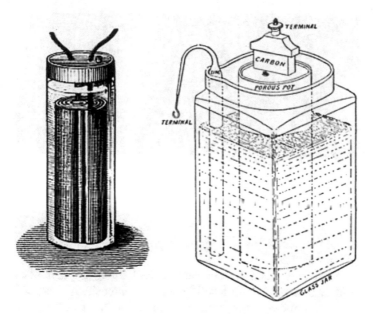

Fig. 12.2. Nineteenth-century illustration of Planté's original lead–acid cell, Leclanché cell

It consists of lead–acid cell which contains a lead anode and a lead dioxide cathode immersed in sulfuric acid. By passing a reverse current through the battery, these reactions performed by cell during discharge can be recharge.

Camille Alphonse Faure improved the Plante's version of lead–acid battery that consists of a lead grid lattice into which is pressed a lead oxide paste, forming a plate. These lead–acid batteries are still used in automobiles and other applications.

Georges Leclanché invented a battery that consists of a cell which contains zinc anode and a manganese dioxide cathode wrapped in a porous material, immersed in a jar of ammonium chloride solution. It provided a voltage of 1.4 V. These cells used to power early telephones.

In 1886, Carl Gassner developed a variant of the Leclanché cell which does not liquid electrolyte by using ammonium chloride is mixed with plaster of Paris to create a paste, with a small amount of zinc chloride added in to extend the shelf life. This variant is known as the dry cell. Manganese dioxide is used as a cathode and dipped in this paste, and both are sealed in a zinc shell, which also acts as the anode. It provides a potential of 1.5 V.

Another battery is invented by Waldemar Jungner, the nickel–cadmium battery, which is a rechargeable battery that has nickel and cadmium electrodes immersed in a potassium hydroxide solution; the first battery to use an alkaline electrolyte (Fig. 12.3).

Jungner invented a nickel–iron battery in 1899. Thomas Edison picked up Jungner's nickel–iron battery design, patented it himself, and sold it in 1903. In the late 1950s, the zinc–carbon battery become primary cell battery. The

Fig. 12.3 Nickel-iron batteries and Lithium-ion battery

nickel–hydrogen battery is used as an energy-storage subsystem for commercial communication satellites.

In 1997, Sony and Asahi Kasei released the lithium polymer battery. These batteries hold their electrolyte in a solid polymer composite instead of in a liquid solvent, and the electrodes and separators are laminated to each other. In 2019, John B. Goodenough, M. Stanley Whittingham, and Akira Yoshino were awarded the Nobel Prize in Chemistry 2019, for their development of lithium-ion batteries.

12.4   Need for Virtual Battery

Main advantage of using virtual battery is to extend battery lifetime and improve the user experience. Operating Systems (OSs) implement low-power operating modes that change system behaviors based on the battery status, such as the charging/discharging condition and remaining energy. Alongside, to protect user and critical system data, systems automatically start saving their content in the permanent storage devices suspending their operations until the battery systems are recharged. This state of the battery is called hibernation.

As the importance of battery management in mobile cyber-physical systems is ever increasing, such battery-related software components are becoming more powerful, complicated, and error-prone. Today, Li-ion is the choice of battery technology due to its good energy density, good power rating, and charge/discharge efficiency. But Li-ion is also very sensitive to overcharge, and this may damage the battery or can even cause hazardous situations. Another charging and discharging of the battery is not always in synchronous. As battery is a combination of cells, so few cells get charge early or discharge quickly then how to decide when to stop charging and discharging. Below are few points which shows the advantages of using virtual battery:

1. **Time and Money**

   a. With the help of virtual battery, we can shorten development cycle and which reduces cost.
   b. It can accurately replicate real-world battery performance.
   c. It allows critical evaluation of and feedback in battery design process.

2. **Accurate and Reliable**

   a. It provides fast and accurate control
   b. It is designed to withstand harsh environmental conditions

3. **Easy to use**

   a. It helps battery software developers and embedded system developers to monitor battery conditions and make decisions.

The importance of virtual battery is very important in IoT devices. Many IoT devices are deployed on remote locations and work on battery. With the help of virtual battery, devices can be programmed to make decision based on environment and signals when to go into hibernation to conserve energy. It can is also very useful in power grid where power demand that varies over a baseline can be handled using the virtual battery. Battery current, voltage, and temperature are the critical inputs for battery protection and determining State of Charge (SoC), State-of-Health (SoH), and State-of-Function (SoF) estimation.

12.5 Parameters Impacting a Virtual Battery

With the help of virtual battery (i.e., a framework) we are able to emulate the dynamic changes in better parameters and inputs and investigate their impacts on battery performance. Performance of battery depends on both internal parameters like State of Charge (SoC), State of Discharge (SoD), impedance and design parameter and external parameters like environmental factor (temperature, vibration) and user behavior.

Thomas et al. [9] performed experiments to find the effects of aging time, temperature, and SOC on the power degradation performance of Li-ion battery and shows that at 100 % SOC, the power degradation is more critical and depends on different temperature.

Venugopal [10] studied the effect of temperature cut-off mechanism by studying the impedance-temperature, Open-Circuit Voltage (OCV), temperature behaviors and shows that increase in cell impedance due to the PTC device occurred gradually over the temperature range 60–125 °C. The rate of change in impedance increased as a function of temperature.

Bloom [11] studies the effects of temperature, time, SOC, and changes in SOC on both the calendar and cycle life of Li-ion battery and concluded that useful cell life was affected by temperature and time. With the temperature accelerated, it results in a cell performance degradation.

Shim et al. [12] found in that during high-temperature (e.g., 60 C) cell loses 65% of its initial capacity at 60 °C as compared to only a 4% loss at room temperature and cell impedance increased significantly with an increase in temperature cycling, resulting in some of loss of capacity.

Chaturvedi [13] studied the design to battery management system to study the aging and power degradation mechanisms from a control perspective.

With the help of these studies, we found that the main factors of interests are time, temperature, SOC, and SOD, with respect to the battery performance. SOC and SOD indicate the current capacity of battery cells. Many models for battery simulation have been proposed to study the relationship between different factors (e.g., time, temperature, SOC, etc.) impacting battery performances.

12.6 Designing Virtual Battery

In theory, a virtual battery is used to keep supply and demand in balance, but existing battery technologies offer no cost savings over power production but making the power producers to trust that virtual battery, however, requires rigorously quantifying its capacity and charge and discharge rates. In this section, we will discuss the basic design of the virtual battery. Resistances and capacitances are functions of the SoC, temperature, current, and voltage.

According to model, the voltage of the battery depends on its capacity of cell, which is decided by the status of battery, charging or discharging power and temperature. Therefore, the ratio of usable cell capacity at time t, $C(t)$, is a function of temperature $T(t)$ and external current $i(t)$ (Fig. 12.4):

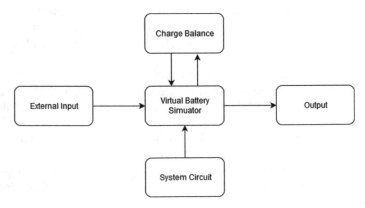

Fig. 12.4 Model of the Virtual Battery

$$C(i(t), T(t), t) = C_0 - \int_0^t i(t)K_1(T(t))dt, Discharge$$

$$C(i(t), T(t), t) = C_0 + \int_0^t i(t)K_2(T(t))dt, charge$$

where C_0 is the initial capacity ratio when charge and discharge start, and $K_1(T(t)$ and $K_2(T(t))$ are the charge and discharge factors which are functions of temperature T.

The cell voltage at any given time t is a function of cell capacity:

$$V(i(t), T(t), t) = V_0 + F(C(i(t), T(t), t))$$

where V_0 is the initial voltage when charge starts.

Here, we have shown some behavior of different characteristics based on different modes of charging. All these figures are derived from Ju et al. [9]. simulation framework.

Virtual battery has four components which we would discuss below

1. External Inputs: There are two types of external inputs: charging and temperature. First, charging which is classified into three types:

a. Charge: The charging power is high for fast charge and less for slow charge.
b. Discharge: It is the process of system discharging when its in running mode.
c. Idle: System stops and there is no charging.

In the slow charge mode, we assume the initial capacity of each cell before charging is set to be empty. Charging is done through a constant current source with small value of current. Cell capacity linearly increasing with respect to the time and which will show the increase in battery voltage (Fig. 12.5).

In the fast charge method, battery is supposed to be charged in shorter period of time using larger power. With the doubling the current amount from slow charge, the charging is faster (Fig. 12.6).

In case of discharge, we supposed the battery is fully charged and no external charging source is needed. Therefore, the charging current is zero. In case of the battery temperature is kept constant, the resistance and impedance from the system determine the discharging speed of the battery. Cell capacity and battery voltage decrease with a constant temperature (Fig. 12.7).

In case of the battery temperature is kept constant, the resistance and impedence from the system determine the discharging speed of the battery. Cell capacity and battery voltage decrease with a constant temperature (Fig. 12.8).

One of the important points is simultaneously charge and discharge. In a few embedded systems when cell capacity goes below a certain threshold, there is a backup charging mechanism which starts the charging of the cells. If the charging

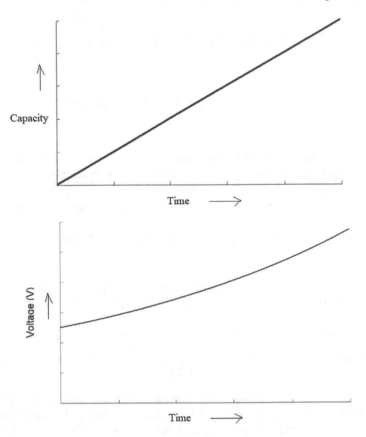

Fig. 12.5 Behavior of slow charge mode on cell capacity and battery voltage

speed of the cell is higher than the discharging speed then the cell capacity should increase, along with the voltage of the battery increase, along with the voltage of the battery (Fig. 12.9).

Second is temperature which is a combination of both external and internal system temperature and affects the rate of charging and discharging which is already been proven by many researchers. The difference in temperature is due to the heating and cooling of the system. Temperature impacts the battery lifetime and working efficiency. Rise in temperature with a certain quantity speed up the charging and discharging process both. With different temperatures, the cell capacity and battery voltage change at various speeds.

2. Output: The outputs of the virtual battery are the voltage, current of battery pack, and the capacity of each cell, etc. In the system, capacity is represented by the state of charge and state of discharge, corresponding to charging and discharging status.

3. Charge Balance: A single block connected to battery cells in a close loop represents the cell balance module. The objective of charge balancing is to ensure the

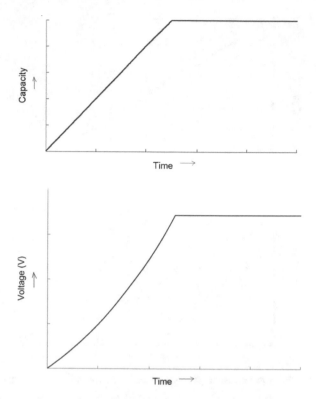

Fig. 12.6 Behavior of fast charge mode on cell capacity and battery voltage

SOCs of all the cells reaching the minimum or maximum level in parallel, thus it can utilize a larger portion of the usable capacity and increase the life of the battery pack. The charge balance module takes the system of charges of all the cells as input and modifies the resistor of cells, thus controlling the function of each battery cell. The charge balance keeps taking the value of the SOC of each cell and controls the cell with the lowest state of charge to discharge in as low speed compared to other cells. It can be attained by setting the equivalent register value of the cell to be significantly very large, i.e., hundred times larger than the value of other registers. This will reduce the discharging rate of the specific cell controlled by charge balancing. The cell can be charged simultaneously with the increase in state of charge.

4. System Circuit: Resistance and impedance present in the external circuit of the system also impact the charging and discharging rate of the battery.

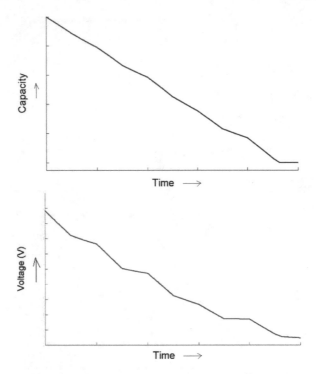

Fig. 12.7 Behavior of working discharge mode on cell capacity and battery voltage

Algorithm: The cell balancing algorithm
    Input: Current SOC of all the cells
    Output: Control signals for each cell
    1 All control signals = 1;
    2 **foreach** *sample time* **do**
    3   **foreach** *cell i* **do**
    4   **if SOC(i)** *is the smallest* **then**
    5   Control signal(i) = 1.1;
    6   All other control signals = 1;
    7   **end**
    8   **end**
    9 **end**

12.7 Conclusion and Future Work

Today, most of the products work on batteries and companies are trying to increase the capacity of these batteries and reduce their cost. But as we go toward, we need to start using renewable resources of energy and in my opinion virtual battery will be very

Fig. 12.8 Behavior of
working discharge mode on
cell capacity and battery
voltage influenced
temperature

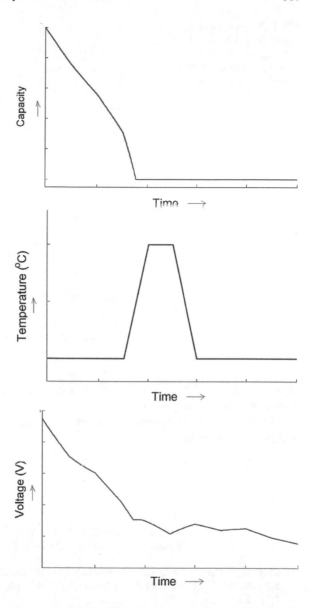

helpful in implementing that. We should also need to focus on designing batteries
which are environment friendly. Research and implementation of the concept of
virtual battery for big power plants and what other factors may impact the functioning
of virtual battery based on different applications is still a focus of research. Many
private companies and government are working on making this dream a reality. As
many IoT devices running around today, which use battery and contains processor
from 4-bit to 64-bit, we should find out how virtual battery can be implemented

Fig. 12.9 Behavior of
simultaneous charge and
discharge mode on cell
capacity and battery voltage
influenced temperature

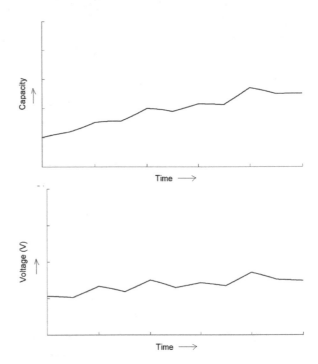

for these embedded devices and how it will impact the functionality and cost of the
device from user's perspective.

Virtual battery is the future of batteries to provide clean and cheaper energy and
will help in managing energy usage.

References

1. Ju F, Wang J, Li J, Xiao G, Biller S (2013) Virtual battery: a battery simulation framework for
   electric vehicles. IEEE Trans Autom Sci Eng 10(1):5–15. https://doi.org/10.1109/TASE.2012.
   2203121
2. Hughes JT, Domínguez-García AD, Poolla K (2016) Identification of virtual battery models
   for flexible loads. IEEE Trans Power Syst 31(6):4660–4669. https://doi.org/10.1109/TPWRS.
   2015.2505645
3. Hewlett-Packard Application Note 1088. Designing the Virtual battery
4. Hentinen (2019) Virtual Battery- Simulation Tools for Engineering. ECV National Seminar,
   Espoo. Sep. 24, 2019
5. Wu D (2016) Virtual batteries. Department of Energy (Office of Energy Efficiency and
   Renewable Energy)
6. Boukhal MA, Lagrat I, Elbannay O (2019) Implementation of a lithium-ion battery state of
   charge estimation algorithm for BMS on the real time target NI myRio. In: 2019 international
   conference on wireless technologies, embedded and intelligent systems (WITS), Fez, Morocco,
   pp 1–5

7. Brandl M et al (2012) Batteries and battery management systems for electric vehicles. In: 2012 design, automation & test in Europe conference & exhibition (DATE), Dresden, 2012, pp 971–976. doi: https://doi.org/10.1109/DATE.2012.6176637
8. Abdelraheem MS, Hameedi S, Abdelfattah M, Peroulis D (2019) A flexible virtual battery: a wearable wireless energy harvester. IEEE Microwave Mag 20(1): 62–69. doi:https://doi.org/10.1109/MMM.2018.2875629
9. Thomas EV, Case HL, Doughty DH, Jungst RG, Nagasubramanian G, Roth EP (2003) Accelerated power degradation of Li-ion cells. J Power Sourc 124:254–260
10. Venugopal G (2001) Characterization of thermal cut-off mechanisms in prismatic Lithium-ion batteries. J Power Sourc 101:231–237
11. Bloom, Cole BW, Sohn JJ, Jones SA, Polzin EG, Battaglia EG, Henriksen GL, Motloch C, Richardson R, Unkelhaeuser T, Ingersoll D, Case HL (2001) An accelerated calendar and cycle life study of Li-ion cells. J Power Sourc 101:238–247
12. Sllllll J, Rustucki D, Richardson T, Song X, Striebel KA (2002) Electrochemical analysis for cycle performance and capacity fading of a Lithium-ion battery cycled at elevated temperature. J Power Sourc 112:222–230
13. Chaturvedi NA, Klein R, Christensen J, Ahmed J, Kojic A (2010) Modeling, estimation, and control challenges for Lithium-ion batteries. In: Proceedings of the American Control Conference, Palo Alto, CA, pp 1997–2002

Mayank Singh is currently working as a Post-Doctoral Fellow, Department of Electrical, Electronic and Computer Engineering, at University of KwaZulu-Natal, Durban, South Africa since September 2017. Prior, he worked as Professor and Head, Department of Computer Science and Engineering at Krishna Engineering College, Ghaziabad. He has 15+ years of extensive experience in IT industry and Academics in India. He completed his Ph.D. in Computer Science and Engineering from Uttarakhand Technical University in 2011. He obtained his M.E. (2007) in Software Engineering from Thapar University, Patiala and B.Tech. (2004) in Information Technology from Uttar Pradesh Technical University, Lucknow. Dr. Singh has published around 46 research papers in peer reviewed Transaction, Journals and Conferences of National and International repute and has supervised 12 M.Tech and 3 Ph.D. students in the areas of Cloud Computing, Software Testing, and Ontology and supervising 5 Ph.D. and 2 M.Tech. students. He has awarded two patents and 1 submitted. He is currently also serving as a Life Member of Computer Society of India (CSI), Life Member of Indian Science Congress Association (ISCA), Member of IEEE, Senior Member of ACM, Senior Member of IACSIT, and Member of Indian Society for Technical Education (ISTE). He is also serving for IEEE Committees as a TPC Member and designated Reviewer for various IEEE Conferences. Dr. Singh is currently serving as Reviewer for IET Software, IEEE Communications, IEEE TENCON, IEEE Sensors, The Journal of Engineering, IET Wireless Sensor Networks, and Program Committee Member of innumerable IEEE & Springer International Journals and Conferences Worldwide.

Hemant Gupta received the Bachelor of Engineering degree in Electronics and Communication Engineering from Birla Institute of Technology, Mesra, Ranchi, India in 2012 and first post graduate degree, Master of Technology in Software Engineering from Birla Institute of Technology and Science, Pilani, India, in 2015 and second Master of Science degree at Computer Science from Carleton University, Canada, in 2019. He is currently pursuing the Ph.D. degree in Computer Science at the same Carleton University. From 2012 to 2017, he worked as a Software Developer in CISCO-ODC at Aricent (currently known as Altran), Gurgaon and Ciena, Gurgaon. His interest includes embedded system, Internet of Things, system security, networking, cryptography, robotics. Mr. Gupta's awards and honors include Graduate Honor Award, Indira Gandhi Fellowship Award and Few team awards and Individual Excellence Award for delivering quality product to CISCO.

Chapter 13
An Architecture for Low Overhead Grid, HPC, and Cloud-Based Simulation as a Service

Scalable Experimentation, Optimization, and Machine Learning for Modeling and Simulation

Matthew T. McMahon, Brian M. Wickham, and Ernest H. Page

Abstract This chapter describes the development of the MITRE Elastic Goal-Directed simulation framework (MEG), designed to provide modelers and analysts with access to (1) grid, cloud, and HPC computing support, (2) a wide range of Design of Experiments (DOE) methods, and (3) robust data processing and visualization. We review the motivation and use cases for MEG, the architecture and functionality of the current framework, discuss recent examples of its use, address its current challenges and future development, and finally consider its applicability and advantages in the context of cyber-physical systems.

13.1 Introduction

In the modeling and simulation life cycle described by Balci [1], the experimental design and simulation execution phase typically requires executing independent instances of a given simulation over a range of values for multiple inputs, which quickly becomes computationally expensive. Along with long run times for a single instance of a simulation model, the need to make multiple simulation runs can become a bottleneck in the life cycle, particularly when a timely analysis is required. For example, the number of independent runs in a full factorial DOE increases combinatorically for each input parameter, quickly resulting in an intractable number of

M. T. McMahon (✉) · B. M. Wickham · E. H. Page
The MITRE Corporation, Mclean, VA, USA
e-mail: mcmahon@mitre.org

B. M. Wickham
e-mail: bwickham@mitre.org

E. H. Page
e-mail: epage@mitre.org

© Springer Nature Switzerland AG 2020
J. L. Risco Martín et al. (eds.), *Simulation for Cyber-Physical Systems Engineering*,
Simulation Foundations, Methods and Applications,
https://doi.org/10.1007/978-3-030-51909-4_13

simulation runs to complete the experiment. Fortunately, executing these independent simulation runs on a parallel computer or compute cloud is a coarse-grained parallel computation, effectively providing linear speedup: 1000 independent runs of a parameterized model with a run time of one hour can all be executed in one hour on 1000 processors (this property is typically referred to as "embarrassingly parallel" granularity). This and other approaches to improving the speed of running extremely long simulation experiments have been well studied. As we described in [GRID], these include the following:

- Parallel Decomposition can decrease the run time of a single large model by subdividing the model into smaller parts, executing those parts concurrently on a parallel computer, and assembling the partial results into a single model-scale solution. Approaches to this are discussed in detail in Fujimoto [2]. In contrast to embarrassingly parallel granularity, this is a much more finely grained approach requiring considerable effort to refactor a single simulation to span multiple processors.
- Statistical approaches can be used to diminish the total number of replications required for a given study [3], or by minimizing the total number of runs needed to derive an estimate for a system variable, regardless of the number of system variables of interest, their interrelationships, or underlying distributions [4, 5].
- In parallel simulation experimentation, where many independent runs of a single model must be executed and compared, each replication of the model can be executed independently on a separate processor in a grid or HPC system.
- A simulation metamodel of the model (in effect, a model of the model) can be created, trading some accuracy for an extremely fast run time once the model has been trained (e.g., via a neural network) using multiple samples from the original model (reviewed in [6].

MEG design focuses on the latter two of these, to expedite the computationally expensive portions of the Simulation Modeling loop, with a usability focus on (1) facilitating access to distributed high-performance and cloud-computing resources, (2) enabling the rapid design and execution of simulation experiments, and (3) managing and visualizing large volumes of results.

The development of MEG has been inspired by research in Design of Experiments (DOE) methodologies, Simulation Optimization, Multidisciplinary Design Optimization, and high-performance computing. Early versions of MEG were based on design concepts in the Unified Search Framework [7], with extensions added to support computational grids, cloud computing and containerized environments; to support the addition of new Simulation Optimization tools; and to manage large-scale experiments. The Applications section of this chapter reviews relevant work supporting these extensions, and the DOE capabilities implemented in MEG including classical full- and partial-factorial parameter sweeps, simulation optimization via evolutionary algorithm, and factor screening for neural network metamodeling. We review a representative subset of the MEG DOE algorithms, practical considerations in experiment and algorithm choice, trade-offs in computational performance for these choices, and discuss published and in-development examples using MEG.

The Design and Architecture section explores the evolution, design goals, and implementation of the technical components of MEG including a complete enumeration of DOE capabilities. We also provide a detailed example of how a parameterized model is developed in MEG.

Finally, the Conclusions section summarizes the work to date, current status, emerging research and future directions for MEG, along with a brief discussion of potential touchpoints for using MEG in the context of Cyber-Physical Systems modeling and simulation.

13.2   Applications in Experimental Design and Simulation Optimization

Simulation Experiments to understand the relationship of model parameters (input variables[1]) to model outputs (response variables[2]) can be broadly categorized as Design of Experiments (DOE) and Simulation Optimization methods. DOE methods are largely focused on relating the influence of model input variables to the response variables by directly specifying ranges and combinations of values for the input variables, running the simulation, and observing the output for each parameter setting. Simulation Optimization—reviewed in Amaran et al. [8]—focuses on algorithmically searching for the input variable settings that yield the best response, for a given simulation and scenario. This category spans many methods depending on the types of variables in the model, the randomness in the model, the number of response variables, and whether the model is mathematically closed-form or complex. Amaran et al. [8] discuss the state-of-the-art algorithms in the field, compare and contrast the different approaches, review some of the diverse applications that have benefited from the use of these methods, and speculate on future directions in the field. In this section, we describe some of the algorithms implemented in MEG for both the classical DOEs and Simulation categories, observations on algorithm selection, and computational considerations for HPC and cloud environments.

13.2.1   Classical Design of Experiments in MEG

Classical DOEs comprise an array of statistical methods for understanding the influence of input variables on response variables. For the types of models MEG is typically used for, we focus on factorial experiments, where multiple inputs are methodically varied to achieve orthogonality and minimal confounding between interaction effects. The model is executed for each unique set of inputs (as described in [9]. The MEG full-factorial algorithm provides for setting input variable type, range, and

---

[1] Also referred to as independent variables.

[2] Also referred to as independent variables.

resolution (by minimal step-size increments allowed for variations in a variable), along with a desired model output of interest.

While a full-factorial DOE exhaustively searches the specified input variable space, it is computationally expensive in that the number of independent model runs grows combinatorically with the number of variables. MEG also provides a spreadsheet-based DOE via its Web Services API, which allows for analyst-specified partial-factorial designs, "what-if" explorations of subsets of variables, and scenario testing.

An example of the use of the Full Factorial DOE in MEG is found in Bookstaber et al. [10], where MEG was used to conduct a full-factorial DOE using an Agent-Based model of cascading financial crisis, spanning on the order of ten million independent model runs.

13.2.2 Simulation Optimization in MEG

A simulation model is a descriptive tool for analyzing real-world systems. Simulation Optimization refers to finding the input variable settings for that model that maximize or minimize the model's output. Techniques include random search, ranking and selection, response surface methodology, gradient search, and evolutionary algorithms (these and the breadth of additional approaches are reviewed in [11]).

For the types of models with which MEG is typically used, we focus here on Evolutionary Algorithms (EAs). The benefit of EAs in the context of simulation optimization is that they don't require continuously valued input parameters or gradient optimization. Thus, this class of global search is suited to mixed input variable types and does not typically require information about the underlying model in order to be useful (hence the term "black box optimization" commonly applied to EAs). This is useful in a multidisciplinary environment where models and scenarios change frequently, or where analysts are not familiar with the details of a model's implementation. Furthermore, EAs and Genetic Algorithms (GAs) are implicitly parallel, and are thus amenable to computational speedup and efficiency on compute clusters and clouds. As with the Classical DOE techniques, MEG provides for mixed input variable types (e.g., input comprising boolean, integers, continuous, and discrete categorical variables), allowing for optimization of most executable models.

Guided by the respective works of Holland [12], Goldberg [13], and Bäck [14], the MEG standard generalized Genetic Algorithm was implemented for maximum flexibility in GA configuration and algorithmic choice, with functionality added to support a wide variety of simulation types, input types, and optimization goals. An example of its use is in Barry et al. [15], where the MEG Genetic Algorithm was used to optimize the placement of sensors in a public venue defense scenario, simultaneously maximizing detection of targets and minimizing casualties.

Genetic Algorithms operate iteratively over populations of designs, and are inherently parallel, because all simulations in a population can be run independently and concurrently. However, unlike a factorial experiment where all required input values are defined a priori, the scalability of that parallelism is limited by population size: for a GA population of size N, only N simulations can be run concurrently before an iterative step is required to configure the next generation of simulations. This inspired the extension of the standard GA to an Island Model GA, based on Skolicki and De Jong [16], and was well-reviewed from the performance perspective in Abdelhafez et al. [17]. Abdelhafez, et al. [17] review not only the performance benefits of using GAs for optimization, but also the further advantages afforded by fine-grained parallelization and the exchange of information that are exploitable in distributed Genetic Algorithms.

Further extensions of the EA paradigm are implemented in MEG to add flexibility for optimizing simulations with multiple competing objectives (e.g., maximize performance and minimize cost). These include the Nondominated Sorting Genetic Algorithm (NSGA-II) and the Strength Pareto Evolutionary Algorithm (SPEA 2), respectively, described in Deb et al. [18], and Zitzler and Thiele [19]. Each of these supports exploring optimal values for multiple output objectives, allowing for analysis of trade-offs between input variables. In the above cost/performance example, maximum performance may come at an exorbitant cost, while minimal cost may yield an unacceptably low performance. While both of these algorithms are useful for multiobjective optimization, in practice, SPEA 2 has a broader coverage of the input parameter space (i.e., less clustering), while NSGA-II tends to provide a broader range of unique solutions.

Table 13.1 summarizes the algorithms discussed in this section, high-level guidance on when to choose each type, and computational performance considerations, in accordance with the No Free Lunch Theorems elucidated by Wolpert and Macready [20].[3]

13.2.3 Response Surface Approximations in MEG

As the size and complexity of simulations increase, the time to execute MEG simulation experiments grows concomitantly. This motivates the use of metamodeling to reduce the run-time complexity of the models. Early work with MEG used data from full-factorial and GA optimization runs to train neural network metamodels, with promising results in reduced run time, at the cost of a small loss of precision relative to the full model [21]. Based on these results, factor screening DOEs were added to MEG to produce inputs specifically tailored for the training of metamodels [22].

---

[3] Wolpert and Macready's No Free Lunch Theorems in Optimization establish that for any optimization algorithm, elevated performance over one class of problems is offset by diminished performance over another class. There is no "shortcut" in algorithm choice—i.e., no one optimization solution is ideal across all problem types in a class of problems, and thus algorithm choice is determined based on model type, performance needs, timeliness of solution requirements, etc.

Table 13.1 Representative MEG algorithms from design of experiments approaches, evolutionary algorithms, and factor screening

Algorithm	MEG implementation	Application	Computational performance notes
Evolutionary Algorithms (EAs)	Standard generalized Genetic Algorithm (GA)	Input parameter optimization for a single simulation output objective scalar or utility function	Scales with the size of a single GA population and with the number of iterations (generations) until convergence [practical limit]
	Island Model GA	Input parameter optimization where there are multiple local optima in the solution space	Scales with the number of subpopulations (islands) and the number of generations, potentially spanning the number of available processors
	Multiobjective nondominated sorting Genetic Algorithm (NSGA-II)	Input parameter optimization for multiple competing objectives	Scales similar to the standard GA
	Multiobjective Strength Pareto Evolutionary Algorithm (SPEA 2)	Input parameter optimization for multiple competing objectives	Scales similar to the standard GA
Factor screening	Full-factorial design	Exhaustive parameter sweeps, exploratory and trade studies	Scales linear with the number of available processors
	Analyst-specified design via spreadsheet or API	Focal parameter sweeps, partial factorial design, "what-if" and counterfactual analysis	Scales linear with the number of available processors
	Latin hypercube sampling (LHS)	Determining input variables' impact on model output, sensitivity analysis	Useful for downstream development of neural network-based metamodels
	Controlled sequential bifurcation (CSB)	Determining input variables' impact on model output, sensitivity analysis	

These techniques have the computational advantage of minimizing the number of full-scale model runs required to achieve acceptable precision in the resulting meta-models. The advantage of a trained metamodel is reduced computational complexity versus the original model, e.g., the running time of the complex iterative Agent-Based model in Rosen et al. [21] was reduced from three minutes for the full model to the computational complexity of a neural network invocation, running in milliseconds.

13.3   MEG Design and Architecture

The design and concept of operations for MEG are described in detail by Page et al. [23]. This section summarizes those principles, and details developments since its publication—namely the complete refactoring of the source code to a service-oriented framework, the introduction of new simulation optimization algorithms, and the development of new algorithms to support neural network-based metamodeling.

Figure 13.1 depicts the high-level architecture of MEG, comprising four principal sets of services: (1) user interface, (2) job scheduling (replication management), (3) DOE support, and (4) data management and visualization. We briefly describe each of these below and provide additional detail in the rest of this section:

(1)   User Interface: The UI has two roles: it (1) provides a programming interface (API) for using MEG, insulating the user from developing and manipulating the varying syntax and litany of scripts associated with typical grid and cloud computing software, and (2) provides a persistent workspace for experiment design and monitoring.

Fig. 13.1 The goal of MEG is to facilitate and expedite the computationally expensive experimental design and model execution part of the Simulation Modeling life cycle. For a given experimental design ("Outer Loop"), the model is executed multiple times for varying input parameters ("Inner Loop")

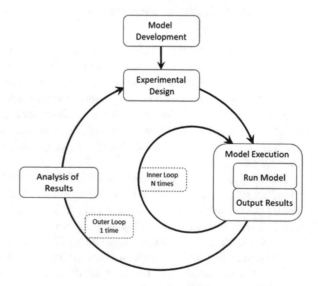

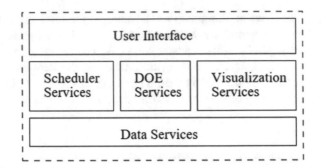

Fig. 13.2 High-level MEG Architecture, highlighting the four key services

(2) Scheduler Services: The objective of the MEG scheduler is to support the widest range of grid and cloud schedulers available—with the ability to extend to new schedulers as they emerge—and to permit a user to submit jobs to MEG without having to select a particular target cluster for execution.

(3) DOE Services: These services are used for selecting, configuring, and orchestrating the various DOE and optimization algorithms in MEG. Sections 13.3.2 and 13.3.3 provide details on these services.

(4) Visualization Services: They are discussed in Sect. 13.5 (Fig. 13.2).

Originally, 20,000 lines of Java code running as a single monolithic service, MEG has been refactored into separate, loosely coupled components running as Web Services. Figure 13.3 depicts the details of the refactored MEG Web Services.

13.3.1 Architectural Enhancement

Lessons learned from the development of the initial version of MEG, and its predecessor USF [7], provided key architectural challenges to address for MEG. A major weakness of previous architectures was the amount of knowledge required to develop a new DOE designer module, representing a new search heuristic, or implementing an interface to a new execution environment. A developer had to implement and clearly understand the entire code base to accomplish these simple tasks. In addition, a developer was required to be an experienced Java programmer for the implementation of the new DOE plugins.

Defining a loosely coupled software architecture while providing a developer-friendly experience was essential to the success of MEG's architecture. MEG architecture design work was completed circa 2012, when Platform as a Service (PaaS) and Infrastructure as a Service (IaaS) paradigms were in their infancy (development of the PaaS standard is described in [24]). MEG is referred to as Simulation as a Service (SaaS) since it successfully integrated the execution of simulation models with the cloud infrastructures mentioned above. Service Oriented Architecture (SOA) was fairly mature, and in particular, web services development using

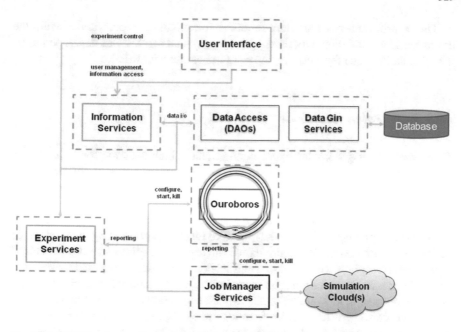

Fig. 13.3 MEG has been refactored as a Web Services framework, depicted in detail here

Simple Object Access Protocol (SOAP) with Web Service Definition Language (WSDL) was extremely popular for the development of loosely coupled software architectures (see [25, 26] for an overview of these technologies). A key benefit of using SOAP/WSDLs is the removal of the requirement for a developer to implement with a specific development language to interface with a software application.

MEG web services were developed as a collection of components representing DOE functionality. The architecture consists of five primary services including the experiment, ouroboros, job manager, user interface, and data services, described below.

13.3.2   Experiment Service

The experiment service provides an external interface with the interconnected set of MEG services. DOE functionality to control the life cycle of a simulation experiment is provided by this service. A user can create, execute, monitor, and delete simulation experiments from the experiment service interface. Job results can be retrieved for experiments that have completed or are currently in process. In addition, running experiments can be terminated. The experiment service is persistent meaning that the service is always running.

The experiment service provides functions to configure, execute, and monitor the execution of a simulation experiment. The experiment service was developed to be the external interface for a user or software to interact with MEG.

13.3.3   Ouroboros Service

The ouroboros service implements a configurable search heuristic for a simulation experiment. The search heuristic controls how the model inputs are modified as the execution progresses from one run to the next. As simulation jobs are completed, the outputs are pushed back to the ouroboros service. A value for the response variable, objective, or scoring function is reported and this value is stored, and compared with that of the other completed jobs. The objective function, specified by the modeler or analyst, evaluates the output of a simulation run to generate a fitness value or vector of values, describing how well the associated set of model inputs performed for this run. The ouroboros implementation determines how to use the returned objective function in the selection of model inputs for subsequent simulation runs to be launched.

The ouroboros service provides functions to configure, start, and stop an implemented search heuristic. In addition, there are functions for configuring links between the job manager service and experiment service to set up callbacks. The ouroboros interface is an internal MEG interface, external interaction is not required.

13.3.3.1   Ouroboros Designers

There are several ouroboros designer implementations (referred to as "designers") that have been developed and are available with MEG. However, MEG is not limited to these implementations. The plug-and-play architecture and interfaces provide for the ability to seamlessly add new ouroboros implementations for search heuristics not currently available in MEG.

Table 13.2[4] enumerates the ouroboroi implemented in the current version of MEG.

13.3.4   Job Manager Service

The job manager service provides a common interface to configure, launch, and monitor a simulation job, with a given backend execution environment. This service provides the ability to verify a simulation experiment in a local execution environment

---

[4]The MOEA Framework is a free and open-source Java library for developing and experimenting with multiobjective evolutionary algorithms (MOEAs) and other general-purpose single and multiobjective optimization algorithms. Accessible at http://moeaframework.org/.

Table 13.2 Ouroboros Designers implemented in the Web-Services version of MEG

Name	Description
Random	Selects a random set of simulation model inputs based on each simulation job
Full factorial	Launches a simulation job for every possible combination of simulation model inputs represented by the model input space, within specified ranges and step size for each variable. Requires discrete model input definitions.
Genetic Algorithm	Implemented from the generalized Genetic Algorithm presented in Goldberg [13], Holland [12], and in Skolicki and De Jong [16], later extended to include Evolutionary Algorithm features from Bäck [14]
Latin hypercube	Creates a collection of simulation model input/output combinations required to train a metamodeling representation of the simulation model (in [22])
Controlled sequential bifurcation	Evaluates simulation model input sensitivity. Results are used to determine which model inputs to vary for simulation optimization experiments [22]
Industrial strength compass	Implemented the ISC algorithm described in Hong and Nelson [39], and in Xu et al. [40]. The ISC algorithm is written in C++, and SOAP was selected for integration with MEG's web services
Parallel empirical stochastic branch and bound (PESBB)	Implemented the PESBB algorithm from Taghiyeh and Xu's early unpublished work, Xu and Nelson [41], and refinements detailed in Rosen et al. [42]. The PESBB algorithm is written in C++, and SOAP was selected for integration with MEG's web services
Multiobjective Evolutionary Algorithm (MOEA) Library	Integrated the open-source MOEA Java library into MEG. A user can select an MOEA library algorithm implementation and execute it in MEG. MOEA keeps track of the objective values and marks the solutions that exist on the Pareto front

while seamlessly scaling to an HPC cluster or cloud environment for production-level runs.

The job manager service also contains functions to configure, start, and stop a simulation job. In addition, there are functions for configuring links between the ouroboros service and experiment service to set up callbacks between running simulations and the storage service. The ouroboros interface is an internal MEG interface; no user interaction is required.

13.3.4.1 Job Manager Interfaces

There are several job manager implementations that have been developed and are available with MEG out of the box. However, a MEG user is not limited to this list. As with the Ouroboros service, MEG's extensible architecture provides for the ability to seamlessly add additional job managers for a user-defined implementation.

Table 13.3 gives a list of job managers currently implemented in MEG.

13.3.5 User Interface [UI] Service and Data Visualization

The user interface service is the external web service functions utilized for experiment data configuration and display. It provides the user with a portal to interact with MEG. MEG's open web service architecture provides simulation model developers access to MEG functionality without the additional requirement of an analyst learning a new interface. An analyst can define a set of model runs with a simulation model's Graphical User Interface and execute the backend runs through MEG. In addition, as the simulation model runs complete, job results can be pulled via the experiment service to display to the analyst.

Table 13.3 Default MEG job manager implementations

Name	Description
Local	Launches and monitors simulation jobs in the local command line environment. The thread pool size for launching parallel processes is configured based on the total number of CPUs in the local execution environment
Condor	Launches and monitors simulation jobs through the Local Resource Manager [LRM] HTCondor
GridWay	Launches and monitors simulation jobs using the meta-scheduler GridWay. GridWay interfaces with Globus Toolkit to launch and monitor jobs across many disparate LRM-controlled clusters such as Sun Grid Engine [SGE], Torque, PBS, and HTCondor
Moab	Launches and monitors simulation jobs using LRM Moab which interfaces with Torque
SLURM	Launches and monitors simulation jobs using LRM SLURM
Docker Compose	Launches and monitors single and co-operative simulation jobs using Docker Compose. Makes use of docker images to ensure execution environment consistency across runs
OpenShift	Launches and monitors simulation jobs using OpenShift executing on top of Kubernetes, a docker orchestration software package. Makes use of docker images to ensure execution environment consistency across runs. OpenShift provides a template concept which fits nicely into MEG's notion of templated inputs for a simulation job and great features for auto-scaling and load balancing of software applications

The use of third-party data visualization tools to analyze and display simulation model data is easy to accomplish with the open architecture embraced by MEG. Multiple models in the past have successfully employed data visualization software or developed model dashboards to display and/or analyze model data generated by MEG, using MATLAB, Python, Apache Superset, and similar visualization tools.

13.3.6   Data Services

The experiment service employs a MySQL database to store configuration, monitoring, and result objects for a simulation experiment. The data schema for the experiment service defines tables for an experiment, execution, and job instance. There is a hierarchical relationship between the aforementioned objects: an experiment has a one-to-many relationship with executions and an execution has a one-to-many relationship with jobs. A job defines a single model instance run of a simulation model with a concrete set of model inputs and a resulting set of model outputs. Figures 13.4 and 13.5 provide a simplified view along with a UML class diagram representing the relational hierarchy of the stored experiment objects.

The experiment table defines all the experiment-level configuration data required to create a running instance. Columns saved for an experiment include name, description, universally unique identifier (UUID), and the Web Services URL locations and types for the ouroboros and job manager to run for the experiment. In addition, there are tables to store the configuration for the ouroboros and job manager configurations for an experiment. These tables use the experiment UUID as a foreign key to link rows to the experiment table.

The execution table defines a running instance of a row in the experiment table. Multiple executions can be associated with a single experiment row in the experiment

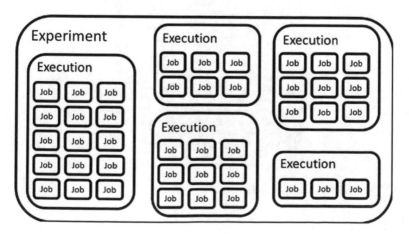

Fig. 13.4   MEG Object Hierarchy for a simulation experiment

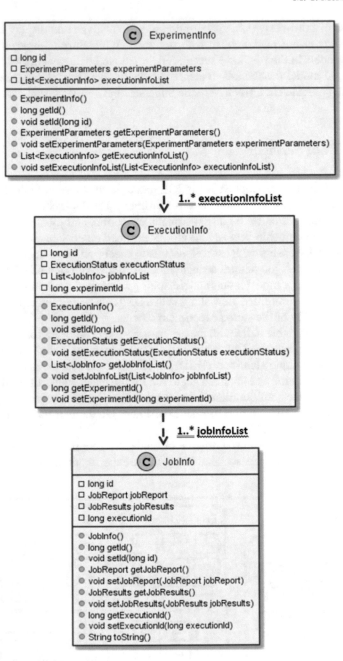

Fig. 13.5 MEG Object Hierarchy UML diagram

table. Columns saved for an execution include execution UUID, foreign key link to experiment UUID, start and end times, success, error code, and error reason.

The job table is divided into a monitoring table and a results table. The monitoring table contains columns that report on the status of a job such as where the jobs were executing and the status of a job. The results table contains columns to capture job inputs, outputs, and ids. Columns included in the job monitoring table include job UUID, foreign key link to execution UUID, execution host, execution time, start and finish times, job name, owner, priority, state, and job type. Columns included in the job results table include job UUID, foreign key link to execution UUID, inputs, outputs, and Pareto front. The Pareto front column is populated with the MOEA ouroboros when multiple objects are defined (e.g., the NSGA and SPEA multiobjective algorithms described in the preceding section).

13.3.7   MEG Experiment Primer—Ackley Function

In this section, an example MEG experiment is described highlighting the steps required to configure, execute, and monitor a simulation optimization experiment within the framework. The Ackley function provides a good example of a simulation response surface.

The benchmark defined by Ackley [27] is a non-convex function, with many local minima and maxima (Fig. 13.6). It provides for an N-dimensional input space and returns a single output. This benchmark is frequently utilized for performance testing of optimization search heuristics.

13.3.7.1   MEG Service Configuration

The experiment, ouroboros, and job manager services need to be configured for a MEG experiment to execute. The input service configuration is specified via the experiment service interface. Below is an excerpt of a Java properties file to highlight the types of data that are defined for a MEG experiment. Note that the generation of a Java properties file is not required. The file format is displayed here for reference to the experiment configuration. These properties can be sent directly to the experiment service via its Web Interface.

The experiment configuration properties (see example in Table 13.4) define the service implementation types and experiment host location. An experiment name, definition, and owner are defined here to capture metadata on the experiment.

The ouroboros configuration properties configure the search heuristic and the host location for the ouroboros service execution. Ouroboros properties are dependent on the implementation type. There is a function to query ouroboros' properties prior to configuration. In the example in Table 13.5, the ouroboros properties file defines the implementation of a genetic algorithm.

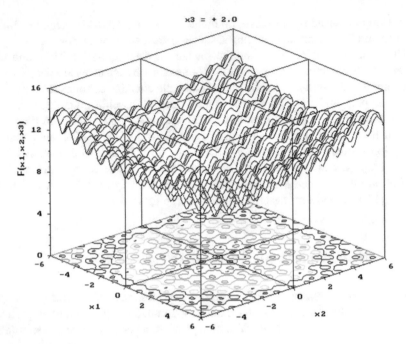

Fig. 13.6 A plot of the Ackley function with 3-dimensional input

Table 13.4 An example MEG experiment configuration

```
    experiment.owner=default
    experiment.name=Experiment 1
    experiment.description=Smoke   test   for   the   Experiment
service and service factory.
    experiment.ouroboros.type=ouroboros.gasimple
    experiment.jobmanager.type=jobmanager.standard
    experi-
ment.resourcemanagerdelegate.type=resourcemanagerdelegate
.local
    experiment.endCount=4

    experiment.service.url=http://localhost:8080/megsf-
experiment
```

Table 13.5 Ouroboros configuration

```
ouroboros.gasimple.nGenerations=1
ouroboros.gasimple.populationSize=100
ouroboros.gasimple.nKeep=2
ouroboros.gasimple.tournamentSize=2
ouroboros.gasimple.crossoverRate=0.5
ouroboros.gasimple.mutationType=1
ouroboros.gasimple.mutationSD=1
ouroboros.gasimple.forceUnique=false
ouroboros.gasimple.maximize=false
ouroboros.gasimple.mutationRate=0.5

ourobo-
ros.factory.service.url=http://localhost:8002/megsf-
service-factory/serviceFactoryService
```

The job manager configuration properties (Table 13.6) define the job name, simulation execution script, filenames for stdout/stderr, job hardware requirements, and query period. The query period specifies the time to wait between job status queries, with a time unit of milliseconds. The host location for the job manager service is also specified.

Below the job manager section is the definition of the input parameter space for a simulation model. This defines the dimensions of the input parameter space that an ouroboros will search.

Model input parameters in MEG can be specified as integer, double, object enumeration, or boolean. The object enumeration parameter type can be used to create a set of unique objects to select from, such as a set of input files.

Once the experiment, ouroboros, and job manager service properties are defined, an experiment is saved and can be executed by referencing the saved experiment properties.

13.3.7.2 MEG Execution Script

In the experiment properties, there is a value for the filename of the execution script (jobmanager.standard.script in the previous section). This script contains the set of commands that are executed for each simulation job submitted from an ouroboros in MEG. The input parameters can be accessed from this script while a simulation is running, by referencing them as environment variables at run time. In addition, MEG provides additional environment variables to uniquely identify simulation model runs, such as model output values per run.

Table 13.6 Job manager configuration and simulation model input parameter space

```
    jobmanager.standard.name=ackley_test_job
    jobmanager.standard.script=runModel.bat
    jobmanager.standard.workingdirectory=.
    jobmanager.standard.stdout=stdout.log
    jobmanager.standard.stderr=stderr.log
    jobmanager.standard.requirements=
    jobmanager.standard.queryperiod=10000

    jm.factory.service.url=http://localhost:8002/megsf-
  service-factory/serviceFactoryService

    parameter.total=3

    parameter1.type=double
    parameter1.name=x0
    parameter1.description=3 inputs
    parameter1.stepsize=0.01
    parameter1.maximum=32.0
    parameter1.minimum=-32.0

    parameter2.type=double
    parameter2.name=x1
    parameter2.description=3 inputs
    parameter2.stepsize=0.01
    parameter2.maximum=32.0
    parameter2.minimum=-32.0

    parameter3.type=double
    parameter3.name=x2
    parameter3.description=3 inputs
    parameter3.stepsize=0.01
    parameter3.maximum=32.0
    parameter3.minimum=-32.0
```

MEG utilizes the stdout I/O stream from the execution script to capture the scoring objective values. Model outputs are defined by a <key>=<value> per line. For example, if a single objective value is being passed back, the resulting output line might be

```
objectiveValue=42.0
```

The contents of the execution script used for this MEG experiment are provided in Table 13.7.

Table 13.7 An example execution script used in the MEG experiment described in this section

```
setlocal enabledelayedexpansion

benchmarks\ackley.py %x0% %x1% %x2% -s 0.1 -r 10
> objective.txt

set /P OBJECTIVE_VALUE=< objective.txt
echo objectiveValue=%OBJECTIVE_VALUE%
```

13.3.7.3 MEG Experiment Results

As simulation jobs complete and model outputs are collected from the execution script, these job results are pushed back to the ouroboros and experiment service. Returned job results are processed by the ouroboros service to determine which set of model inputs to launch for the next set of runs. The experiment service stores the job results in the MySQL database to persist the information after a job completes.

Job results for a MEG experiment can be viewed in a variety of ways. They can be accessed directly in the database, or a third-party visualization tool can be connected to the database for post data analysis. The choice of data visualization and analysis tools for MEG experiment results is selected by the user.

13.4 MEG and Cyber-Physical Systems

In this section, we discuss elements of Cyber-Physical Systems (CPS) that are amenable to Modeling and Simulation, and how the DOE and optimization capabilities implemented in MEG can potentially complement cyber-physical system research, design, and development. As in many multidisciplinary domains, Modeling and Simulation are used to design, demonstrate, and calibrate CPS. Examples of each of these facets are found in Oksa et al. [28]; Giaimo et al. [29]; and in Canadas et al. [30]. While a fully software-based simulation of CPS is complicated by the tightly coupled mixture of real-time systems, embedded systems, and human interaction, there are several touchpoints that can benefit from the formal DOE research and experimentation tools described in this chapter.

Giaimo et al. [29] describe the experimentation challenges unique to CPS systems comprising varying mixes of hardware types and human interaction. In their research, they have found that data collected in fielded systems is an underutilized resource. The automated data collection, storage, post-processing, and visualization tools used with MEG could potentially facilitate better use of data collected from fielded systems. Automated experimentation tools which require full algorithmic control of a simulation and control of input variables (e.g., an optimization experiment) are likely out of scope for a fielded mixed system as a whole. However, CPS systems where simulation is a functional component of the larger whole are amenable to MEG and similar tools and techniques. An example of this latter case is described by Mittal et al. [31], where an "HPC-in-the-loop" was described for a real-time CPS. For this simulation component, machine-directed optimization was implemented using a GAMS model of a home energy management system, and solving via mixed-integer linear programming (MILP) solver every 15 min. MEG can similarly provide this capability, through integration with the broader CPS architecture. This scenario is part and parcel of MEG's cardinal use case (optimization of a fully constructive computational simulation), and MEG's modular service-oriented design.

Oksa et al. [28] describe cyber-physical demonstrator systems, which are fully software-based simulations of the system as a whole, intended to prototype, roughly model, or portably demonstrate a CPS in development. Koulamas and Kalogeras [32] describe the use of a fully software-based digital twin to simulate the environment, behavior, and interactions of a planned system during the design phase. Gabor et al. [33] define some of the challenges in—and a reference architecture for—creating fully virtual digital twins to represent CPS. In contrast to the constraints of experimentation with fielded systems, these simulations of the CPS as a whole are likely most amenable to using the MEG framework, and could potentially benefit immensely from ongoing and emerging research in Simulation Optimization, experimentation, metamodeling, and data analytics, along with developments in automated experimentation at scale using the SaaS concepts embodied by the MEG architecture.

13.5   Summary and Future Work

MEG has been in continual development over the last decade in service of our three original design goals: access to (1) grid, cloud, and HPC computing support, (2) a wide range of Design of Experiments (DOE) methods, and (3) robust data processing and visualization. Here, we discuss MEG's current status and research challenges, avenues for future research and development, and how MEG can potentially support Modeling and Simulation in Cyber-Physical Systems.

13.5.1   MEG Summary and Current Status

In Sect. 13.3, we described MEG's current design—evolved from early support for multicore workstations, then extending to multi-server compute clusters, to computational grids, and finally to compute clouds and containers. Along the way, we have implemented and refined new DOE and Simulation Optimization algorithms, starting with early full factorial designs experiments, then adding simulation optimization with evolutionary algorithms, refinement of those to multiobjective optimization, and more recently implementing response surface algorithms for metamodeling, as described in Sects. 13.2 and 13.3.

As we've addressed real-world simulation challenges in the Defense, Aviation, Finance, and Healthcare realms, the organization of large amounts of data, experimental designs, and results have evolved to fit larger problems and more complex models. As new needs arise, we anticipate this ongoing development will continue, and there are unsurprisingly many new developments in each of our four focus areas.

13.5.1.1   High-Performance and Cloud Computing

The emergence of Exascale computing [34] promises additional computational power to support more sophisticated simulation experiments. At the same time, this means that the current Big Data challenge will become an Even Bigger Data challenge [35]. Extending the MEG architecture to efficiently support new high-performance computing capabilities will continue to be a design and development challenge.

13.5.1.2   DOE and Simulation Optimization

As we've seen herein, DOE methodologies and Simulation Optimization are also continually evolving. While refinement to MEG's existing capabilities is continual, there are several emerging trends that offer opportunities for new capabilities. Neuroevolution, the merging of Evolutionary computation with Neural network analysis, is increasingly being used for simulation metamodeling (see the review by [36]). There is continuing research in modern simulation optimization approaches previously unimplemented in MEG, including Particle Swarm Optimization, Tabu Search, and Simulated Annealing. As with neuroevolution, implementing, exploiting, and combining these techniques with each other and with Genetic Algorithms is a recent trend (see Ahmadian et al. [37] for a discussion of many of these algorithms and their hybridized use in the context of electrical distribution system planning and optimization). Finally, implementing and extending these approaches for use in a parallel computing environment remains a continual challenge for MEG development.

13.5.1.3   Usability, Data Processing, and Visualization

MEG has progressed from a purely research-oriented platform to validation and use in runs of record for larger real-world applications. In support of these projects, elements supporting usability, post-processing, and visualization have been added along the way. For example, we have had promising results using the Apache Superset[5] visualization tools, coupled with the MEG MySQL database, to quickly visualize experiment results. Future work in usability includes bolstering the visualization components of the framework to support these and additional open, modern, scalable visualization tools. In addition, the MEG pipeline detailed in Sect. 13.3.3 supports facile post-processing using Python or other standard data analytics tool stacks. Recent additions to MEG supporting containerization via the OpenShift[6] platform promise to lower barriers to use by non-technical analysts. Finally, there is emerging research in explanatory AI that the MEG team has heretofore not investigated deeply, but which can improve the effectiveness of human–simulation interaction and the understanding of results. For a review of explanatory AI in the context of DARPA's AINext[7] 3rd-wave AI initiative, see Mueller et al. [38]. All told, these emerging approaches hold the promise of improved usability, better guiding of algorithmic choice, managing, and understanding Simulation DOEs.

As mentioned, MEG has been rewritten as a service-oriented framework using open-source tools, libraries, and languages. In meeting MEG's three overarching design goals, we see a clear need to not only broaden access to MEG by a broader user base but also enable contributions from the larger Modeling and Simulation research community. Thus, MEG itself is in the process of being released under an open-source license.

Acknowledgements The authors wish to thank the many people who have contributed to the conceptualization, development, research, and refinement of the MEG Simulation Framework.

Glenn Roberts provided early support and guidance in developing the vision for MEG. Phil Barry, Brian Tivnan, Tobin Bergen-Hill, and Matt Koehler contributed to early prototypes and to the ongoing development of MEG as a simulation optimization tool. Laurie Litwin, Sheng Liu, and Elizabeth Chang contributed to the design and development of interfaces, visualization tools, and the database structure for MEG. Billy Baden and Pete Kuzminski contributed to algorithm development, and to the deployment and testing of MEG in a production environment. Christine Harvey, Scott Rosen, Pete Salemi, and Ashley Williams contributed to the development of MEG DOE algorithms, parallel processing algorithms, and response surface algorithms. We also thank Scott Rosen and David Prochnow for their valuable insight in reviewing and refining this chapter.

---

[5]Apache Superset is an open-source, Web-based, enterprise-ready business intelligence and visualization tool released as an Apache Incubator project (https://superset.apache.org/).

[6]Red Hat OpenShift is an enterprise-ready Kubernetes container platform with full-stack automated operations to manage hybrid cloud and multicloud deployments. (https://www.redhat.com/en/tec hnologies/cloud-computing/openshift).

[7]DARPA's AINext, announced in 2018, is accessible at https://www.darpa.mil/work-with-us/ai-next-campaign.

References

1. Balci O (2011) A life cycle for modeling and simulation. In: Proceedings of the 2011 winter simulation conference. Phoenix, AZ, pp 176–182
2. Fujimoto RM (2000) Parallel and distributed simulation systems. Wiley
3. Alexopoulos C, Seila AF (1988) Output data analysis. In: Banks J (ed) Handbook of simulation. Wiley, pp. 225–272
4. Balci O, Sargent RG (1984) Validation of simulation models via simultaneous confidence intervals. Am J Math Manage Sci 4(3–4):375–406
5. Juneja S, Shahabuddin P (2006) Rare event simulation techniques: an introduction and recent advances. In: Henderson SG, Nelson BL (eds) Handbooks in operations research and management science: simulation, vol 13, ch 11. North Holland, pp. 291–350
6. Wang GG, Shan S (2007) Review of metamodeling techniques in support of engineering design optimization. ASME J Mech Des 129(4):370–380
7. Ye T, Kaur HT, Kalyanaraman S, Yuksel M (2008) Large-scale network parameter configuration using an on-line simulation framework. IEEE/ACM Trans Netw 16(4):777–790
8. Amaran S, Sahinidis NV, Sharda B, Bury SJ (2016) Simulation optimization: a review of algorithms and applications. Ann Oper Res 240:351–380
9. Law A (2014) Wintersim Proceedings of the 2014 Winter Simulation Conference. In: Tolk A, Diallo SY, Ryzhov IO, Yilmaz L, Buckley S, Miller JA (eds) A tutorial on design of experiments for simulation modeling
10. Bookstaber R, Paddrick M, Tivnan B (2014) An Agent-based model for financial vulnerability. Office of financial research working paper, https://www.financialresearch.gov/working-papers/2014/07/29/an-agent-based-model-for-financial-vulnerability/. Accessed 15 Sept 2019
11. Tekin E, Sabancuoglu I (2004) Simulation optimization: a comprehensive review on theory and applications. IIE Trans 36(11):1067–1081
12. Holland J (1992) Genetic algorithms. Sci Am 267:66–72
13. Goldberg DE (1989) Genetic algorithms in search, optimization, and machine learning, 1st edn. Addison-Wesley Longman Publishing Co., Inc, Boston, MA, USA
14. Bäck T (1996) Evolutionary algorithms in theory and practice: evolution strategies, evolutionary programming, genetic algorithms. Oxford University Press Inc, New York, NY, USA
15. Barry PS, Koehler MTT, McMahon MT, Tivnan BT (2009) Agent-directed simulation for systems engineering: applications to the design of venue defense and the oversight of financial markets. Int J Intell Control Syst 14(1):20–32
16. Skolicki Z, De Jong K (2004) Improving evolutionary algorithms with multi-representation island models. In: Proceedings of 8th international conference on parallel problem solving from nature. PPSN VIII, Birmingham, UK
17. Abdelhafez A, Alba E, Luque G (2019) Performance analysis of synchronous and asynchronous distributed genetic algorithms on multiprocessors. Swarm Evol Comput 49:147–157
18. Deb K, Pratap A, Agarwal S, Meyarivan T (2002) A fast and elitist multiobjective genetic algorithm: NSGA-II. IEEE Trans Evol Comput:182—197
19. Zitzler E, Thiele L (1999) Multiobjective evolutionary algorithms: a comparative case study and the strength pareto approach. IEEE Trans Evol Comput 3(4):257–271
20. Wolpert DH, Macready WG (1997) No free lunch theorems for optimization. IEEE Trans Evol Comput 1:67–82
21. Rosen SL, Saunders C, Guharay S (2012) Metamodeling Of Time Series Inputs And Outputs. In: Laroque C, Pasupathy R, Himmelspach J (eds) Proceedings of the 2012 winter simulation conference. Institute of Electrical and Electronics Engineers, Inc, Piscataway, New Jersey, pp 2366–2377
22. Rosen SL, Guharay SK (2013) A case study examining the impact of factor screening for neural network metamodels. In: Pasupathy R, Kim S-H, Tolk A, Hill R, Kuhl ME (eds) Proceedings of the 2013 winter simulation conference, Washington, DC, pp. 486–496

23. Page EH, Litwin L, McMahon MT, Wickham BM, Shadid M, Chang E (2012) Goal-directed grid-enabled computing for legacy simulations. In: Proceedings of the 12th IEEE/ACM international symposium on cluster, cloud and grid computing. IEEE, Ottowa, Ontario, Albgerta CAN, pp 873–879

24. CSSC Cloud Standards Customer Council (CSCC) (2015) Practical guide to platform-as-a-service version 1.0. Accessed 19 Nov 2019. http://www.cloud-council.org/CSCC-Practical-Guide-to-PaaS.pdf

25. Chinnici R, Moreau J, Ryman A, Weerawarana S (2004) Web services description language (WSDL) version 2.0 part 1: core language. W3C working draft. 26. Accessed 19 Nov 2019. https://www.w3.org/TR/2007/REC-wsdl20-20070626/

26. Lewis G (2010) Getting started with service oriented architecture (SOA) terminology. Software Engineering Institute, Carnegie Mellon. Accessed 19 Nov 2019. https://resources.sei.cmu.edu/asset_files/WhitePaper/2010_019_001_30118.pdf

27. Ackley DH (1987) A connectionist machine for genetic hillclimbing. Kluwer Academic Publishers, Boston MA

28. Oksa SJ, Jalowskia M, Fritzschea A, Mösleinab KM (2019) Cyber-physical modeling and simulation: a reference architecture for designing demonstrators for industrial cyber-physical systems. In: 29th CIRP Design Conference May 2019, Póvoa de Varzim, Portgal

29. Giaimo F, Yin H, Berger C, Crnkovic I (2016) Continuous experimentation on cyber-physical systems: challenges and opportunities. In: Proceedings of the scientific workshop proceedings of XP2016. ACM, Edinburgh Scotland, UK

30. Canadas N, Machado J, Soares F, Barros C, Varela L (2018) Simulation of cyber physical systems behaviour using timed plant models. Mechatronics 54:175–185

31. Mittal S, Ruth M, Pratt A, Lunacek M, Krishnamurthy D, Jones W (2015) A system-of-systems approach for integrated energy systems modeling and simulation. In: Proceedings of the conference on summer computer simulation, summer Sim '15. Society for Computer Simulation International, San Diego, CA, USA, pp 1–10

32. Koulamas C, Kalogeras A (2018) Cyber-physical systems and digital twins in the industrial internet of things. Computer 51(11):95–98

33. Gabor T, Belzner L, Kiermeier M, Beck MT, Neitz A (2016) A simulation-based architecture for smart cyber-physical systems. In: 2016 IEEE international conference on autonomous computation (ICAC), pp 374–379

34. Roe R (2019) Expect exascale. Sci Comput World 165:4. Accessed 5 Dec 2019. https://www.scientific-computing.com/feature/expect-exascale

35. Varghese B, Buyya R (2018) Next generation cloud computing: new trends and research directions. Future Gen Comput Syst 79(3):849–861

36. Sloss AN, Gustafson S (2019) Evolutionary algorithms review, Cornell computing research archive. Accessed 17 June 2019. https://arxiv.org/abs/1906.08870

37. Ahmadian A, Elkamel A, Mazouz A (2019) An improved hybrid particle swarm optimization and Tabu search algorithm for expansion planning of large dimension electric distribution network. Energies12:16

38. Mueller ST, Hoffman RR, Clancey WJ, Emrey A, Klein G (2019) Explanation in human-ai systems: a literature meta-review, synopsis of key ideas and publications, and bibliography for explainable AI. In CoRR https://arxiv.org/abs/1902.01876

39. Hong LJ, Nelson BL (2006) Discrete optimization via simulation using COMPASS. Oper Res 54:115–129

40. Xu J, Hong LJ, Nelson BL (2010) Industrial strength COMPASS: a comprehensive algorithm and software for optimization via simulation. ACM TOMACS 20:1–29

41. Xu WL, Nelson BL (2013) Empirical stochastic branch-and-bound for optimization via simulation. IIE Trans 45(7):685–698

42. Rosen SL, Salemi P, Wickham B, Williams A, Harvey C, Catlett E, Taghiyeh S, Xu J (2016) Parallel empirical stochastic branch and bound for large-scale discrete optimization via simulation. In: Proceedings of the 2016 winter simulation conference (WSC), Washington, DC, USA, 11–14 Dec 2016, pp 626–637

Matthew T. McMahon is a Principal Simulation & Modeling Engineer at the MITRE Corporation. He received a MS in Computer Science from Virginia Tech and a BE in Electrical Engineering from Auburn University. He has published in the areas of high-performance computing, engineering optimization, and complex systems.

Brian M. Wickham is a Lead Simulation & Modeling Engineer at the MITRE Corporation. He received a MS in Computer Science from Johns Hopkins University and BS in Computer Science and Mathematics from Rose-Hulman Institute of Technology. His research interests include optimization via simulation, simulation interoperability, and high-performance computing.

Ernest H. Page is the DARPA Portfolio Manager at the MITRE Corporation. He received the Ph.D. in Computer Science from Virginia Tech. He has published over 50 articles in the areas of simulation modeling methodology and distributed computing.

Part IV
Reliability Issues

Chapter 14
Cloud-Based Simulation Platform for Quantifying Cyber-Physical Systems Resilience

Md. Ariful Haque, Sarada Prasad Gochhayat, Sachin Shetty, and Bheshaj Krishnappa

Abstract Cyber-Physical Systems (CPS) often involve trans-disciplinary approaches, merging theories of different scientific domains, such as cybernetics, control systems, and process design. Advances in CPS expand the horizons of these critical systems and at the same time, bring the concerns regarding safety, security, and resiliency. To minimize the operating costs and maximize the scalability, often time, it is preferable to use the cloud environment for deploying the CPS computation processes and simulation environments. With the expanding uses of the CPS and cloud computing, major cybersecurity concerns are also growing around these systems. The cloud itself has security and privacy issues. This chapter focuses on a cloud-based simulation platform for deriving the cyber resilience metrics for the CPS. First, it presents a detailed analysis of the modeling of the resilience metrics by mapping them with cloud security concerns. Then, it covers modeling and simulation (M&S) challenges in developing simulation platforms in the cloud environment and discusses a way forward. Overall, we aim to discuss resilience metrics modeling and automation using the proposed simulation platform for the CPS in the cloud environment.

Md. A. Haque
Computational Modeling and Simulation Engineering, Old Dominion University,
5115 Hampton Blvd, Norfolk, VA 23529, USA
e-mail: mhaqu001@odu.edu

S. P. Gochhayat · S. Shetty (✉)
Virginia Modeling Analysis and Simulation Center, Old Dominion University,
1030 University Blvd, Suffolk, VA 23435, USA
e-mail: sshetty@odu.edu

S. P. Gochhayat
e-mail: sarada1987@gmail.com

B. Krishnappa
Risk Analysis and Mitigation, ReliabilityFirst Corporation, 3 Summit Park Drive, Suite 600,
Cleveland, OH 44131, USA
e-mail: bheshaj.krishnappa@rfirst.org

© Springer Nature Switzerland AG 2020
J. L. Risco Martín et al. (eds.), *Simulation for Cyber-Physical Systems Engineering*,
Simulation Foundations, Methods and Applications,
https://doi.org/10.1007/978-3-030-51909-4_14

14.1   Introduction

Cyber-physical systems (CPS) are engineered systems built from the integration of computation, networking, and physical processes. CPS have put their marks on modern societies by providing efficient and reliable operation of systems and services necessary for our daily life. Critical infrastructures, such as energy delivery systems, oil and gas systems, healthcare systems, industrial plants, transportation systems, autonomous vehicle industry, etc. heavily depend on the CPS.

Researchers often generalize CPS as the integrated system of cyber and physical systems, where we use embedded computers and networks to compute, communicate, and control the physical processes. In a broader sense, CPS are a combination of two technologies: the cyber layer (information technology (IT)) and the physical layer (operational technology (OT)). The cyber layer consists of servers, database systems, hosts, etc. which are necessary for the business operations. The physical layer consists of sensors, actuators, control functions, feedback systems, etc. which are responsible for handling the production facilities according to the intended design of the system. The risks of cyber-attacks come from the integration of the cyber and physical domains, in other terms, integration of IT and OT domains.

Recently, there is a growing trend (Kim[1]) to utilize the cloud service platform for deploying the CPS control and computation processes. Cloud computing is a convenient service platform for the on-demand access to a shared pool of configurable computing resources (Mell et al. [2]). The availability of high-capacity systems, low-cost computers, and storage devices, as well as the widespread adoption of hardware virtualization, service-oriented architecture, and autonomic computing, have led the growth in cloud computing in an exponential manner. With cloud computing, there is no need to make significant upfront investments in hardware and software. The cloud service platform can be rapidly provisioned and released with minimal management effort or service provider interaction. Hence, the crucial benefits that the cloud services provide over the traditional fixed infrastructures are agility, reduction in cost, elasticity, scalability, fast deployment, location independence, easy maintenance, improved productivity, etc.

Generally, people perceive that security in cloud computing is as good as or even better than other fixed network systems. Because service providers can devote resources to solve security issues that many customers cannot afford to tackle because of not having the required technical skills and resources. But like many other systems, there are cybersecurity concerns, such as the fear of loss of control over sensitive data and processes. The complexity of security significantly increases when data are distributed over a wide geographic area, as well as in multi-tenant systems shared by unrelated users.

Traditionally in the CPS and cloud domain, deployment of intrusion detection systems (IDS) and intrusion prevention systems (IPS) help in protecting these systems from the potential cyber adversary. Although the security measures taken by IDS and IPS are of utmost necessity, we need cyber-resilient assessment processes for enhancing security. As the cloud services are comparatively inexpensive and provide flexibility, deployment of simulation platforms in the cloud is attracting interests from the research communities. We consider addressing the security concerns and developing the resilience assessment metrics as complementary to each other. In this chapter, we relate the resilience of the CPS from the modeling and simulation (M&S) perspective to the cloud computing and leverage the cloud services to propose a simulation platform for the assessment of the resilience metrics for the CPS. Overall the chapter addresses the following critical issues:

- A detailed discussion on the CPS threats and vulnerabilities, including cloud security threats;
- Mathematical modeling for deriving quantitative cyber resilience metrics for the CPS using system critical functionality;
- A proposed simulation platform for cyber resilience assessment for CPS in the cloud; and
- Complexities, challenges, and a way forward for implementing the CPS simulation platform in the cloud environment.

We organize the rest of the chapter as follows. Section 14.2 presents details on the CPS. Section 14.3 discusses cloud computing service models and cloud architectures. Section 14.4 illustrates CPS and cloud threats and security issues. Section 14.5 discusses the resilient CPS characteristics, and a formal method to compute the cyber resilience metrics. Section 14.6 provides the architecture of the proposed quantitative simulation platform in the cloud. This section also covers a use case using a qualitative simulation platform for cyber resilience assessment for the CPS that we have already implemented utilizing the Amazon Web Service (AWS). Section 14.7 highlights the complexities of implementing the CPS simulation methodologies in the cloud platform and ways to move forward. Finally, Sect. 14.8 concludes the chapter.

14.2   Cyber-Physical Systems (CPS)

Cyber-physical systems represent a complex class of systems consisting of a robust combination of computational and physical components. The National Institute of Standards and Technology (NIST) CPS Public Working Group (PWG) defines CPS as *smart systems that include engineered interacting networks of physical and computational components* (Griffor et al. [3]). CPS are indispensable in implementing most modern technologies, such as wireless sensor networks (WSN), Internet of Things (IoT), industrial Internet, industrial control systems (ICS), machine-to-machine (M2M) communication, smart devices, etc.

"Cyber-Physical Systems (CPS) comprise interacting digital, analog, physical, and human components engineered for function through integrated physics and logic. These systems will provide the foundation of our critical infrastructure, form the basis of emerging and future smart services, and improve our quality of life in many areas (Griffor et al. [3])."

CPS consist of control and feedback systems, which are highly interconnected and heterogeneous. The control systems include physical processes, such as sensors and actuators, which operate in real time. These systems can be networked or distributed. There may be humans and environmental interactions involved in the process. A typical example of the CPS is the power systems network. We provide a generic CPS architecture in Fig. 14.1 utilizing the NIST recommended defense-in-depth security model (DHS CSSP [4]) to address the security concerns in CPS. There are three broad layers in the defense-in-depth security architecture: cyber layer, control layer, and physical layer. We explain here the architecture using a bottom-up approach for facilitating a smooth transition to the details coming later on.

The physical layer is the place where the production facilities are installed and expected to behave according to the intended functional specifications. According to the NIST defense-in-depth security architecture, this layer divides itself into three separate segments. The lowest section, denoted as level zero, contains the field devices and sensors. For example, for the case of the power system, this layer consists of generators, transformers, field buses, sensors, and transducers (e.g., temperature sensors, proximity sensors, pressure sensors, capacitive level sensors, ultrasonic level sensors, smoke detectors, etc.).

The segment marked as level one in the physical layer consists of the field controllers and actuators, such as protective relays, circuit breakers, hydraulic actuators, etc. These devices drive the lower layer field devices based on the commands from the human–machine interface (HMI) and other input/output (I/O) devices from layer two (Macaulay and Singer [5]). The level two consists of phasor measurement units (PMU), intelligent electronic devices (IED), remote terminal units (RTU), etc. These devices exchange information about the operational status of the field devices as found from the sensors in the level zero. Level one and two are parts of supervisory control and data acquisition systems (SCADA), distributed control systems (DCS), or hybrid systems. These systems have some redundancy in their implementation either by using the redundant devices or physical links (e.g., electric bus systems) to ensure the functional continuation of the services.

The next layer is the control layer, which segments itself into control demilitarized zone (DMZ) and the control local area network (LAN). These levels contain domain controllers, HMI, application servers, historians, file servers, patch management servers, engineering workstations, etc. The primary function of these layers is to monitor the performance of the field devices based on the collected data from the physical device networks. If necessary, the operators can issue control commands for maintaining the service functionalities. According to NIST defense-in-depth secu-

Fig. 14.1 A generic
defense-in-depth
architectural view of
cyber-physical systems
security depicting the cyber,
control, and physical layers.
We have adapted the figure
from US-CERT
defense-in-depth architecture
(DHS CSSP [4]) to illustrate
the significant components
of CPS and ways to address
security concerns

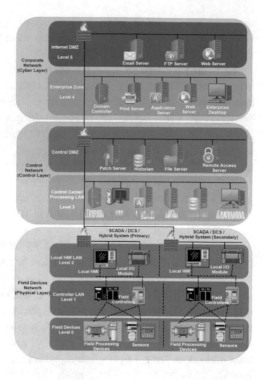

rity model, there is a need to have some barriers (e.g., firewalls) between the control
and the physical layers to protect the intended information flow and manage security
issues efficiently. The topmost layer is the corporate or business layer, which contains application servers, email servers, web servers, enterprise desktops, etc. This
layer is essential for the regular business operations of the organization or entities.
NIST suggests using DMZ to isolate the part of the corporate network open to the
Internet. In the next subsection, we discuss the scope of CPS operations in different
domains of systems before going into the deep drive of the security assessment platform in the cloud. Within the chapter, we consider CPS as a plural term representing
cyber-physical systems.

14.2.1 CPS and Other Related Fields

As CPS overlap with other emerging fields and applications, we aim to provide here a
brief idea about the scope of operations of CPS. Here, we present a concise discussion
on the relation and distinction of CPS to the emerging applications to highlight the
area of operational scope that falls under CPS.

CPS and Embedded Systems: An embedded system is part of an extensive process that incorporates elements of control logic in a limited scope from functionality

and resources point of view. An embedded system typically confines to a single device. On the other hand, CPS operate at a much larger scale, including many embedded systems or other devices.

CPS and System of Systems (SoS): A system of systems (SoS) is a system composed of elements, which are independent systems in their domains. The constituent systems which are operationally independent, physically distributed, and continuously evolving collaborate to produce a global behavior that is not attainable individually. CPS comprise separate constituents, and, like SoS, CPS also tackle challenges of coping with dependable emergence, evolution, and distribution (Kopetz et al. [6]). Although we consider CPS constituent systems are independent, it is not an essential requirement because most of the underlying physical processes and systems are interdependent. Similarly, although it is often the case that SoS do incorporate elements of computation as well as real-world interaction, this is not a defining property of an SoS.

CPS and Internet of Things (IoT): CPS and IoT have significant overlaps in their conceptual definitions. The IoT is a vision of the future which considers millions of devices connected over the Internet. These devices allow them to collect information about the real world remotely and share it with other systems. Some consider IoT as a subset of CPS, and others think the opposite. The CPS and IoT concepts emerged from different communities, with CPS primarily emerging from a system engineering and control perspective. The IoT concept emerged mainly from a networking and information technology perspective, which envisioned integrating the digital realm into the physical world (Greer et al. [7]). IoT has a strong emphasis on uniquely identifiable and internet-connected devices and embedded systems. On the other hand, CPS engineering has a strong focus on the relationship between computation and physical processes.

"The IoT emphasizes the networking, and is aimed at interconnecting all the things in the physical world, thus it is an open network platform and infrastructure; the CPS emphasizes the information exchange and feedback, where the system should give feedback and control the physical world in addition to sensing the physical world, forming a closed-loop system (Ma [8])."

CPS and Industry 4.0: Industry 4.0 refers to a new phase in the industrial revolution that focuses heavily on interconnectivity and automation. It includes cyber-physical systems, Internet of Things, cloud computing, cognitive computing, smart digital technology, machine learning, big data, etc. Industry 4.0 fosters the idea of "smart factory." Within modular structured smart factories, cyber-physical systems monitor physical processes, create a virtual copy of the physical world, and make decentralized decisions. Over the Internet of Things, cyber-physical systems communicate and cooperate with humans, in real time, both internally and across organizational services offered and used by participants of the value chain (Hermann et al. [9]). Figure 14.2 illustrates a generic diagram for industry 4.0, including the cyber-physical systems and the Internet of Things only. The figure illustrates that

CPS and IoT work in their self-domain, but Industry 4.0 comprises both CPS and IoT in terms of operations and scope.

With the vast scope of applications, CPS have various challenges. For example, there are complexities in modeling physical processes and real-time behavior. There are challenges of interconnectivity and interoperability because of the heterogeneity. Challenges also lie in the secure integration of various components. From the M&S perspective, CPS need to handle the analysis of specification, development of design methodologies, scalability and complexity, and overall verification and validation (Mittal et al. [10]). Because of the integration of the cyber and physical domains, there is a significant amount of cyber risk that CPS need to handle. In this chapter, we aim to focus on those challenges and propose a simulation platform to facilitate the security and resilience assessment for the CPS in the cloud environment. To have a better understanding of the cloud computing platform, we present the service models and a generic architecture of cloud computing in Sect. 14.3.

14.3   Cloud Computing Environment

NIST defines cloud computing as *a model for enabling convenient, on-demand net-work access to a shared pool of configurable computing resources (e.g., networks, servers, storage, applications, and services) that can be rapidly provisioned and released with minimal management effort or cloud provider interaction* (Mell et al.

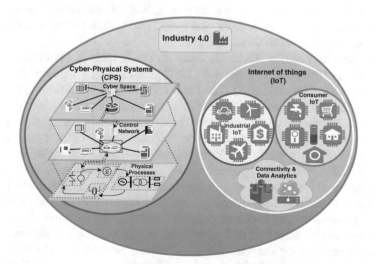

Fig. 14.2 Relation among the scope of operations of cyber-physical systems, Internet of Things, and Industry 4.0. IoT consists of industrial and consumer IoT, where connectivity, cloud computing, and data analytics play a significant role. CPS consist of physical, control, and cyber components. Both CPS and IoT fall under the broad umbrella of Industry 4.0

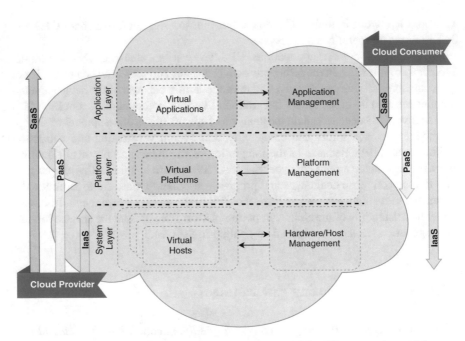

Fig. 14.3 Cloud provider and consumer scope of concentration for different service models

[2]). In this section, we discuss the generic cloud computing platform and its service and deployment models to help the readers to gain a detailed understanding of the cloud computing services. This discussion would also facilitate in the security threat analysis, which we cover in Sect. 14.4.2.

Cloud Service Models: There are three service models in the cloud environment. These are Software as a Service (SaaS), Platform as a Service (PaaS), and Infrastructure as a Service (IaaS). The SaaS enables the users to use applications running on the cloud infrastructure. The consumer does not manage or control the underlying infrastructure, including network, servers, operating systems, storage, or even individual application capabilities. The PaaS is the service model that allows its consumers to deploy required applications created using programming languages, libraries, services, and tools supported by the cloud provider. Again, the consumer does not manage or control the underlying cloud infrastructure but has control over the deployed applications and possibly configuration settings for the application-hosting environment. The IaaS service model provides the consumer to provision processing, storage, networks, and other fundamental computing resources where the consumer can deploy and run arbitrary software. The consumer has control over operating systems, storage, and deployed applications; and possibly limited control over selected networking components (e.g., host firewalls).

Figure 14.3 illustrates the differences in scope and control between the cloud consumer and cloud provider for each of the service models discussed above. The arrows at the left and right of the diagram denote the approximate range of the cloud providers and consumers' scope and control over the cloud environment for each service model. In the cloud architecture, the system layer is the lowest layer providing services through hosts and networks. This layer contains hypervisors, virtual machines, virtual data storage, and virtual network components, which are required to realize the infrastructure upon which a computing platform may establish. One example of this layer is the Amazon Elastic Compute Cloud (EC2) service. The platform layer is the second layer of cloud architecture. This layer consists of virtualized operating systems, application programming interfaces (APIs), compilers, libraries, middleware, and other software tools needed to implement and deploy applications. A few examples of this layer are AWS Elastic Beanstalk, Windows Azure, etc. The application layer is the top layer of the cloud architecture and provides virtual applications. Google Apps, Microsoft Office 365 are some examples of this layer.

Cloud Deployment Models: On the deployment side, the cloud has four types of deployment models. The consumers should have an excellent overall idea to deploy their system using either of these models. The models are private, community, public, and hybrid cloud. In the private cloud deployment model, we provision the cloud infrastructure for exclusive use by a single organization comprising multiple consumers (such as different business units of a system). In the community cloud deployment model, the cloud providers provide services for use by a specific community of consumers from organizations that have shared concerns. It may be owned, managed, and operated by one or more of the organizations in the community, a third party, or some combination of them. In the public cloud deployment model, the cloud providers make the infrastructure open for use by the public, and it exists on the premises of the cloud provider. In the hybrid cloud deployment model, the cloud infrastructure is a composition of two or more distinct cloud infrastructures but are bound together by standardized or proprietary technology.

To use the cloud platform for CPS simulation platform development, we prefer to use the public cloud to reduce costs and promote the ubiquitous use of the tool. We would illustrate the ways to handle security issues while deploying the simulation platform by utilizing the public cloud deployment model in Sect. 14.7. In Sect. 14.3, we discuss a detailed analysis of the CPS threats and cloud security concerns before presenting the resilience quantification methodologies in Sect. 14.5. We suppose having sound knowledge on the CPS and cloud security threats and attack vectors would help the audiences to understand the rationale of the resilience quantification methodologies used in Sect. 14.5.

14.4 CPS and Cloud Security Concerns

Due to the complex operational nature of CPS, the systems are prone to various cyber threats and vulnerabilities. To shift the business and control operations in the cloud platform, it also needs to consider cloud security and privacy issues (e.g., data confidentiality, integrity, and availability). There requires a significant amount of technical and managerial capabilities to ensure the safety and resiliency of the systems operations in the cloud environment. In this section, we aim to provide a detailed analysis of CPS cyber threats and challenges to handle from security perspectives. We also discuss cloud security issues. These discussions would give the reader a solid basis for the CPS resilience assessment methodology development leveraging the cloud platform. The study would also help the readers to realize the resilience modeling and simulation platform discussions in Sects. 14.5 and 14.6, respectively.

14.4.1   CPS Security Threats

CPS face threats from adversaries in the physical, network, and cyber layer. In the physical layer, the availability of the services and functionalities provided by the field controllers and sensors are of utmost concern. There is also a risk of information alteration by modifying the embedded codes of the physical devices (e.g., PLC logic codes) (Basnight et al. [11]). In the network layer, most attacks are taking place, such as a distributed denial of service (DDoS), eavesdropping (man-in-the-middle attack), jamming, and selective forwarding. Most of these attacks are possible to encounter by implementing intrusion detection and intrusion prevention systems, and regular updating of system patches. Threats in the cyber layer can lead to leaking of confidentiality, stealing of credentials, unauthorized access to the system, social engineering, etc. We classify some of the threats according to the area of origination by analyzing the discussion provided by Ginter [12] and Cardenas et al. [13]. We provide the categorization in Fig. 14.4. Here sometimes, we alternately use the terms ICS instead of CPS to focus on the specific components that may get impacted by the cyberattacks.

External Threats: These threats arise from adversaries such as nation sponsored hackers, terrorist groups, and industrial competitors through espionage activities. Cyber intruders may launch an APT attack, where the goal is to steal some valuable information on the network's assets without getting detected. One example of such an attack in recent times is the Stuxnet attack (see Chen and Abu-Nimeh [14]) on the Iranian nuclear centrifuges.

Internal Threats: Today, as the work processes in any industry are segmented and done by third-party contractors, CPS companies need to share system access information to outside business partners. That makes the CPS system vulnerable to potential cyber threats. There exist insider threats from the employees of the ICS support providers who have authorized access to the ICS network. We call them credentialed ICS insider.

Technology Threats: Many of the ICS networks run on legacy technology, where security is not the utmost priority. Instead, the availability of services and the continuation of the protocol-level communication among different ICS products without interruptions or disturbances are of utmost necessity. Thus, many of them are lacking strong authentication or encryption mechanism as pointed by Laing [16]. The weak

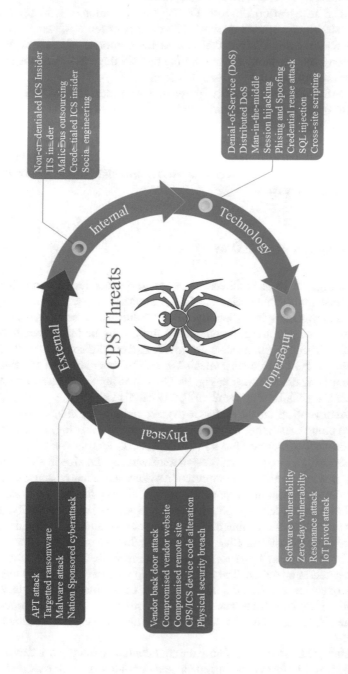

Fig. 14.4 Cyber threats and attack vectors in cyber-physical systems. We have adapted the categorization by following Haque et al. [15]

security mechanisms (e.g., insecure password, default user accounts, etc.) are not enough to protect the system from smart adversaries.

ICS and ITS Integration Threats: Due to the integration of field devices with the control system and corporate network, ICS devices become vulnerable to cyber-attacks. The reason behind this is that part of the corporate network is open for communication over the Internet. Only putting the ICS devices behind the firewalls do not necessarily safeguard the valued ICS assets.

Physical Infrastructure Security Threats: Sometimes, the absence of proper access control or infrastructure security to the ICS devices poses severe threats to the ICS network. One such example of poor physical security is sharing of the floor space of ICS devices with the ITS devices (e.g., employees' computers, routers, switches, etc.) and thus, providing easy access to the ICS devices (e.g., PLC, IED, RTU, PMU, etc.). Other sorts of threats may arise from compromised vendor websites or compromised remote sites.

14.4.2 Cloud Security Issues

Cloud computing, due to its architectural design and characteristics, provides several security benefits, such as centralization of security, data and process segmentation, redundancy, and high availability. Many traditional security risks are possible to counter effectively in cloud computing. But due to the infrastructure control and ownership sharing in different service models, there are some distinct security threats present in the cloud environment. We present here some of the threats in the cloud computing environment in Fig. 14.5 by following the discussions given by Zissis and Lekkas [17], Roschke et al. [18], and Modi et al. [19].

Multi-tenancy: Multi-tenancy refers to the resource sharing characteristics of the cloud. As the cloud computing business model is resource sharing, the cloud users share resources such as memory, program, storage, hosts, and networks. With a multi-tenant architecture, service providers design a software application in a way so that it can virtually partition its data and configuration so that each client organization works with a customized application instance. Because of sharing the same resources with virtual partitions, multi-tenancy poses several privacy and confidentiality threats. There exists the risk of data confidentiality breach due to remanence (Sindhiya et al. [20]). Data remanence is the residue of digital data that remains even after there are attempts to erase the data. This residue may come from data being left intact by a nominal file deletion operation. The physical media may have properties that allow the recovery of previously erased data. As a result, data remanence may lead to the unwilling disclosure of private data and, thus, poses the threats of confidentiality breach of data. Hardware and network security are also concerns in the cloud due to multi-tenancy.

Data Security: Data security refers to the confidentiality, integrity, and availability of data. The cloud platform maintains data confidentiality through user authentication. Lack of a robust authentication mechanism may lead to unauthorized access to

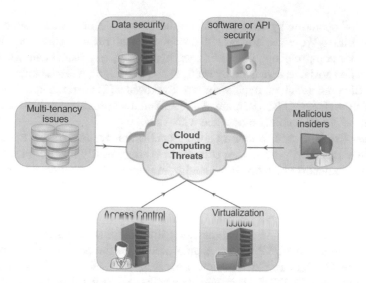

Fig. 14.5 Cybersecurity threats and concerns in the cloud platform. The major cloud threats come from the data breach (with possible loss of confidentiality and integrity), software and API security, hosting multiple user applications on the same hardware (multi-tenancy), access control including security policy, virtualization (i.e., use of VM images and hypervisor), and malicious insiders

the users' accounts, which may ultimately lead to a privacy breach. Data integrity refers to protecting data from unauthorized deletion, modification, or fabrication. The cloud computing environment has the threats of alteration of data at rest and in transit and data interruption or removal, including sophisticated insider attacks. Data availability is also a concern because of the heavy reliance on the cloud provider's ability to maintain the uninterrupted continuation of the services.

Software Security: Software or the APIs running on the cloud platform are also facing confidentiality and integrity threats. In the cloud, unauthorized access to software applications is possible through the exploitation of the vulnerability associated with the apps or due to a lack of robust security measures. In the case of web applications, Cross Site-Scripting (XSS) is a typical class of attacks. An attacker may insert JavaScript code into the web page to steal user data by utilizing a session cookie, as pointed out by Roschke et al. [18]. Cloud computing providers implement APIs for users to manage and interact with cloud services. These APIs also pose severe threats to software integrity as an unauthorized user gaining control of them may cause significant damages to the applications and consumer data.

Virtualization Issues: The cloud platform depends heavily on the virtualization techniques as it uses virtual machines (VMs) for logical separation of users utilizing the same platform and resources. A hypervisor or virtual machine monitor (VMM) is computer software, firmware or hardware that creates and runs the virtual machines. The hypervisor presents the guest operating systems with a virtual operating platform and manages the execution of the guest operating systems. Multiple instances of a variety of operating systems may share the virtualized hardware resources. For example, Linux, Windows, and macOS instances can all run on a single physical x86 machine. By compromising the lower layer hypervisor, an attacker can gain control over installed VMs, as discussed by Modi et al. [19]. Along with the known attacks such as Blue pill attack on hypervisor (see Rutkowska [21]), direct kernel structure manipulation (DKSM) (see Bahram et al. [22]) on the virtual layer, there are threats from zero-day vulnerability associated with VMs.

Access Control: Access control and authentication are two means to keep data away from unauthorized users. Access controls are typically identity based, which makes authentication of the user's identity an essential issue in cloud computing. Some cloud providers use the Security Assertion Markup Language (SAML) standard to authenticate users before providing access to applications and data. SAML messages use Simple Object Access Protocol (SOAP), whose format is eXtensible Markup Language (XML). Some cases illustrate that XML wrapping attacks by manipulating the SOAP messages are possible on the Amazon EC2 instances, as shown by Gajek et al. [23]. In some other examples, eXtensible Access Control Markup Language (XACML) is in use to control access to cloud resources. Messages transmitted between XACML entities are also susceptible to malicious attacks, as explained by Keleta et al. [24].

Malicious Insider: There always exist the risks of insider attacks on cloud and information resources, whether it is information technology systems (ITS) or a cloud environment. Attackers often utilize social engineering techniques (see Ghafir et al. [25]) to get the required access to the system or platform.

In the next section, we present the formal modeling approach for quantifying the cyber resilience metrics for CPS using the vulnerability graph. We utilize the model in the design of the simulation platform.

14.5 Modeling CPS Cyber Resilience Metrics

In this section, we present the definitions of cyber resilience from scholarly articles. We offer here a mathematical formulation for deriving quantitative cyber resilience metrics using the vulnerability graph model. Section 14.5.1 presents the related definitions of cyber resilience; Sect. 14.5.2 discusses background information of common vulnerability scoring systems. Section 14.5.3 explains the vulnerability graph model and some preliminary graph theories that we utilize in the simulation platform. Finally, Sect. 14.5.4 proposes the mathematical formulations of cyber resilience metrics using graph properties and critical system functionality.

14.5.1   Cyber Resilience: Definition and Characteristics

The National Academy of Science (NAS) defined resilience as *the ability to prepare and plan for, absorb, recover from, or more successfully adapt to actual or potential adverse events* (Cutter et al. [26]). Bruneau et al. [27] proposed a conceptual framework initially to define seismic resilience, and later, Tierney and Bruneau [28] introduced the R4 framework for disaster resilience. The R4 framework comprises four components. These are *robustness* (the ability of systems to function under degraded performance), *redundancy* (identification of substitute elements that satisfy functional requirements in the event of significant performance degradation), *resourcefulness* (initiate solutions by identifying resources based on prioritization of problems), and *rapidity* (ability to restore functionality in a timely manner).

Several current research works focus on the resilience analysis of CPS and ICS. We mention a few of them here. The National Institute of Standards and Technology (NIST) provides a framework (see Sedgewick [29]) for improving the cybersecurity and resilience of critical infrastructures that are supported by both ITS and ICS. The NIST framework identifies five functions that organize cybersecurity at the highest levels. These are *identify* (develop understanding of and manage risk to systems, assets, data, and capabilities), *protect* (develop and implement appropriate safeguards to ensure delivery of critical infrastructure services), *detect* (identify the occurrence of a cybersecurity event), *respond* (take action regarding a detected cybersecurity event), and *recover* (maintain plans for resilience and to restore any capabilities or services that are impaired due to a cybersecurity event) (for details see Sedgewick [29]).

Stouffer et al. [30] provide detailed guidelines for ICS system security. Haque et al. [31] illustrate the gap in resilience analysis and propose a comprehensive cyber resilience framework to quantify resilience metrics. The proposed framework for assessing the cyber resilience of ICS considers the physical, technical, and organizational domains of cyber operations. Koutsoukos et al. [32] present a modeling and simulation integrated platform for the evaluation of cyber-physical system resilience with an application to the transportation systems. Clark and Zonouz [33] present intrusion resilience metrics for cyber-physical systems. Most of the above works handle resilience from intrusion detection and prevention perspectives. In this chapter, we focus our limit on the quantification of resilience metrics for the CPS and implementation of the CPS simulation platform in the cloud environment.

Resilient CPS Characteristics: Wei and Ji [34] present the resilient industrial control system (RICS) model where the authors have identified the following three vital characteristics of the ICS system to be cyber-resilient:

- Ability to minimize the undesirable consequence of an incident.
- Ability to mitigate most of the undesirable incidents.
- Ability to restore to normal operation within a short time.

All the above characteristics are some form of repetition of the four broad resilience metrics (i.e., robustness, resourcefulness, redundancy, and rapidity), as illustrated by

Tierney and Bruneau [28]. As CPS or ICS work closely with the field devices with having interfaces to the control centers, a wide range of efforts spanning the system level and organizational level are necessary to make the CPS cyber-resilient.

Need for Simulation Platform to Quantify Resilience Metrics: As pointed out before, a lot of research works are going on the development of standard practices and guidelines to make the CPS cyber-resilient, which have a lack of specific quantitative cyber resilience metrics. The availability of quantitative cyber resilience metrics would assist the concerned industry operators in assessing and evaluating the CPS. One of the objectives of this chapter is to derive quantitative resilience metrics. The other aim is to develop a simulation platform to automate the process of quantification. There is a high need for resilience metrics automation across various industries. Thus, an approach to quantify the resilience metrics and development of a simulation platform to automate the metrics generation process is timely research. Also, the inclusion of the modeling and simulation paradigm in the CPS resilience study and leveraging the cloud platform are crucial research aspects to consider.

The next subsections present the required preliminaries for the mathematical formulation of the cyber resilience quantification process. Sections 14.5.2 and 14.5.3 discuss the common vulnerability scoring system (CVSS) and vulnerability graph model that we utilize in the formal modeling in Sect. 14.5.4.

14.5.2 Common Vulnerability Scoring System (CVSS)

We present a brief description of the common vulnerability scoring system (CVSS) here to facilitate the audiences with a smooth migration to the formal modeling approach. CVSS proposed by Mell et al. [35, 36] provides a way to capture the principal characteristics of vulnerability and produce a numerical score reflecting its severity. CVSS attempts to assign severity scores to vulnerabilities, allowing responders to prioritize responses and resources according to the threat. Scores are calculated based on a formula that depends on several metrics that approximate the two factors: ease of exploit and the impact of exploit. Scores range from 0 to 10, with 10 being the most severe. The current version of CVSS (CVSSv3.1) was released in June 2019. The earlier most popular version was CVSSv2.

CVSS scores are composite scores of the following three categories of metrics as described by Mell et al. [36]:

- **Base metrics**: This group represents the static properties of a vulnerability that do not change over time, such as access complexity, access vector, the degree to which the vulnerability compromises the confidentiality, integrity, and availability of the system, and the requirement for authentication to the system.
- **Temporal metrics**: This group measures the properties of a vulnerability that do change over time, such as the existence of an official patch or functional exploit code.

Fig. 14.6 A sample vulnerability graph with arbitrary edge weights. Edge weight represents important quantitative scores of the exploitability of that edge or the impact on the system by exploiting the edge. The number of nodes used in this illustration is ten. The edge weights are within the range of 0~10 to keep similar to CVSS scores

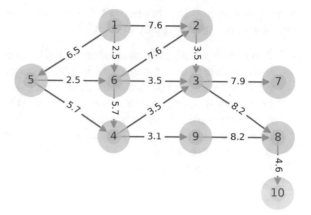

- **Environmental metrics**: This group measures the properties of a vulnerability that are representative of users' IT environments, such as the prevalence of affected systems and potential for loss.

We encourage readers to explore the details of the CVSS scores in the articles by Mell et al. [35, 36].

14.5.3   Vulnerability Graph Model

A vulnerability graph is a directed weighted graph $G = (N, E, W)$ where N is the finite set of nodes (or vertices), $E \subseteq N \times N$ is the set of graph links or edges, and W is the weight matrix of the graph. If an edge $e = (i, j)$ connects two nodes i and j, then the nodes i and j are said to be adjacent to each other. A path in a graph is a walk from a source node to a destination node without repeated nodes. An adjacency matrix A of a graph $G = (N, E, W)$ with $|N| = n$ is an $n \times n$ matrix, where $A_{ij} = W_{ij}$, if $(i, j) \in E$ and $A_{ij} = 0$ otherwise. The weights W_{ij} between the edge (i, j) is coming from the CVSS exploitability score (see Mell et al. [35]) of the node j. That is why we call the graph G a vulnerability graph. Figure 14.6 presents a sample vulnerability graph with arbitrary edge weights.

In the following paragraphs, we discuss different components of the graph model and network structure to provide the readers with the necessary background information about our resilience quantification approach. We frequently refer to SCADA systems for illustration purposes as we formulate the mathematical models by keeping in mind the energy delivery systems (EDS) as an example of CPS. Readers may consider SCADA as a monitoring and control system for the physical field devices. We find that researchers commonly refer to cyber-physical power systems (CPPS) (see Shi et al. [37], Zhang et al. [38]) when it comes to the discussion of energy systems cybersecurity. That is why we take the case of the power systems to illustrate the model that applies equally to other CPS.

(1) **Network Topology and Connectivity**: In a CPS network, the network design follows specific system architecture, and the security policies (e.g., firewall rulesets). In the CPS, as per the NIST guidelines (Stouffer et al. [30]), the message or protocol-level communications among SCADA and field devices are done through ICS firewalls having specific rulesets. We utilize the adjacency matrix as defined above to represent the network topology and connectivity in our vulnerability graph model.

(2) **Control Function**: We define a control function as a logical connection that carries (or, transmit) the data from the field devices to SCADA and control commands from SCADA to the field devices to perform the specific task (such as voltage regulation, phase angle adjustment, etc.). Mathematically, we define a control function $C_f(i, j)$ between node i & j as $\{C_f(i, j) = e(i, j) \mid \exists\, e(i, j) \in E,\ A_{ij} \neq 0\ \&\ W_{ij} > 0\}$, and the importance of the control function $C_f(i, j)$ is determined by its weight W_{ij}. As we utilize the CVSS base scores, this importance indicates the easy of exploit and impact of exploiting the particular control function. We do not consider the degree of operability of the control functions in this model, because that brings a different research question of modeling and incorporating the functional dependencies in the cyber resilience assessment.

(3) **Base, Exploitability, and Impact Metrics**: CVSS (Mell et al. [35]) defines the exploitability and impact metrics for every known vulnerability. The national vulnerability database (NIST [39]) provides the CVSS scores for all the reported (i.e., known) vulnerabilities. The exploitability metric is composed of three base metrics: Access Vector A_V, Access Complexity A_C, and Access Authentication A_U. Similarly, the impact metric is also composed of three base metrics: Confidentiality Impact I_C, Integrity Impact I_I, and Availability Impact I_A. CVSS computes the exploitability E_i and impact I_i of a vulnerability i using below equations.

$$E_i = 20 \times A_V^i \times A_C^i \times A_U^i$$

$$I_i = 10.41 \times (1 - (1 - I_C^i)(1 - I_I^i)(1 - I_A^i))$$

The exploitability score is on a scale of 0–10, and the higher value indicates higher exploit capability by a cyber attacker. Similarly, the higher the impact score, the higher is the possible damage an attacker may cause upon exploiting the vulnerability. To define the base score, CVSS define a impact function as given below:

$$f(I_i) = \begin{cases} 0 & \text{if } I_i = 0 \\ 1.176 & \text{otherwise} \end{cases}$$

Finally, CVSS computes the base score (BS) of vulnerability i using the below equation (see Mell et al. [35]):

$$BS_i = \text{roundTo1Decimal} \left(\left((0.6 \times I_i) + (0.4 \times E_i) - 1.5 \right) \times f(I_i) \right) \quad (14.1)$$

(4) **Multi-Edge to Single Edge Transformation**: In a network, if a node has multiple vulnerabilities, the graph becomes a multi-digraph, where the number of paths from source to the destination increases exponentially and thus creates scalability problems for large networks. To avoid this, we transform the multi-edged directed vulnerability graph to a single-edged directed graph (simple graph) using the composite exploitability score. As the severity of the exploitability and impact are different for different vulnerabilities, we use a severity-based weight approach (see Table 3 of Haque et al. [40] to incorporate the severity level of the vulnerability). The composite exploitability score (ES), impact score (IS), and base score (BS) for node j, having vulnerabilities $i = 1 \sim n$ is defined in Eqs. (14.2), (14.3), and (14.4).

$$ES_j = \frac{\sum_{i=1}^{n} w_i^j \times E_i^j}{\sum_{i=1}^{n} w_i^j} \quad (14.2)$$

$$IS_j = \frac{\sum_{i=1}^{n} w_i^j \times I_i^j}{\sum_{i=1}^{n} w_i^j} \quad (14.3)$$

$$BS_j = \frac{\sum_{i=1}^{n} w_i^j \times BS_i^j}{\sum_{i=1}^{n} w_i^j} \quad (14.4)$$

Here, w_i^j, E_i^j, I_i^j, and BS_i^j are the severity weights, exploitability score, impact score, and base score of vulnerability i of node j. We find BS_i^j from NVD database [39] or using Eq. (14.1), and we compute BS_j using Eq. (14.4) which refers to the composite base score of node j.

(5) **Edge Weight Computation**: The edge weights of the vulnerability graph are coming from the composite base scores as we compute using Eq. (14.4). This way we take into consideration of both the exploitability and impact of a vulnerability in our edge weight. We define the weight matrix as below.

$$W_{ij} = \begin{cases} BS_j & \text{if } (i, j) \in E \\ 0 & \text{otherwise, i.e., if} (i, j) \notin E \end{cases}$$

(6) **Betweenness Centrality (BC)**: BC quantifies the number of times a node acts as a bridge along the shortest path between two other nodes. BC is a crucial graph-theoretic metric that indicates the possible criticality of a node, i.e., the possibility of attack progression through a node. The betweenness centrality of a node n, B_n is the fraction of the shortest paths going through n and is given by

$$B_n = \sum_{s \neq n \neq t} \frac{\sigma_{st}(n)}{\sigma_{st}} \tag{14.5}$$

where σ_{st} is the total number of shortest paths from node s to node t, and $\sigma_{st}(n)$ is the number of those paths that pass-through node n.

(7) **Katz Centrality (KC)**: KC measures the number of all nodes that are connected through a path, while we penalize the contributions of distant nodes. Haque et al. [40] (for details, please see Chap. 4, Sect. 4.2 of Haque [41]) define the asset value of a node by the importance of the information contained by the network component, which is also dependent on the predecessor nodes' importance. Thus, the asset value that is addressed by Haque et al. [40] to rank critical nodes is suitable to formalize by KC. Mathematically, the KC of node i is defined in Eq. (14.6), where α is an attenuation factor and $0 \leq \alpha \leq 1$.

$$C_{Katz}(i) = \sum_{k=1}^{\infty} \sum_{j=1}^{n} \alpha^k (A^k)_{ji} \tag{14.6}$$

14.5.4 Resilience Metrics Formulation

In this section, we present the mathematical formulations of critical system functionality, cyber resilience, and network criticality metrics by following a top-down approach, as shown by Haque et al. [42]. We utilize the critical system functionality to derive the cyber resilience metrics, and network criticality to formulate the critical functionality.

14.5.4.1 Critical System Functionality (CSF)

Arghandeh et al. [43] illustrate resilience as a multi-dimensional property of the system, which requires managing disturbances originating from physical component failures, cyber component malfunctions, and cyberattacks. The authors also describe critical system functionality (CSF) as maintaining the essential functionality of the system in the presence of unexpected extreme disturbances. Bharali and Baruah [44] define the average network functionality using the network criticality metric and considering random failures. We extend the analysis of Bharali and Baruah [44] for the case of random cyber-attacks on the CPS. We consider the removal of an edge in the vulnerability graph as making a control function or service unavailable by removing the logical connection. Here we treat the average network functionality metric as the CSF, which is the level of functionality maintained by the CPS in case of an adverse incident (i.e., after deactivation of some control functions).

Let us consider G be the original graph, and $G \backslash e$ be the graph obtained by removing the edge e, then τ and τ_e be the network criticality of G and $G \backslash e$. Then we define the critical system functionality by Eq. (14.7).

$$\eta = 1 - \frac{1}{m} \sum_{e \in E} \left[H^+(\tau_e - \tau) \frac{\tau}{\tau_e} + H^-(\tau_e - \tau) \frac{\tau}{\tau_e + \frac{2n}{\mu}} \right] \qquad (14.7)$$

where m is the number of edges in G, μ is the smallest non-zero eigenvalue of G, $H^+(x) = 1$ if $x \geq 0$ and 0 otherwise, and $H^-(x) = 1$ if $x < 0$ and 0 otherwise. For a connected graph G, $\mu = \mu_1$ which is the algebraic connectivity of G and $0 \leq \eta \leq 1$. Thus, η indicates the system functionality of the CPS under adverse cyber events, and a higher value of η means a higher degree of system functionality is maintained. We discuss the computation process of the network criticality τ in subsection 14.5.4.3.

14.5.4.2   Cyber Resilience Metric

Roberson et al. [45] define bulk power system resilience as the safeguarding of the critical system functionality when subject to perturbations and restoration after outages. We estimate the cyber resilience for the CPS by utilizing the system performance curve, as given in Fig. 14.7 and using critical system functionality. The nature of the recovery behavior of a system during an adverse event is typically non-linear and is a function of the system under consideration (S), duration of recovery (T), recovery rate (r), time (t), and the functionality level (η) maintained. Using the functional notation, we can express the recovery behavior as $Q_r(t) = f(S, \eta, r, T, t)$. Zobel [46] addresses the recovery behavior and proposes several functional forms to model the recovery over time. In this work, we utilize the inverted exponential functional form of the recovery curve from Zobel [46], which seems suitable to model the resilience for the CPS by addressing the non-linearity. We model the time-dependent system recovery behavior $Q_r(t)$ by following the Eq. (6) of Zobel [46] to demonstrate quantitative resilience metric under adverse events where loss of performance $= 1 - \eta$ and $0 \leq \eta \leq 1$.

$$Q_r(t) = (1 - \eta) \left(1 - e^{\left(-\frac{\left(T - (t - t_i^{ri}) \right) ln(n)}{T} \right)} + \frac{\left(T - (t - t_i^{ri}) \right)}{nT} \right) \qquad (14.8)$$

Here, t_i^{ri} = time of recovery initiate, and t_i^{cr} = time of complete recovery of system functions for attack incident i. The period of recovery is $T = t_i^{cr} - t_i^{ri}$. The parameter n defines the level of concavity inherent in the inverted exponential curve.

In Fig. 14.7, T^* is the system-specific maximum allowable time for the recovery to occur which is selected by the decision-makers as the acceptable time within which the system must recover to be considered as operational (i.e., not abandoned). The area formed by the points e-a-d is the amount of system functionality losses over time due to the cyber attack incident i. The area enclosed by the points a-b-c'-d' is the area of system resilience. To compute the resilience metrics, we first compute the area enclosed by the points e-a-d, which is as follows:

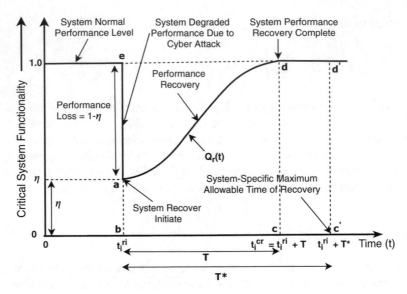

Fig. 14.7 System performance curve during a cyber-attack incident i on the CPS. We use the graph from Haque et al. [42], which is a modified form of the resilience graph as presented by Wei and Ji [34]

$$A_{e-a-d} = (1-\eta) \int_{t_i^{ri}}^{t_i^{ri}+T} \left(1 - e^{\left(-\frac{\left(T-(t-t_i^{ri})\right)ln(n)}{T}\right)} + \frac{\left(T-(t-t_i^{ri})\right)}{nT}\right)dt$$

Simplifying the above equation, we find the following reduced form as in Eq. (14.9).

$$A_{e-a-d} = (1-\eta)T\left[1 - \frac{n-1}{nln(n)} + \frac{1}{2n}\right] \tag{14.9}$$

From Fig. 14.7, we find the area of e-b-c'-d' is $1 * T^* = T^*$ and the area of e-a-d is defined by Eq. (14.9). Thus, the cyber resilience of the CPS system is the area under the curve enclosed by points a-b-c'-d' over period T^* as given in Eq. (14.10).

$$\xi = \frac{1}{T^*}\left[T^* - (1-\eta)T\left(1 - \frac{n-1}{nln(n)} + \frac{1}{2n}\right)\right] \tag{14.10}$$

The term $\left(1 - \frac{n-1}{nln(n)} + \frac{1}{2n}\right)$ is a constant term for specific n, and is denoted by β. Thus, Eq. (14.10) becomes $\xi = \frac{1}{T^*}\left[T^* - (1-\eta)T\beta\right]$.

14.5.4.3   Network Criticality

Bharali and Baruah [44] and Tizghadam and Leon-Garcia [47] proposed graph-based network criticality metric. We apply the same metric here to measure the criticality of the overall CPS OT network. We use the Moore–Penrose inverse of the Laplacian matrix L to compute the network criticality τ. As we are using the directed weighted graph, the Laplacian matrix L is defined as per Chung [48] as below where P is the transition matrix of the graph, Φ is a matrix with the Perron vector of P in the diagonal and zeros elsewhere.

$$L = I - \left(\Phi^{\frac{1}{2}} P \Phi^{\frac{-1}{2}} + \Phi^{\frac{-1}{2}} P^T \Phi^{\frac{1}{2}} \right) / 2 \tag{14.11}$$

Another way to derive L is by using the normalized graph Laplacians L_{sym} and random walk Laplacian L_{rw}, as below.

$$L_{sym} = D^{\frac{-1}{2}} L D^{\frac{-1}{2}} = I - D^{\frac{-1}{2}} W D^{\frac{-1}{2}}$$

$$L_{rw} = D^{\frac{-1}{2}} L_{sym} D^{\frac{1}{2}}$$

where D is a diagonal matrix formed by the degree of the nodes in the vulnerability graph and defined as $D = diag(d_1, d_2, ..., d_m)$. Here $d_i = \sum_{j=1}^{m} W_{ij}$. The Moore–Penrose inverse of the Laplacian matrix (L) L^+ as computed by Bernstein [49] is given in Eq. (14.12).

$$L^+ = \left(L + \frac{J}{n} \right)^{-1} - \frac{J}{n} \tag{14.12}$$

where J is an $n \times n$ matrix whose entries are all equal to 1. The network criticality τ is defined by Eq. (14.13).

$$\tau = 2n * trace(L^+) \tag{14.13}$$

Here, n is the number of nodes, L^+ is the Moore–Penrose inverse of the Laplacian matrix L, and $trace(L^+) = \sum_{i=1}^{n} (L^+)_{ii}$. The larger the value of τ indicates the more vulnerable is the network from the exploitability perspective. The normalized network criticality is found by Eq. (14.14).

$$\hat{\tau} = \frac{2 * trace(L^+)}{n(n-1)} \tag{14.14}$$

We utilize the above vulnerability graph-based resilience metric derivation methodologies in the analytical engine of our simulation platform, as we discuss in Sect. 14.6. There are other mathematical analyses for critical devices, attack paths, and links identification, which are available in Haque et al. [42]. We omit those here to restrict our focus only on the resilience metrics.

14.6 Cloud-Based Simulation Platform

In this section, we present the proposed simulation platform for the CPS resilience
assessment in the cloud environment. Section 14.6.1 presents the architecture of the
tool. Section 14.6.2 presents the deployment plans for the simulation tool in the cloud
platform. The proposed simulation platform is still under development stage. That is
why we present here another simulation platform based on a qualitative assessment,
which we have already deployed in Amazon AWS using an EC2 instance, as we
illustrate in Sect. 14.6.3.

14.6.1 Simulation Platform Architecture

As we have explained before, one of our aims in this chapter is to present a quantitative simulation platform for cyber resilience assessment for the CPS. Figure 14.8
offers a very high-level architecture of the tool that we propose in this chapter. The
deployment model may differ based on the user's preferences to use private or public
clouds. We prefer to utilize the public cloud instances to implement the simulation tool because of reduced cost and dynamic configuration management as well
as flexible handling of scalability issues. For example, we can utilize the Amazon
Elastic Cloud Computing (EC2) instances for deploying the tool. It is also possible
to implement the simulation platform in the private cloud using VMs. We present
a brief discussion on the major functional blocks of the simulation platform in the
following paragraphs.

Interface Management System (IMS): The IMS of the quantitative simulation
platform has two interfaces: the admin interface and the user interface. The admin
interface can make changes in the simulation platform (e.g., the creation of new user
accounts, modification in the input systems, etc.). The user interface is not allowed
to create new users, but it can make changes in the network topology as required. The
network topology is to set up the network architecture for the simulation. The users
can directly build the network manually by specifying the network elements and their
configurations at different layers of the defense-in-depth architecture. Also, we will
keep provisions to use the network scanning results (found from tools like Nessus,
OpenVAS, etc.) as an input to the system. The simulation engine will produce a graph-
theoretic network based on the user inputs of the network components, software, and
other applications as specified.

Database Management System (DMS): DMS is the local repository of the commonly known vulnerabilities. It downloads the vulnerability information provided
by the National Vulnerability Database (NVD) [39]. We use a local database to store
the data. The local database contains necessary vulnerability information and their
CVSS exploitability and impact metrics, as illustrated by Mell et al. [35]. We use

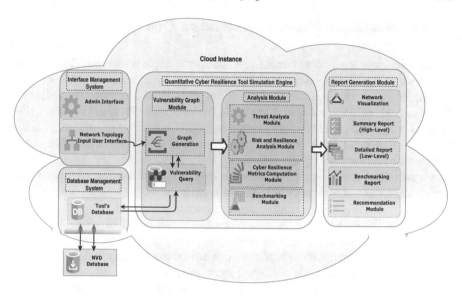

Fig. 14.8 Cloud-based CPS resilience assessment simulation platform architecture

these metrics to compute the edge weights in the graph-theoretical models. The only component that requires communication outside the cloud is the local database and the NVD database to keep the vulnerability information synchronized.

Quantitative Simulation Engine: There are two major parts in the simulation engine. One is the vulnerability graph generation module, and the other is the analytical module. The vulnerability graph module takes the network topology as an input from the user, extracts the corresponding vulnerabilities for the network nodes by communicating with the vulnerability repository in DMS. It then generates the graph model for the simulated network. The simulation engine can only handle the known vulnerabilities, and it cannot control the zero-day vulnerabilities (Levy [50]) yet. The analysis module analyzes the threats, computes the risk and resilience metrics, and performs the benchmarking for the resilience assessment of the CPS. We utilize the mathematical models presented in Sect. 14.5 to compute the quantitative resilience metrics. The tool would be able to provide critical network elements, easily exploitable paths, and other relevant security metrics.

Report Generation Module: The report generation module provides the network visualization, high-level summary report, a detailed low-level report, and benchmarking reports. The tool would be able to provide cost-effective remediation strategies and essential recommendations for resilience improvement.

14.6.2 Simulation Platform Deployment Plan

In the following, we discuss our plan regarding the deployment model and implementation of the simulation platform in the cloud environment in brief.

Implementation: To implement the above simulation tool, we may utilize the Amazon AWS EC2 instance using an elastic IP. An Elastic IP address is a static IPv4 address designed for dynamic cloud computing. An Elastic IP address is an associated user's AWS account. With an Elastic IP address, it is easy to mask the failure of an instance or software by rapidly remapping the address to another instance in the user's account. As we want to provide the simulation platform to different utility companies to assess their network risk and resilience, we prefer the multi-instance model for scalability and security.

Scalability: For handling the scalability issue, we prefer to utilize the multi-instance model as we describe in the next topic. Although, from the cost perspective, the multi-tenancy model is the best if we can ensure the security and privacy of the data from multiple users, we prefer the multi-instance model. The reason behind choosing the multi-instance model is that different types of utilities would use the simulation platform, and it would be needed to use as many times as required by the client. The multi-instance model would give dedicated instances per client and thus ensure enhanced security for the network and other configuration related data.

Multi-tenancy Versus Multi-instance Model: The multi-tenancy commingles the data and processing for multiple clients in a single application instance. In contrast, a multi-instance architecture uses one application instance per client. With the multi-instance model, organizations need to allocate their time to efficiently create and manage multiple application instances. On the other hand, organizations that opt for multi-tenancy often need to invest in application code to prevent the exposure of data from one client to another. The multi-instance model maintains the privacy and confidentiality of data more than the multi-tenancy model. Thus, we prefer to utilize the multi-instance model for the final deployment of our simulation platform for multiple users.

14.6.3 Use Case: An AWS-Based Qualitative Simulation Platform for Resilience Assessment

The quantitative simulation platform that we present in this chapter is still under development. But we have already implemented a qualitative simulation platform for the cyber resilience assessment for the CPS, which is similar to the proposed quantitative simulation platform in terms of resilience assessment. The qualitative approach is complementary to the quantitative approach in the sense that it also provides meaningful insights into the resilience assessment by using the subject matter experts' judgment. The details of the qualitative simulation platform and mathematical modeling are available in [51]. Here we present a use case to assess the cyber resilience metrics for EDS. We also give the deployment details and a few simulation results generated from the tool.

Use Case Description: We have presented a generic CPS architecture in Sect. 14.2 using the defense-in-depth security strategy, which includes the cyber, control, and physical layers. In this section, we assume a similar architecture for the EDS system. We compute the cyber resilience metrics by using the following system properties: robustness, redundancy, resourcefulness, and rapidity. Each of those sub-areas subdivides itself into three domains: physical, organizational, and technical. We have used a set of questionnaires to assess the resilience metric and its underlying sub metrics.

For this use case, we want to generate cyber resilience metrics for an EDS by performing a survey and utilizing the tool that we have deployed in the cloud platform. We employ a qualitative assessment method, as illustrated by Haque et al. [51]. The idea of the qualitative assessment is to perform a comprehensive survey on the system's overall security. We expect that the system administrators and operators working closely in the IT and OT domains would participate in the research survey. For this use case, we randomly select fifteen system-level experts working in the EDS sector to participate in the study. We have sent them the survey request to their email addresses through the qualitative simulation tool that we have deployed in the AWS cloud. We use the simulation platform to collect the responses. The simulation platform then processes the responses and generates the aggregated resilience metrics based on the users' responses. The tool also creates different reports, which contains detailed resilience metrics as well as recommendations for improving the overall resilience posture.

Deployment: Figure 14.9 shows an overview of the tool deployment in the Amazon AWS environment. The web application has two interfaces: admin and user interfaces. Admin interface can create new survey requests, release survey, and view or download the comprehensive reports from the web application. The users can not create a new survey. Users can only participate in the study by using their respective login details only if they receive a survey request from the survey system of the tool. The tool utilizes MongoDB databases to store the responses. It uses several python APIs for mathematical computation and utilizes several supporting JavaScript libraries for rendering the simulation results in graph format. For the application server, we use t2.xlarge type instance with four vCPUs (virtual CPU) and 16GB memory. For the debug server, we have used one t2.large type instance with two vCPUs and 8GB memory. For secure communication, we use the Hypertext Transfer Protocol Secure (https) to ensure the security of the communications from user responses to database storage.

Simulation Results: We present two simulation results from the qualitative simulation platform, as presented in Figs. 14.10 and 14.11. Figure 14.10 presents the broad domain metrics of cyber resilience based on the assessment performed using the tool. The scores here are on a Likert scale from 1.0~5, with 5.0 being the highest performance and 1 be the lowest performance. We can see that based on the automatic assessment, the EDS system is performing well in physical and organizational robustness with score 5.0, and slight underperformance in technical robustness cri-

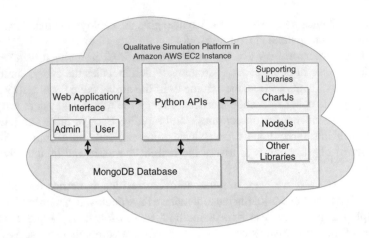

Fig. 14.9 Cloud-based qualitative simulation platform for CPS resilience assessment. We utilize Amazon AWS EC2 instance for the deployment of the qualitative tool

teria. Again the illustrations and definitions of those metrics can be found in Haque et al. [51]. Figure 14.11 shows the detailed underlying metrics that constitute the upper layer metrics in the resilience assessment. By utilizing the simulation results generated from the tool, the network administrator can analyze the overall resilience posture and find out the areas that need improvements or actions. The tool also presents automatic reports and provide necessary recommendations. We consult the standards and industry best practices to derive the recommendations.

14.7 Challenges of Cloud-Based CPS Simulation Platform and Way Forward

In this chapter, we are proposing a simulation platform to quantify cyber resilience that we could deploy and operate in the cloud environment. We are not focusing on putting part of CPS operations in the cloud. We are also not covering the deployment of the IDS/IPS system for CPS in the cloud. There are a lot of research works already available in that specific area. Thus the challenges we discuss in this section are limited to only the complexities that we may face to model, develop, and set up a simulation platform for quantifying security and resilience analytics in the cloud environment.

Complexities in Capturing CPS: CPS is an active research field with numerous complexities and challenges from the cybersecurity perspective. The complications arise because of the complex nature of the system design, components heterogeneity, critical interconnections, lack of overall visibility, and the tradeoff between security and reliability of physical processes, etc. as illustrated by Haque et al. [15] and Mittal et al. [52]. When it comes to deriving security and resilience analytics for CPS, we

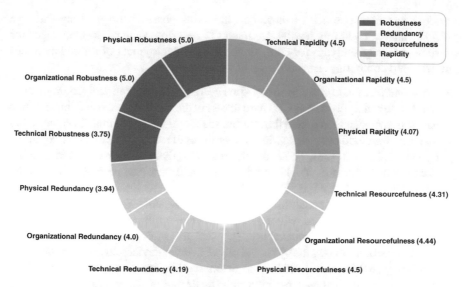

Fig. 14.10 A sample simulation result generated from the qualitative simulation tool illustrating the broad resilience domain metrics. The color legends on the top right corner represent the four broad resilience metrics: robustness, redundancy, resourcefulness, and rapidity

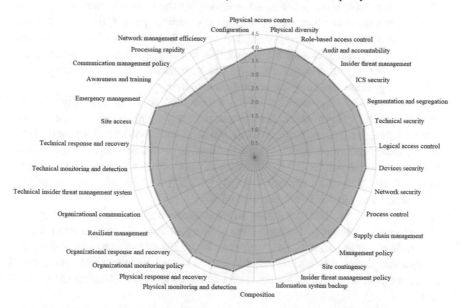

Fig. 14.11 A sample simulation result generated from the qualitative simulation tool illustrating the detailed underlying metrics which contribute to the overall cyber resilience metric

need to consider as much of those constraints as possible to get comprehensive security analytics. In this section, we intend to describe some of the challenges for deriving security analytics for CPS, along with the deployment of a simulation tool in the cloud computing platform.

- *Real-timeliness nature*: CPS are real-time systems. The operational requirement of real-time availability makes CPS hard to assess the security threat and implement a preventive mechanism in real time within the tolerable delay limit. Cryptographic mechanisms could cause delays in the operations of real-time devices, as explained by Humayed et al. [53]. To handle the real-time dynamism, we plan to create a socket communication which would take the real-time scanning results from the vulnerability scanners (i.e., Nessus, OpenVAS, etc.) as input. Keeping the other processes remain the same; the tool would then be able to capture the real-time dynamism if it can be fed with the real-time vulnerability details.
- *Heterogeneity and interconnectivity*: The CPS are heterogeneous having complex interconnections among the cyber and physical layer devices. The proprietary protocols (e.g., Modbus and DNP3 in ICS or smart grids) are not free of vulnerabilities due to the isolation consideration during the design of those protocols. It is hard to capture every interconnection in a simulation platform. We plan to provide a detailed network design as an input in the tool to obtain the interconnections. There would have been provisions for setting up the network manually with the connections using boolean logic (1-means connection present, 0-means no direct relationship between two components). Again, with thousands of devices and tens of thousands of links, it is nearly impossible to capture everything within a simulation environment. A way forward would be to assess part of the network at a time and continue repeating the process as required.
- *Underlying physical processes*: CPS are designed to operate jointly with the embedded physical processes. The complex underlying processes reduce the visibility of the overall security of the CPS, and thus, we should consider the underlying processes and dependencies in the resilient network topology design process. To capture the dependence of different business processes, we plan to utilize a more robust form of vulnerability graph, which would include the dependency relationships in our future analytical model.
- *Secure integration*: The integration of new components with the existing CPS should perform security testing before putting online. There is a need for the vulnerability assessment of the devices to be added to the current systems. The simulation platform would be helpful in this regard to assess the security and resilience posture by including the components under test in the simulation mode only.
- *Understanding the consequences of an attack*: Often, it is difficult to visualize the effects of an attack in the ICS or CPS. It is essential to perform penetration tests and assess the impacts of a cyberattack on the CPS. Also, the prediction of the attacker strategy is crucial in defending the CPS network. The simulation platform would be beneficial in this regard by pointing to the most critical attack paths and

vulnerable devices. Also, it is possible to identify the control functions, which, upon exploitation, could create the highest impact on the overall system.

Complexities in Cloud Deployment: Cloud computing is full of potentials and offers several benefits to the users. Among the benefits that the cloud services provide are reduced cost for IT resources, scalability, flexible deployment, and access to up to date systems with automatic updates, etc. But like many other systems, the cloud also suffers from cybersecurity risks and, thus, need to adopt stringent measures for the security and privacy of the user data. In this subsection, we focus on some of the challenges in deploying a simulation tool for the CPS security analysis in the cloud environment.

- *Governance and compliance*: Governance implies control and oversight by the organization over policies and standards for application development and information technology service acquisition. The authority should also span to the design, implementation, testing, use, and monitoring of deployed or engaged services. On the other hand, compliance refers to the responsibility of the organization to operate in agreement with established laws, regulations, standards, and specifications. Most of the compliance issues fall on the cloud providers' side. But when we need to deploy our simulation platform in the cloud, part of the responsibility falls on our part as well. We need to make sure that we develop the tools following the require compliance and standard practices. Often there could arise a potential conflict on the governance and compliance issues if a well-articulated service level agreement (SLA) between the cloud provider and user is absent.
- *Trust management*: In cloud computing, cloud users have to keep trust in the security and protection provided by the cloud provider because the users do not have direct control over the security measures deployed to the infrastructure or applications. Thus, there could be insider access to sensitive data and network information. Also, data ownership management is a big concern when utilizing cloud-based simulation platforms. Another matter of trust is composite services. Sometimes, cloud services use nesting and layering with other cloud services. For example, a SaaS provider could build its services upon those of an IaaS cloud. Thus maintaining trust and security throughout the entire service platform is a significant concern.
- *Data security and privacy*: In the cloud, the application or infrastructure spans in a diverse geographical location which may create issues in data security. Also, sharing the shared environment or utilizing a multi-tenancy model may risk the security and privacy of the data. Complexities in hypervisors or virtual machine monitors may make it complex to protect data security. While deploying a simulation platform, there are also concerns because the tool would be fed with network configuration related data to provide security and resilience analytics as accurately as possible.
- *Tools architecture*: The deployed tool's architecture itself may create potential attack vectors for cyber intruders. If there are additional application programming interfaces or sockets, it may facilitate an attacker to cause damage to the system by

utilizing those APIs or sockets. Handling security in virtual machine environments can also be challenging because of the complexity.

To handle some of the challenges discussed above, we utilize a separate database system that would handle user access control and identification. The network configuration details would only be used for the analytical engine to compute the metrics. It would not store the information permanently. Once the administrator generates the assessment report, he/she could remove the network configuration or scanning files permanently if there is no plan to reuse the same configuration. As it would be a web-based application, it is possible to disable the cookies as well. And, as we plan to utilize the multi-instance model for deploying for multiple clients, we safeguard the risk and security concerns of a shared environment.

14.8 Conclusion

Cyber-physical systems play a vital role in critical infrastructures, industrial control systems, and many other applications. The complex nature of CPS, in conjunction with the lack of clear visibility due to the integration of different cyber, cyber-physical, and physical components, make it challenging to handle the security issues. To address the gap in resilience assessment and automation, in this chapter, we present a mathematical model for deriving the cyber resilience metrics for the CPS and propose a cloud-based simulation platform to help in the automation of cyber resilience analytics.

We provide a detailed discussion on the CPS and cloud threats and security concerns to facilitate the understandings of these two domains. It would help to leverage the cloud services for the potential deployment of automation and simulation tools. Our proposed analytical model for resilience assessment utilizing the critical system functionality would provide some decisive insights to the CPS operator. We address details of the proposed simulation platform architecture and deployment plan in the cloud environment. We discuss the complexities and ways forward for implementing such simulation tools in the cloud. Overall, we offer a comprehensive analysis of the CPS resilience analytics formulation and automation using the established M&S methodologies and leveraging the cloud service opportunities. Thus, the chapter would provide the readers with an in-depth idea about CPS resilience, M&S challenges, simulation platforms for security assessment, and use of cloud services for deployment of security assessment tools.

Acknowledgements This material is based upon work supported by the Department of Energy under Award Number DE-OE0000780.

Disclaimer This report was prepared as an account of work sponsored by an agency of the United States Government. Neither the United States Government nor any agency thereof, nor any of their employees makes any warranty, express or implied, or assumes any legal liability or responsibility for the accuracy, completeness, or usefulness of any information, apparatus, product, or process disclosed, or represents that its use would not infringe privately owned rights. Reference herein to

References

1. Kim JH (2017) A review of cyber-physical system research relevant to the emerging it trends: industry 4.0, IoT, big data, and cloud computing. J Ind Integr Manag 2(3):1750011
2. Mell P, Grance T et al (2011) The NIST definition of cloud computing
3. Griffor ER, Greer C, Wollman DA, Burns MJ (2017) Framework for cyber-physical systems: volume 1, overview. Technical report
4. DHS CSSP (2009) Recommended practice: improving industrial control systems cybersecurity with defense-in-depth strategies. US-CERT Defense in Depth
5. Macaulay T, Singer BL (2016) Cybersecurity for industrial control systems: SCADA, DCS, PLC, HMI, and SIS. Auerbach Publications
6. Kopetz H, Bondavalli A, Brancati F, Frömel B, Höftberger O, Iacob S (2016) Emergence in cyber-physical systems-of-systems (CPSOSS). Cyber-physical systems of systems, pp 73–96. Springer
7. Greer C, Burns M, Wollman D, Griffor E (2019) Cyber-physical systems & the internet of things. NIST Special Publ 1900:202
8. Ma H-D (2011) Internet of things: objectives and scientific challenges. J Comput Sci Technol 26(6):919–924
9. Hermann M, Pentek T, Otto B (2016) Design principles for industrie 4.0 scenarios. In: 2016 49th Hawaii international conference on system sciences (HICSS), pp 3928–3937. IEEE
10. Mittal S, Zeigler BP, Tolk A, Ören T (2017) Theory and practice of M&S in cyber environments. The profession of modeling and simulation: discipline, ethics, education, vocation, societies and economics
11. Basnight Z, Butts J, Lopez Jr J, Dube T (2013) Firmware modification attacks on programmable logic controllers. Int J Crit Infrastruct Prot 6(2):76–84
12. Ginter A (2017) The top 20 cyber attacks against industrial control systems. White Paper, Waterfall Security Solutions
13. Cardenas A, Amin S, Sinopoli B, Giani A, Perrig A, Sastry S et al (2009) Challenges for securing cyber physical systems. In: Workshop on future directions in cyber-physical systems security, vol 5
14. Chen T, Abu-Nimeh S (2011) Lessons from stuxnet. Computer 44(4):91–93
15. Haque MdA, Shetty S, Krishnappa B (2019) Cyber-physical system resilience. Complexity challenges in cyber physical systems: using modeling and simulation (M&S) to support intelligence, adaptation and autonomy
16. Laing C (2012) Securing critical infrastructures and critical control systems: approaches for threat protection: approaches for threat protection. IGI Global
17. Zissis D, Lekkas D (2012) Addressing cloud computing security issues. Future Gen Comput Syst 28(3):583–592
18. Roschke S, Cheng F, Meinel C (2009) Intrusion detection in the cloud. In: 2009 eighth IEEE international conference on dependable, autonomic and secure computing, pp 729–734. IEEE
19. Modi C, Patel D, Borisaniya B, Patel H, Patel A, Rajarajan M (2013) A survey of intrusion detection techniques in cloud. J Network Comput Appl 36(1):42–57
20. Sindhiya V, Navaneetha Krishnan M, Ravi R (2016) Analyzing and improving the security of cryptographic algorithm against side channel attack
21. Rutkowska J. Subverting VistaTM kernel for fun and profit. In: Black hat briefings

22. Bahram S, Jiang X, Wang Z, Grace M, Li J, Srinivasan D, Rhee J, Xu D (2010) DKSM: subverting virtual machine introspection for fun and profit. In: 2010 29th IEEE symposium on reliable distributed systems, pp 82–91. IEEE
23. Gajek S, Jensen M, Liao L, Schwenk J (2009) Analysis of signature wrapping attacks and countermeasures. In: 2009 IEEE international conference on web services, pp 575–582. IEEE
24. Keleta Y, Eloff JHP, Venter HS (2005) Proposing a secure XACML architecture ensuring privacy and trust. Research in progress paper, University of Pretoria
25. Ghafir I, Prenosil V, Alhejailan A, Hammoudeh M (2016) Social engineering attack strategies and defence approaches. In: 2016 IEEE 4th international conference on future internet of things and cloud (FiCloud), pp 145–149. IEEE
26. Cutter SL, Ahearn JA, Amadei B, Crawford P, Eide EA, Galloway GE, Goodchild MF, Kunreuther HC, Li-Vollmer M, Schoch-Spana M et al (2013) Disaster resilience: a national imperative. Environ Sci Policy Sustain Dev 55(2):25–29
27. Bruneau M, Chang SE, Eguchi RT, Lee GC, O'Rourke TD, Reinhorn AM, Shinozuka M, Tierney K, Wallace WA, Von Winterfeldt D (2003) A framework to quantitatively assess and enhance the seismic resilience of communities. Earthq Spectra 19(4):733–752
28. Tierney K, Bruneau M (2007) Conceptualizing and measuring resilience: a key to disaster loss reduction. TR News (250)
29. Sedgewick A (2014) Framework for improving critical infrastructure cybersecurity, version 1.0. Technical report
30. Stouffer K, Falco J, Scarfone K (2011) Guide to industrial control systems (ICS) security. NIST Special Publ 800(82):16
31. Haque MdA, De Teyou GK, Shetty S, Krishnappa B (2018) Cyber resilience framework for industrial control systems: concepts, metrics, and insights. In: 2018 IEEE international conference on intelligence and security informatics (ISI), pp 25–30. IEEE
32. Koutsoukos X, Karsai G, Laszka A, Neema H, Potteiger B, Volgyesi P, Vorobeychik Y, Sztipanovits J (2017) Sure: a modeling and simulation integration platform for evaluation of secure and resilient cyber-physical systems. Proc IEEE 106(1):93–112
33. Clark A, Zonouz S (2017) Cyber-physical resilience: definition and assessment metric. IEEE Trans Smart Grid 10(2):1671–1684
34. Wei D, Ji K (2010) Resilient industrial control system (RICS): concepts, formulation, metrics, and insights. In: 2010 3rd international symposium on resilient control systems, pp 15–22. IEEE
35. Mell P, Scarfone K, Romanosky S (2007) A complete guide to the common vulnerability scoring system version 2.0. In: Published by FIRST-forum of incident response and security teams, vol 1, p 23
36. Mell P, Scarfone K, Romanosky S (2006) Common vulnerability scoring system. IEEE Secur Priv 4(6):85–89
37. Shi L, Dai Q, Ni Y (2018) Cyber-physical interactions in power systems: a review of models, methods, and applications. Electric Power Syst Res 163:396–412
38. Zhang T, Wang Y, Liang X, Zhuang Z, Xu W (2017) Cyber attacks in cyber-physical power systems: a case study with GPRS-based SCADA systems. In: 2017 29th Chinese control and decision conference (CCDC), pp 6847–6852. IEEE
39. NIST (2020) National vulnerability database. https://nvd.nist.gov/vuln/data-feeds. Accessed 14 Jan 2020
40. Haque MdA, Shetty S, Kamdem G (2018) Improving bulk power system resilience by ranking critical nodes in the vulnerability graph. In: Proceedings of the annual simulation symposium, p 8. Society for Computer Simulation International
41. Haque MdA (2018) Analysis of bulk power system resilience using vulnerability graph
42. Haque MdA, Shetty S, Krishnappa B (2019) Modeling cyber resilience for energy delivery systems using critical system functionality. In: 2019 resilience week (RWS), pp 33–41. IEEE
43. Arghandeh R, Von Meier A, Mehrmanesh L, Mili L (2016) On the definition of cyber-physical resilience in power systems. Renew Sustain Energy Rev 58:1060–1069

44. Bharali A, Baruah D (2019) On network criticality in robustness analysis of a network structure. Malaya J Mat (MJM) 7(2):223–229
45. Roberson D, Kim HC, Chen B, Page C, Nuqui R, Valdes A, Macwan R, Johnson BK (2019) Improving grid resilience using high-voltage DC: strengthening the security of power system stability. IEEE Power Energy Mag 17(3):38–47
46. Zobel CW (2014) Quantitatively representing nonlinear disaster recovery. Decis Sci 45(6):1053–1082
47. Tizghadam A, Leon-Garcia A (2008) On robust traffic engineering in transport networks. In: IEEE GLOBECOM 2008-2008 IEEE global telecommunications conference, pp 1–6. IEEE
48. Chung F (2005) Laplacians and the cheeger inequality for directed graphs. Ann Comb 9(1):1–19
49. Bernstein DS (2018) Scalar, vector, and matrix mathematics: theory, facts, and formulas-revised and expanded edition. Princeton University Press
50. Levy E (2004) Approaching zero. IEEE Secur Priv 4:65–66
51. Haque MdA, Shetty S, Krishnappa B (2019) ICS-CRAT: a cyber resilience assessment tool for industrial control systems. In: 2019 IEEE 6th international conference on big data security on cloud (BigDataSecurity), IEEE international conference on high performance and smart computing, (HPSC) and IEEE international conference on intelligent data and security (IDS), pp 273–281. IEEE
52. Mittal S, Cane SA, Schmidt RBC (2019) Taming complexity and risk in internet of things (IoT) ecosystem using system entity structure (SES) modeling. Complexity challenges in cyber physical systems: using modeling and simulation (M&S) to support intelligence, adaptation and autonomy
53. Humayed A, Lin J, Li F, Luo B (2017) Cyber-physical systems security–a survey. IEEE Internet Things J 4(6):1802–1831

Md. Ariful Haque is a Ph.D. candidate in the Department of Computational Modeling and Simulation Engineering at Old Dominion University. He is currently working as a graduate research assistant in the Virginia Modeling Analysis and Simulation Center. Mr. Haque has earned Master of Science (M.S.) in Modeling Simulation and Visualization Engineering from Old Dominion University (ODU) in 2018. He has received Master of Business Administration (M.B.A) degree from the Institute of Business Administration (IBA), University of Dhaka in 2016. He holds a B.S. degree in Electrical and Electronic Engineering from Bangladesh University of Engineering and Technology in 2006. Before joining Old Dominion University as a graduate student, he has worked in the Telecommunication industry for around seven years. His research interests include but not limited to cyber-physical system security, cloud computing, machine learning, and Big data analytics. Mr. Haque has authored couple of ACM/IEEE conference papers and a book chapter in Wiley.

Sarada Prasad Gochhayat received M.Tech. (in Signal Processing) from the IIT Guwahati, India and Ph.D. (in Communication Networking) from the IISc, Bangalore, India. Currently, he is working as a PostDoc Fellow in the Virginia Modeling, Analysis and Simulation Center, ODU, USA. Where, he is working on developing the theoretical foundation for the Blockchain simulator. Before joining ODU, he was an assistant professor at Manipal University Jaipur and a postdoctoral research fellow at the Department of Mathematics, University of Padova, Italy. Where, he was part of the Security and PRIvacy Through Zeal (SPRITZ) research group. He has published 20 reputed journals and conferences which includes only IEEE Transactions, Elsevier and Springer.

Sachin Shetty is an Associate Professor in the Virginia Modeling, Analysis and Simulation Center at Old Dominion University. He holds a joint appointment with the Department of Modeling, Simulation and Visualization Engineering and the Center for Cybersecurity Education and Research. Sachin Shetty received his Ph.D. in Modeling and Simulation from the Old Dominion University in 2007. He is the site lead on the DoD Cyber Security Center of Excellence, the Department of

Homeland Security National Center of Excellence, the Critical Infrastructure Resilience Institute (CIRI), and Department of Energy, Cyber Resilient Energy Delivery Consortium (CREDC). He has authored and coauthored over 150 research articles in journals and conference proceedings and two books. He has served on the technical program committee for ACM CCS, IEEE INFO-COM, IEEE ICDCN, and IEEE ICCCN. He is an Associate Editor for International Journal of Computer Networks.

Bheshaj Krishnappa is currently working as a Principal at ReliabilityFirst Corporation. Mr. Krishnappa is responsible for risk analysis and mitigation of threats to bulk power system reliability and security across a large geographic area in U.S. He has over 22 years of professional experience working for large and mid-sized companies in senior roles implementing and managing information technology, security, and business solutions to achieve organizational objectives. He is a business graduate knowledgeable in sustainable business practices that contribute to the triple bottom line of social, environmental and economic performance. He is motivated to apply his vast knowledge to achieve individual and organizational goals in creating a sustainable positive impact.

Chapter 15
Reliability Analysis of Cyber-Physical Systems

Sanja Lazarova-Molnar and Nader Mohamed

Abstract Cyber-Physical Systems (CPS) are marking our time and they are characterized by the smooth integration of cyber and physical parts. This integration carries along both challenges and new opportunities. The combination of software and hardware elements implies more complex systems that are prone to intricate interdependencies that affect the overall reliability. To this, usually, we need to add the human-computer interaction that is also a vital aspect of the functioning of CPS, which further complicates the reliability calculations. Unreliable systems can mean huge losses, both financially as well as in human lives. On a positive note, CPS have data as a central element of their operation. The availability and prevalence of data present a new opportunity to transform the ways in which reliability assessment has been traditionally performed. The goal of this contribution is to provide a holistic overview of the reliability analysis of CPS, as well as identify the impact that data and new data infrastructures may have on it. We, furthermore, illustrate the key points through two well-known cases of CPS, smart buildings and smart factories.

15.1 Introduction

Cyber-Physical Systems (CPS) are marking our time and they are characterized by the flawless integration of the cyber and physical parts. This integration carries along both challenges and new opportunities. The combination of software and hardware elements implies more complex systems that are prone to intricate interdependencies that affect the overall reliability. Reliability, which quantifies the probability that a system operates as expected for a predefined duration of time, while very well defined in hardware, is not as clearly defined in software. To this, we typically

S. Lazarova-Molnar (✉)
University of Southern Denmark, Odense, Denmark
e-mail: slmo@mmmi.sdu.dk

N. Mohamed
California University of Pennsylvania, Harrisburg, USA
e-mail: nader@middleware-tech.net

© Springer Nature Switzerland AG 2020   385
J. L. Risco Martín et al. (eds.), *Simulation for Cyber-Physical Systems Engineering*,
Simulation Foundations, Methods and Applications,
https://doi.org/10.1007/978-3-030-51909-4_15

need to add the interaction with humans that is also a vital aspect of the reliable functioning of CPS as humans do not always exhibit expected behavior. Users, or humans in general, can interact with CPS in numerous ways, either as actual users, or as maintenance or installation staff, for instance. Each of these interactions is both an opportunity and a risk with respect to the reliability of CPS, as has been elaborated in [35]. Therefore, interactions with humans can significantly affect the overall system reliability, and they need to be modeled, analyzed, and addressed in a comprehensive manner. However, including the human interaction aspect further complicates the reliability calculations. Until now, user interaction has been insufficiently addressed in reliability modeling. Even in some efforts to more holistically assess reliability, the human factor has been overseen, e.g., [43, 63, 73].

Unreliable systems translate to large portions of downtime, which can imply significant losses, both financially, as well as in terms of endangering human lives. Furthermore, an unreliable system is much more susceptible to security attacks. On a positive note, CPS have data as a central part of their operation. Availability and prevalence of data present new opportunities to transform the ways in which reliability assessment has been traditionally performed, typically through extensive use of expert knowledge.

The goal of this contribution is to provide a holistic overview of reliability analysis in the context of CPS, as well as identify the impact that data and new data infrastructures may have on it. We elaborate on the three central aspects of CPS reliability: hardware reliability, software reliability, and human interaction's impact on reliability, as well as discuss the potential of data. We further illustrate our key points through examples of two well-known CPS, smart buildings and smart factories. We first begin by providing a background on CPS and reliability analysis.

15.2   On Traditional Reliability in the Context of CPS

CPS are everywhere. Nowadays, when we talk of most systems in use, they are CPS, starting from smartphones, smart watches, smart kitchen appliances, all the way to smart buildings, smart factories, etc. They all have both cyber and physical components that cooperate to achieve certain purposes. On top of the cyber and physical components, most of the CPS extend and achieve their goals through interactions with humans. In the following, we provide a background on the specifics of CPS that relate and impact their reliability, as well as a background on the reliability analysis in a broader sense. This will create a basis for the following sections where we discuss the specific of CPS's reliability analysis.

Reliability is a measure that quantifies the probability that a given system performs as expected during a predefined period of time. As such, it has paramount importance in safety-critical systems, as it is in a tight correlation with safety. That said, an unreliable system is not necessarily unsafe, but an unsafe system is unreliable. The reason for this is that not all faults in a system are safety-critical, i.e., they do not impose danger to humans.

$$R(t) = \Pr\{T > t\} = \int_t^\infty f(x)dx,$$

where is the failure probability density function and is the length of the period of time (which is assumed to start from time zero). Reliability, traditionally, does not consider downtime due to repairs. There are, however, other metrics that capture downtimes, as explained in the following.

Highly related to reliability are the measures of dependability, availability, and performability. Availability differs from reliability in that that it takes into account downtime of equipment due to repairs or maintenance, and it quantifies the ratio of time that a system is operational compared to the overall lifetime. Dependability, on the other hand, is a more complex measure that aims to also encompass all of the other measures that assess the quality of the operation of a system. Performability is a measure that combines both performance and reliability.

To provide a hint of the difference of central reliability concepts with respect to both hardware and software, in Table 15.1, we compare side by side the basic notions related to reliability and availability, i.e., fault, failure, and error. They are similar in their meanings, and we can see that the concept of failure is consistent across both domains. Across both domains, it means a complete halt of the system. Note the distinct difference between a fault and an error. To better illustrate it, let us assume that there could be a fault that does not generate error, for example: a line is physically shortened to 0 (there is a fault); as long as the value on the line is supposed to be 0, there is no error. The main difference in reliability concepts between hardware and software is between the concept of fault, as faults in software typically occur during design, whereas hardware faults happen during operation. Furthermore, they can mutually trigger each other. We can definitely recognize a challenge to integrate both hardware and software faults and failures, and model the manners in which they influence each other, as this would be a very common scenario in CPS. For example,

Table 15.1 Comparison of core reliability concepts between hardware and software

	Software	Hardware
Fault	**Fault** is an incorrect step, process, or data definition in a program [3]. It is a programming error that could lead to an erroneous result. It is commonly known as a software bug	**Fault** is an abnormal condition or defect at the component, equipment, or subsystem level, which may lead to a failure [45]
Error	**Error** is a discrepancy between a computed, observed, or measured value or condition and the true, specified, or theoretically correct value or condition	**Error** is a deviation from correctness or accuracy, usually associated with incorrect values in the system state
Failure	**Failure** is the inability of a system or component to perform its required function within the specified performance requirement [3]. Example: Facebook crash	**Failure** is the state or condition of not meeting a desirable or intended objective, and may be viewed as the opposite of success [5]. Example: Hard drive failure

a fault in a hardware component may cause failure in a software component, or the other way around, i.e., a software fault might instigate a hardware component failure.

15.3 Holistic Reliability of Cyber-Physical Systems

Due to their complexity, CPS represent combinations of many different aspects, supplemented by many external influences. There can be many things that can go wrong, all associated with the different aspects of CPS, and all having their distinct natures. Yet, all different factors need to be viewed in unison as they do influence each other. We can clearly distinguish three key aspects of CPS that impact the reliability of a system: hardware, software, and human interaction. This does not imply that these are the only reliability-relevant aspects; it barely means that these three aspects are essential to capturing the reliability of a system.

15.3.1 Hardware Reliability

Traditionally, reliability has been addressed solely on a physical level. However, throughout time, systems that are in use have evolved to encompass phenomena that are of completely different natures, and their quantifications cannot be performed in the same manner as it is for the physical parts. To illustrate this, in Fig. 15.1 we compare hardware and software reliability. While in hardware most of the components follow the *Bathtub* curve which describes the probability of faults' occurrences, in software one needs a completely different approach, as there is no physical degradation in place. Instead, in software most of the problems are there since the start time. Therefore, the probability of faults' occurrences is related to the probability of using certain functionalities and discovering the problems, and correspondingly addressing them.

Hardware reliability has been very well defined and has a solid theoretical background. Reliability of hardware components is the probability that they perform as expected for a specified duration of time. Typically, Fault Trees [62, 75] can be utilized to capture mutual dependencies of the various components' and systems' faults that could lead to a system failure.

The main reliability measures are defined through the concepts of faults and failures, such as "mean time to failure" or MTTF or "mean time to repair" or MTTR. The assumption for using these two measures is that both repair times and inter-failure times are exponentially distributed, i.e., the rates are constant. These are also the most common reliability features for describing fault models of systems' components. If failures and repairs are non-exponentially distributed, then we use the notions of failure distributions and repair distributions.

The convenience of exponentially distributed failures and repairs is in the number of analytical methods available to tackle this class of problems. As mentioned, one

Fig. 15.1 **a** Bathtub curve for hardware reliability; **b** revised bathtub curve for software reliability [57]

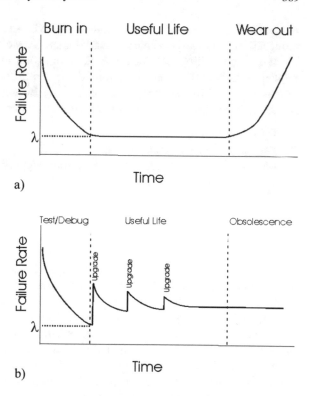

a)

b)

of the most popular reliability modeling approaches that utilizes this assumption is Fault Trees [62, 75]. The assumption for exponentially distributed failures and repair times, however, has often been proven to be an oversimplifying assumption [64, 70], and thus, inadequate for quantifying reliability. Relaxing the assumption of exponentially distributed faults and failure has yielded a number of sophisticated methods [17], such as Dynamic Fault Trees [23] and the Proxel-Based Method [32, 33], or even standard discrete-event simulation. Fault trees have been around for a long time [7, 72] and, thus, they have been utilized in various industries, such as automotive [31] and aircraft [12]. Fault trees are designed by describing system's and components' faults and failures as logical combinations of other components' faults.

15.3.2   Software Reliability

While all reliability-related measures have been solidly defined in hardware, in software they are still somewhat vague, especially in terms of their quantification. Similar to hardware reliability, software reliability is defined as the likelihood that a software system will work without failure under specified conditions and for a specified period of time [54]. Software reliability is considered to be an attribute of software quality, alongside functionality, usability, performance, serviceability, capability, installability, maintainability, and documentation. Software reliability analysis is mainly related to design errors, and, therefore, there is a different approach to it than the one usually used for hardware. As specified by Rosenberg et al. in [61], software reliability is comprised of the following three activities:

1. Error prevention,
2. Fault detection and removal, and
3. Measurements to maximize reliability, thus supporting the first two activities.

   Therefore, to improve software reliability, a large emphasis must be placed from the very beginning of its development, starting even in the requirements specification phase. The earlier the reliability is encompassed, the higher it can become. Typical processes that target improving reliability of software are debugging, early error detection, fast recovery, dynamic and static analysis, and evolution.

   Compared to hardware reliability, software reliability is also linked to many unquantifiable factors, such as programmers' skills or software project management skills. In the case of hardware, the skills of workers that build the hardware also matter, so the design aspect is common for both hardware and software. However, in software there is not wear out. Therefore, most of the reliability in software is tackled through certifications and quality assurance methods [66], and there is very little in terms of widely accepted software reliability assessment methods. One such attempt is presented in [66], where the authors propose systematic software reliability prediction approach based on software metrics.

   The fact that reliability of software is quite different from hardware reliability still does not mean that we cannot learn about software reliability by collecting data relevant to it and model software's failure rates. It often happens that the same software is used in many systems, meaning that data collection on its faults and failures can significantly support future reliability estimates of software of similar types. In fact, software testing can be seen as a time-compacted way of collecting data to both assess and improve the reliability of a software system. The same principle applies to reliability estimates of upgrades. Obtaining more data relevant to their reliability can support discovery of trends and insights, whose extrapolation can only enhance future reliability estimates.

   Moreover, having access to data about software that includes every detail from the very initial planning, and relating this data to the reliability of that same software, could assist extrapolating conclusions and metrics to other new software, as well as making conclusions that could enhance the reliability of newly built software and upgrades. This could further lead to the extraction of important principles that can be utilized in future software development, emphasizing on "design for reliability". A number of studies have been done on the data-driven reliability assessment of software, such as in [76], where hybrid genetic algorithm (GA)-based algorithm is developed which adopts the model mining technique to discover the correlation of failures and to obtain optimal model parameters; [44], where a hybrid approach to software reliability forecasting is proposed that combines both auto regression integrated moving average (ARIMA) and support-vector machines (SVM) models. These efforts are a clear sign that data can significantly enhance the reliability assessment of software [4].

15.3.3   Reliability Related to Human Interaction

The majority of CPS are designed for interaction with humans and, thus, humans have, to a large extent, influence on the operation of these systems. This influence encompasses reliability. There have been numerous research efforts directed at estimating the impact of unexpected human interaction with a wide range of CPS. One example here is smart buildings, where a lot of research has been done on the topic of occupants' behavior [39, 46]. Humans are, however, highly uncertain in their behavior and, thus, modeling of their behavior needs different approaches from the ones applicable to software and hardware aspects. Moreover, the types of interactions that can occur might differ from one system to another. Therefore, it may be needed to provide custom solutions with respect to the type of system and interactions.

Human interaction can, however, be viewed as both an opportunity and a risk, as human interaction can sometimes occur as an indication of a fault, which if attended properly in a timely manner can prevent further problems in the system. A good example of this is a smart building scenario of "opening of a window", which might indicate a faulty sensor that prevents ventilation or cooling systems from operating as intended. If the "opening of a window" is performed by mistake or the window is left open longer inadvertently, it will negatively influence the energy performance of the system, and, if the opening occurs on a repetitive basis, it could lead to shortened lifetime of some components. This simple example indicates how complex the assessment of human interaction events can be. Therefore, human interaction within CPS needs careful investigation and modeling.

We can relate the level of human interaction with a system with the vulnerability of a system to malfunctioning and performing sub-optimally. For instance, smart buildings are CPS that have a high level of interaction with humans. This, in turn, makes them highly vulnerable to both faults and anomalies in their performance. The example with leaving an open window, which is a very common occurrence, can contribute to increased energy consumption.

The relation of reliability to the level of human interaction can be made similar to the classification provided in [58]. This classification can serve in determining the vulnerability of a system to faults due to unexpected human behavior. In that sense, systems defined as "fully manual" are highly susceptible to human misuse, as opposed to those that are "fully automatic", where the interaction is minimal. To exemplify this with our "open window" situation, in a "fully automatic" system, windows would open and close automatically, and, therefore, this type of fault cannot occur. Therefore, the relevance of the inspection and inclusion of faults due to unexpected human behavior or interaction highly depends on the level of automation of a system.

However, as discussed previously, the relation is not straightforward, as an increased level of human interaction is also an increased opportunity to detect faults in a timely manner.

In a similar attempt to study reliability due to human behavior [25], Hollnagel talks about the concept of a "human error" and relates it to, what is termed as, "human reliability". He, furthermore, emphasizes the importance to develop a model to anticipate failures in joint machine-human systems. He concludes that this, however, is a very complex problem, as these failures are of complex nature and they need careful consideration.

Faults due to unexpected human behavior can be of different types, so each type would require a different model. For instance, some faults could be a signal that something is faulty in a system (e.g., opening a window could be a sign that some sensors are faulty, possibly temperature or CO_2), especially if it happens continuously. On the other hand, there could be a repetitive misuse in the manner that some component is being used, and then this would be a different type of model. To summarize, human factor reliability needs to be included for overall CPS reliability modeling. Moreover, it is a very complex issue, which can benefit from the ease of obtaining data nowadays.

As far as approaches to include human uncertainty in the overall reliability assessment of CPS are considered, the efforts are not plentiful. In one of them [10], Bessani et al. present a model to include operator's responsiveness together with machines' faults and failures to evaluate the reliability of a system. Further, more recently, Fan et al. in [21] present a platform and associated methodology to effectively generate accident scenarios by modeling human-machine interaction errors using model-level fault injection, followed by simulation to produce dynamic evolutions of accident scenarios. These are notable efforts toward providing accurate and holistic, and therefore useful, reliability measures of CPS.

15.4   Combined Reliability of CPS

Once we have observed the three different aspects of reliability modeling that are relevant to CPS, the question that pops up is how these three aspects interfere with each other, and is there a way to provide a unified reliability measure that captures them holistically for the system. Then, a further question is what human interaction errors can be prescribed as design errors [11], and thus address them at the design level. One illustrative example is when a machine is interacted in an unexpected way, which results in a software input that is also unexpected, thereby yielding a fault. As a result of this, the system fails as well. In this case, the error could be prescribed as a design error that did not prevent the user to interact with the machine in an improper manner. However, the fault can also be attributed to inadequate training of users if the design is not easily amendable. This example shows how complex error troubleshooting of CPS can be.

The three different aspects that contribute to faulty behavior in CPS have to be combined to provide a comprehensive reliability estimate of CPS. In the following, we review available approaches to holistic reliability evaluation of CPS, as well as list the challenges and opportunities that are associated with it.

15.4.1   CPS Reliability Approaches

Reliability analysis is often an important requirement for CPS since examples of CPS range from safety-critical complex infrastructures to simple but important medical devices [6, 55, 74]. Despite this, the holistic reliability of CPS, such that all contributing elements are considered, has not received an adequate span of attention. There have been a few independent approaches, but it was always far from a systematic and comprehensive reliability approach. The demand for reliability modeling and analysis of CPS, however, has been constantly increasing. To prevent the further deployment of CPS from slowing down, more researchers have recently addressed the reliability of CPS [24, 42, 67].

In [15], reliability of transportation CPS has been discussed by Clarke et al. The work has suggested formal analysis techniques to be used. This work is more focused on software reliability rather than hardware and human factors.

In [19], the reliability of human factors has been addressed by Dragan and Isaic-Maniu. The authors discuss a wide range of modeling frameworks for human reliability modeling and test the models on a flight simulator. The work clearly has only addressed the human factor and missed the software and hardware components.

In [22], Faza et al. have focused on fault injection in studying different faulty situations for the Smart Grid, which is clearly a cyber-physical system. Both software errors and hardware faults have been considered in the study to represent failures in the cyber infrastructure as well as hardware. The paper does not discuss the human factor in the study.

Reliability of CPS power grid ha been discussed by Singh and Sprintson in [68]. The authors classify the state-of-the-art solutions to reliability assurance into three categories: analytical, Monte Carlo simulation-based methods, and hybrid techniques, which are a combination of the first two categories. However, they have concluded that cyber parts in reliability studies of power grid are not addressed properly and the software errors needed to receive more attention.

In [65], a hybrid method which uses fault-tolerant structures with formal verification is proposed by Sha and Meseguer. The presented architecture supports the design of reliable cyber-physical systems. Another example of similar effort is presented by La and Kim in [30], where a service-oriented cyber-physical system with a service-oriented architecture and mobile Internet device is proposed. Again, none of these works study the effect of human uncertainty in reliability.

In one of the related and recent approaches [47], Mitchell et al. present a Petri net-based model that focuses on the effect of intrusion detection and response. The model, however, only focuses on the hardware aspects, and assumes flawless software operation and human interaction, which are oversimplifying.

Furthermore, there has been a lot of research that shows that the link between security and reliability is very strong [9, 26, 59, 69]. It is a fact that a nonsecure system is not reliable, and a non-reliable system cannot be secure [8]. With this respect, in [13] Cardenas et al. discuss survivable CPS through improved security. As survivability is highly linked to reliability, security is not the only aspect that should be considered.

15.4.2 Challenges and Opportunities Associated with Reliability of CPS

Based on our findings throughout relevant literature, we have been able to identify a number of challenges related to the reliability evaluation of CPS. In the following, we elaborate on each of them.

Reliability and security cannot really be observed independently, as both influence each other. A security attack can affect the reliability of a system, just as well as a component defect can also produce a security hole in a system. Therefore, common reliability-security measures need to be introduced that would quantify the combined level of fitness of a CPS system with respect to both measures. This is also shown and discussed by Mitchell and Chen in [47].

Secondly, we discovered that human factors' impact on CPS reliability has not been sufficiently studied. CPS, however, regardless of the level of automation [58], still exhibit a certain level of interaction with humans, be it in the form of typical users of a system, or staff responsible for installation and configuration, or staff that performs maintenance and repairs. This interaction, as is human nature, can be often erroneous and it can have a negative effect on the lifetime of the specific component. In other situations, it can have an impact on the performance of the system, and thus, the system will perform sub-optimally, which, again, in turn, may have an effect on its reliability and lifetime. Modeling human factors' effect on reliability is far from trivial; however, the availability of data that CPS typically feature can enhance modeling processes significantly.

Another challenge with reliability modeling of CPS is modeling of causality across the three paradigms, i.e., modeling how a software fault could instigate a hardware or a system fault or failure, as well as determining what is the actual cause. The same would apply to unexpected human behavior, e.g., opening a window in a smart building could be caused by a faulty temperature sensor, but inadequate handling with a light switch could instigate a fault in the light switch. Collection of relevant data, along with machine learning methods and proper root cause analysis, could be utilized to model relations of faults across paradigms. This, however, will be quite a challenge, as the three paradigms exhibit completely different fault-related behaviors.

To summarize, we have identified the following challenges in reliability modeling for CPS:

- Development of combined reliability-security measures for CPS,
- Development of reliability models that incorporate human factors,
- Development of machine-learning data-based reliability approaches that would take advantage of the available data collection platforms, and
- Utilizing reliability estimates and considering reliability during CPS design processes.

Adequate addressing of these challenges would drastically increase the quality of future CPS. Moreover, these challenges will unquestionably become a necessity in the near future. As CPS mature, reliability requirements will increase, as is typical across all domains. Challenges related to reliability evaluation of CPS need to be addressed in a timely manner, such as to address them while data flow platforms (like the Internet of Things) are still being developed. In addition, a call for "design-for-reliability" in CPS, which would consider reliability from CPS design phases, is vital. Furthermore, reliability needs to be viewed in accordance with other typical CPS performance metrics to better study the effects and trade-offs that they have on each other.

15.5   Data-Driven Reliability Analysis of CPS

Availability of data has become a game changer in the evaluation of systems. With the development of the Internet of Things (IoT), new opportunities for analyzing the reliability of systems are being developed, as well as opportunities to validate the existing approaches. Typically, a lot of expert knowledge is utilized for reliability analysis, and to a large extent, expert knowledge will remain irreplaceable for safety-critical systems, where there is no option for collecting data on faults, as they should not happen at all due to their catastrophic consequences. One example, where a great level of expert knowledge is needed, is the design of Fault Trees for the reliability of aviation systems [56]. However, large portions of systems utilized in manufacturing are not safety-critical, i.e., their faults and failures do not cause damage to the people or the environment. Mainly, the consequences of faults and failures in manufacturing systems are in terms of financial cost.

Therefore, the recent ICT developments and their use in manufacturing facilities create a significant opportunity to effectively gather data on faults and failures of these systems and utilize it to supplement expert knowledge and build more accurate reliability models. The development of Industry 4.0 has yielded a number of new moments and associated opportunities that can change the way in which reliability in manufacturing systems is analyzed and assessed. In particular, we focus on the following aspects:

1. Availability and ease of collection of data,
2. Large portion of systems being non-safety-critical,
3. Same flexible machines being utilized by different manufacturers for different purposes, and
4. New technologies leading to more complex and failure-prone systems.

Each of the four listed aspects provides different benefits and challenges, detailed as follows.

Availability and ease of collection of data have no only yielded collections of new types of data, but also requested the development of new and sophisticated approaches to enable full benefit of the data. For example, typically, the data collected is in the form of time series, without explicitly capturing faults' occurrences. This implies that there is a high necessity of approaches that focus on event detection, such that faults' and other events' occurrences can be extracted from the time series data. Furthermore, accurate root cause analysis methodology will be the next requirement, such as to extract events' dependencies and model them. Once such approaches are sufficiently advanced, reliability analysis of systems can be automated, such as to be automatically performed, based on data from manufacturing machines. The fact that most of the manufacturing machines are not necessarily safety-critical, and faults/failures are relatively common occurrences, makes the data-based approaches very adequate. Collection of data for reliability analysis in safety-critical systems would be impossible, as in those systems the failures can cost human lives. Examples are automotive systems or aircraft. This brings us to the second point that supports the first.

The third fact on our list is the anticipation for flexible machines that can perform different tasks, which implies that highly repetitive use of the same types of machines will be occurring. Thus, a lot more data can be collected for those machines, yielding more accurate models. We have previously emphasized the potential of Collaborative data analytics for both smart buildings and smart factories, and, especially in the case of reliability analysis it has a critical meaning [34, 37, 38]. Namely, data on faults is always sparse, so the collaborative model building can significantly enhance and speed up model building processes. However, as we state by the fourth aspect, the availability of data and the new technologies comes at a cost, which is the increased complexity of the systems. This increased complexity implies vulnerability to faults and failures. For example, CPS' processes often depend on input from sensor data, so a fault in the sensors can lead to significant damaging and costly consequences. This vulnerability only emphasizes the need for accurate and efficient reliability and overall health assessment of the manufacturing systems.

To summarize, in Fig. 15.2, we illustrate the feedback loop that can be enabled by the data-based reliability assessment for smart factories. We begin from the bottom left corner of the figure, where we have the smart factory that collects data from its manufacturing processes. This data then needs to be processed in a way that key reliability-related events are detected (such as fault occurrences, repair starting times and completion times, etc.) and extracted [14, 71]. Next, this data is used for learning fault models, including causality among faults and failures, which is

Fig. 15.2 Data for
supporting reliability
modeling in smart factories

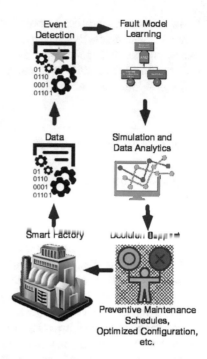

followed by advanced simulation and data analytics. The results of simulation and data analytics are utilized for decision support on improved system configuration and generation of preventive maintenance schedules for increased reliability of the system (i.e., smart factory in this case).

It is evident that machine learning and simulation will play an important role in making the data-based reliability analysis processes possible. Advanced event detection methods will be crucial for gathering data for building reliability-relevant models, and accurate and efficient simulation methods will be needed for evaluating the reliability of the built models. The automatically generated models can then be used for analyzing alternative configurations for given systems, with the purpose of optimizing their reliabilities.

The four described aspects that characterize the latest developments, captured in the Industry 4.0 initiative, are in favor of using data-based approaches for reliability assessment of manufacturing cyber-physical systems. In the following section, we discuss the implications of the data-based reliability assessment of manufacturing systems.

15.6   Illustrative Examples (Case Studies)

In the following, we illustrate the specifics of reliability modeling for two different case studies of CPS, i.e., smart buildings and cyber-physical production systems, also known as smart factories. In both cases, we have systems that are not necessarily safety-critical (unless there are special types of buildings or factories, whose incorrect operation can endanger human lives). This means that we have the opportunity to collect vast amounts of data on faults and, therefore, enable the building of accurate and useful fault models.

15.6.1 Cyber-Physical Production Systems (Smart Factories)

As the competition among manufacturing enterprises grows, there is a necessity for employing more cost-effective and advanced technological approaches to make intelligent decisions for improving and optimizing manufacturing processes and decreasing production costs. One of these technological approaches is employing cyber-physical production systems (CPPS) [53]. CPPS are new innovative manufacturing systems that employ modern computer and communication technology as well as recently innovated manufacturing science and technology to enhance the production processes in manufacturing. These enhancements include improving productivity, quality, reliability, and cost-effectiveness.

In manufacturing, fabricating, or creating practical and final products requires several manufacturing processes. Each of these manufacturing processes participates in the total manufacture cycle to build the needed products [27]. Several manufacturing processes can be attached and incorporated to produce a construction process that can deliver the required final product. Different manufacturing stages and processes can be formed and controlled using a cyber-physical system. Multiple cyber-physical systems belonging to different manufacturing stages and processes are integrated together to from a complete manufacture cycle to build the needed products. Using cyber-physical systems in manufacturing has many advantages including enabling collecting detailed manufacturing data. This data can be utilized to form digital twin models of the real production machines and production processes [29, 60]. The collected data and formed digital twin models can be utilized to provide numerous solutions for improving reliability for production processes and final products by manufacturers.

Reliability levels in manufacturing can be improved by conducting periodic maintenance processes for production machines and facilities. However, maintenance and support can cost between 60 and 75% of the total lifecycle expense of manufacturing processes [20]. This expense can be reduced significantly if the collected data and advanced data analytics are utilized to perform adaptive and strategic maintenance scheduling for production machines and facilities [36]. For example, manufacturing data and advanced analytics models can be utilized to enhance reliability to reduce total energy consumption costs in manufacturing [51, 52]. In addition, the reliability of CPS-based manufacturing can be improved by conducting collaboration data analytics for multiple manufacturing units [38]. This requires using advanced technological infrastructures such as cloud computing, fog computing, and data analytics services that are integrated by advanced middleware platforms [3]. Such infrastructure can provide effective and efficient integration of technologies and processes needed to improve the reliability in manufacturing.

15.6.2   Smart Buildings

Commercial, industrial, and residential buildings exhaust around 49% of the total energy and generate around 47% of the greenhouse gas emissions in the United States [1]. Moreover, commercial, industrial, and residential buildings in Europe exhaust 40% of the total energy used and generate around 36% of the carbon dioxide emissions [16]. Due to these high numbers, both the US Department of Energy and the European Commission aim to gradually reduce the consumption of primary energy in buildings to realize the concept of smart buildings and smart cities.

One of the most important focuses on energy consumption saving is to employ Information and Communication Technology (ICT) to turn regular buildings into smart buildings. This requires deploying and interconnecting sensors, actuators, and subsystems and utilizing smart automation monitoring and control mechanisms to reduce energy needs in buildings. This creates what is known as the building energy management system (BEMS) [18]. With this approach, intelligent controls and computational models can be utilized to automate energy management in buildings. Using BEMS in smart buildings makes them cyber-physical systems as shown in Fig. 15.3. The physical world in these cyber-physical systems involves the buildings' spaces, heating, ventilating, and air-conditioning (HVAC) systems, energy supply, atmosphere conditions such as current lighting level, temperature, ventilation, and building occupants. The cyber world in these cyber-physical systems involves

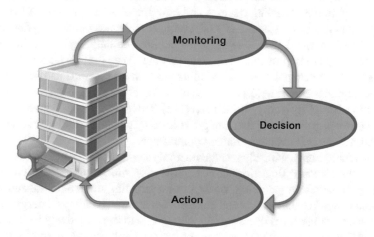

Fig. 15.3 BEMS as a cyber-physical system

the software that runs the building monitoring and control algorithms, sensing and networking, and the hardware systems that perform the control decisions.

One of the main issues with energy efficiency in buildings is faults that may occur in the energy control systems. In some cases, going undetected for some time, faults could in the long run build up a significant energy inefficiency. Moreover, some of these faults are difficult to be discovered and as a result the energy inefficiency could run unobserved for long periods of time. It is projected that faults add between 15 and 30% of the consumed energy [28]. Consequently, Fault Detection and Diagnostics (FDD) is very essential for energy efficiency in smart buildings [40, 41]. Based on some studies, the utilization of automated FDD as part of the BEMS can offer a substantial cut in the operational costs of buildings [28]. The faults that can be occurred in the cyber-physical building systems can be due to software faults, hardware faults, or faults caused due to human interaction with the system. As cyber-physical building systems rely on different software and hardware components such as sensors, actuators, networks, and other devices, faults in these components may create faults in the energy systems. In addition, as humans can interact with these systems in a wrong manner, different faults may occur in such systems. One example of this human behavior is in opening a window while the outside temperature is high [35]. The level of automation in the building energy system can vary on several levels. There are not only buildings that do not permit occupiers to open a window, but also buildings that allow occupiers to deal with a wide range of elements of the building energy management systems. However, the "opening window" case can generate several opportunities for the automated systems to discover the occurrences of some faults in the building energy management system including faults in buildings' sensors, actuators, networks, and other devices [35].

Fortunately, utilizing the cyber-physical system approach for managing energy in buildings can enable many innovative mechanisms for supporting data-driven reliability analysis and solutions for energy management in buildings. Data related to energy consumption, building devices and equipment, smart meters, human needs and behaviors can be collected and analyzed using advanced data analytics models. The collected energy data can be achieved within one building or within multiple similar buildings to improve the data analytics outcomes [34]. In addition, it can be achieved and optimized through all buildings in a smart city [48]. Different middleware technologies can be used to enable collecting energy data from multiple buildings such as SmartCityWare [49] and PsCPS [2]. This data can be effectively analyzed through an energy cloud [50] that can provide advanced computational services and powerful resources for energy-related applications.

These technologies with advanced data analytics can provide services for timely fault and deficiencies' detection and diagnosis for smart buildings. It can deliver efficient procedures for gathering, storing, and evaluating smart building energy data to detect and report energy faults and deficiencies in smart buildings' energy systems. The scalable storage and processing capacity of Cloud can effectively facilitate data-driven fault detection and diagnostics approaches. Faults can be found by applying data mining and other analytics methods to find discrepancies in the data or unusual patterns. This helps in detecting unknown common faults faster among these similar buildings [41]. As a result, the associated increase in consumed energy due to these faults can be avoided earlier.

15.7   Conclusions

Reliability analysis of CPS is of paramount importance to obtain relevant measures of how much a system can meet its expected preset criteria. Providing reliability analysis of CPS has even gained importance now that CPS complexity has increased. Thankfully, CPS' increased complexity comes along with the increased capability of generating and collecting data that captures systems' behaviors. This data can be utilized to automatically derive models for different purposes with a reduced need for experts' knowledge, as we illustrated this by our case studies. For reliability analysis this is a real game changer, as traditionally, reliability models have been developed using extensive expert knowledge.

We provided an analysis of the specifics of assessing the reliability of CPS in the new current reality, where data has become the main currency. Reliability of systems, being one of the paramount performance measures, needs to be accurately estimated such that it can be improved. Unreliable systems can be extremely costly, both in terms of money, as well as in terms of human lives. Furthermore, unreliable systems can also be unsafe and unsecure. Therefore, the opportunity that lies in the availability of data to more accurately assess CPS' reliability needs to be utilized. Thus, new data-driven approaches need to emerge. This is especially important, as traditionally, reliability analysis has been performed relying to a great extent on expert knowledge.

We, furthermore, provided an overview of the advances in the reliability of CPS, with the goal of emphasizing the need for accurate and holistic reliability modeling of the three paradigms involved in CPS: hardware, software, and humans. Each of these paradigms has its own challenges and its specific fault-related behaviors and models. Still, a lot needs to be done with respect to the combined reliability modeling and analysis of CPS, and the ease with which data is being collected nowadays represents a significant opportunity. Finally, to illustrate our key points, we presented two case studies: smart buildings and smart factories.

References

1. 2030 (2011) A challenge for products: critical points. https://docplayer.net/30519472-2030-challenge-for-products-critical-points.html. Accessed 9 Nov, 2019
2. Al-Jaroodi J, Mohamed N (2018) Pscps: a distributed platform for cloud and fog integrated smart cyber-physical systems. IEEE Access 6:41432–41449
3. Al-Jaroodi J, Mohamed N, Jawhar I (2018) A service-oriented middleware framework for manufacturing industry 4.0. ACM SIGBED Rev 15(5):29–36

4. Alsina EF, Chica M, Trawiński K, Regattieri A (2018) On the use of machine learning methods to predict component reliability from data-driven industrial case studies. Int J Adv Manuf Technol 94(5):2419–2433. https://doi.org/10.1007/s00170-017-1039-x
5. Altenbach H, Sadowski T (2014) Failure and damage analysis of advanced materials. Springer
6. Baheti R, Gill H (2011) Cyber-physical systems. Impact Control Technol 12:161–166
7. Barlow RE, Proschan F (1975) Importance of system components and fault tree events. Stocha Processes Appl 3(2):153–173
8. Barnum S, Sastry S, Stankovic JA (2010) Roundtable: reliability of embedded and cyber-physical systems. IEEE Secur Priv 5(8):27–32
9. Bécue A, Cuppens-Boulahia N, Cuppens F, Katsikas S, Lambrinoudakis C (2016) Security of industrial control systems and cyber physical systems. In: First Workshop, CyberICS 2015 and First Workshop, WOS-CPS 2015 Vienna, revised selected papers, vol 9588. Springer, Austria, 21–22 Sept, 2015
10. Bessani M, Fanucchi RZ, Delbem ACC, Maciel CD (2016) Impact of operators' performance in the reliability of cyber-physical power distribution systems. IET Gener Transm Distrib 10(11):2640–2646
11. Booth PA (1991) Errors and theory in human-computer interaction. Acta Physiol (Oxf) 78(1–3):69–96
12. Brière D, Traverse P (1993) AIRBUS A320/A330/A340 electrical flight controls-A family of fault-tolerant systems. In: FTCS-23 The 23rd international symposium on fault-tolerant computing. IEEE, pp 616–623
13. Cardenas AA, Amin S, Sastry S (2008) Secure control: towards survivable cyber-physical systems. In: Distributed computing systems workshops. ICDCS'08. 28th international conference on, 2008. IEEE, pp 495–500
14. Chauhan V, Surgenor BJPM (2015) A comparative study of machine vision based methods for fault detection in an automated assembly machine. 1:416–428
15. Clarke EM, Krogh B, Platzer A, Rajkumar R (2008) Analysis and verification challenges for cyber-physical transportation systems
16. Commission E (2013) Communication from the commission: energy efficiency: delivering the 20% target. Commission of the European Communities, Brussels
17. Distefano S, Trivedi KS (2013) Non-Markovian state-space models in dependability evaluation. Qual Reliab Eng Int 29(2):225–239
18. Dounis AI, Caraiscos C (2009) Advanced control systems engineering for energy and comfort management in a building environment—a review. Renew Sustain Energy Rev 13(6–7):1246–1261
19. Dragan I-M, Isaic-Maniu A (2014) The reliability of the human factor. Procedia Econ Finan 15:1486–1494
20. Efthymiou K, Papakostas N, Mourtzis D, Chryssolouris G (2012) On a predictive maintenance platform for production systems. Procedia CIRP 3:221–226
21. Fan C-F, Chan C-C, Yu H-Y, Yih S (2018) A simulation platform for human-machine interaction safety analysis of cyber-physical systems. Int J Ind Ergon 68:89–100. https://doi.org/10.1016/j.ergon.2018.06.008
22. Faza A, Sedigh S, McMillin B (2010) Integrated cyber-physical fault injection for reliability analysis of the smart grid. In: International conference on computer safety, reliability, and security. Springer, pp 277–290
23. Ge D, Yang Y (2015) Reliability analysis of non-repairable systems modeled by dynamic fault trees with priority AND gates. Appl Stoch Models Bus Ind 31(6):809–822
24. Group CS (2008) Cyber-physical systems executive summary. CPS Summit
25. Hollnagel E (2005) Human reliability assessment in context. Nucl Eng Technol 37(2):159–166
26. Izosimov V, Asvestopoulos A, Blomkvist O (2016) Security-aware development of cyber-physical systems illustrated with automotive case study. In: 2016 Design, automation and test in europe conference and exhibition (DATE). IEEE, pp 818–821
27. Kalpakjian S, Schmid S (2014) Manufacturing engineering and technology. In: Sekar KV ed Prentice hall, pearson education South Asia Pte Ltd., Singapore

28. Katipamula S, Brambley MR (2005) Methods for fault detection, diagnostics, and prognostics for building systems—a review, part I. Hvac&R Res 11(1):3–25
29. Kritzinger W, Karner M, Traar G, Henjes J, Sihn W (2018) Digital twin in manufacturing: a categorical literature review and classification. IFAC-PapersOnLine 51(11):1016–1022
30. La HJ, Kim SD (2010) A service-based approach to designing cyber physical systems. In: 2010 IEEE/ACIS 9th international conference on (ICIS). IEEE, pp 895–900
31. Lambert HE (2004) Use of fault tree analysis for automotive reliability and safety analysis. SAE Tech Pap
32. Lazarova-Molnar S (2005) The proxel-based method: formalisation, analysis and applications. Otto-von-Guericke-Universität Magdeburg, Universitätsbibliothek
33. Lazarova-Molnar S, Horton G (2003) Proxel-based simulation for fault tree analysis. In: 17. Symposium simulationstechnik (ASIM 2003)
34. Lazarova-Molnar S, Mohamed N (2017) Collaborative data analytics for smart buildings: opportunities and models. Cluster Comput 22(1):1065–1077
35. Lazarova-Molnar S, Mohamed N (2017) On the complexity of smart buildings occupant behavior: risks and opportunities. In: Proceedings of the 8th balkan conference in informatics. ACM, p 1
36. Lazarova-Molnar S, Mohamed N (2019) Reliability assessment in the context of Industry 4.0: data as a game changer. Procedia Comput Sci 151:691–698. https://doi.org/10.1016/j.procs.2019.04.092
37. Lazarova-Molnar S, Mohamed N, Al-Jaroodi J (2018) Collaborative data analytics for industry 4.0: challenges, opportunities and models. In: 2018 6th international conference on enterprise systems (ES). IEEE, pp 100–107
38. Lazarova-Molnar S, Mohamed N, Al-Jaroodi J (2019) Data analytics framework for industry 4.0: enabling collaboration for added benefits. IET Collaborative Intell Manufact
39. Lazarova-Molnar S, Shaker HR (2016) A conceptual framework for occupant-centered building management decision support system. In: 12th international conference on intelligent environments. London, United Kingdom, 2016. Ambient intelligence and smart environments, intelligent environments. IOS Press, pp 436–445
40. Lazarova-Molnar S, Shaker HR, Mohamed N (2016) Fault detection and diagnosis for smart buildings: state of the art, trends and challenges. In: 2016 3rd MEC international conference on big data and smart city (ICBDSC). IEEE, pp 1–7
41. Lazarova-Molnar S, Shaker HR, Mohamed N (2016) Reliability of cyber physical systems with focus on building management systems. In: 2016 IEEE 35th international performance computing and communications conference (IPCCC). IEEE, pp 1–6
42. Lee EA (2008) Cyber physical systems: Design challenges. In: 2008 11th IEEE international symposium on object and component-oriented real-time distributed computing (ISORC). IEEE, pp 363–369
43. Li Z, Kang R (2015) Strategy for reliability testing and evaluation of cyber physical systems. In: 2015 IEEE international conference on industrial engineering and engineering management (IEEM) IEEE, pp 1001–1006
44. Lo J-H (2012) A data-driven model for software reliability prediction. In: 2012 IEEE international conference on granular computing. IEEE, pp 326–331
45. Loo AW (2012) Distributed computing innovations for business, engineering, and science. IGI Global
46. Masoso O, Grobler LJ (2010) The dark side of occupants' behaviour on building energy use. Energy Build 42(2):173–177
47. Mitchell R, Chen R (2013) Effect of intrusion detection and response on reliability of cyber physical systems. IEEE Trans Reliab 62(1):199–210
48. Mohamed N, Al-Jaroodi J, Jawhar I (2018) Service-oriented big data analytics for improving buildings energy management in smart cities. In: 2018 14th international wireless communications and mobile computing conference (IWCMC). IEEE, pp 1243–1248
49. Mohamed N, Al-Jaroodi J, Jawhar I, Lazarova-Molnar S, Mahmoud S (2017) SmartCityWare: a service-oriented middleware for cloud and fog enabled smart city services. IEEE Access 5:17576–17588

50. Mohamed N, Al-Jaroodi J, Lazarova-Molnar S (2018) Energy cloud: Services for smart buildings. In: Sustainable cloud and energy services. Springer, pp 117–134
51. Mohamed N, Al-Jaroodi J, Lazarova-Molnar S (2019) Industry 4.0: opportunities for enhancing energy efficiency in smart factories. In: 2019 IEEE international systems conference (SysCon). IEEE, pp 1–7
52. Mohamed N, Al-Jaroodi J, Lazarova-Molnar S (2019) Leveraging the capabilities of industry 4.0 for improving energy efficiency in smart factories. IEEE Access 7:18008–18020
53. Monostori L (2014) Cyber-physical production systems: roots, expectations and R&D challenges. Procedia Cirp 17:9–13
54. Musa JD, Everett WW (1990) Software-reliability engineering: technology for the 1990s. IEEE Softw 7(6):36–43
55. Nannapaneni S, Mahadevan S, Pradhan S, Dubey A (2016) Towards reliability-based decision making in cyber-physical systems. Paper presented at the IEEE international conference on smart computing (SMARTCOMP)
56. Netjasov F, Janic M (2008) A review of research on risk and safety modelling in civil aviation. J Air Transp Manag 14(4):213–220
57. Pan J (1999) Software reliability. Dependable Embedded Syst Carnegie Mellon Univ 18:1–14
58. Parasuraman R, Sheridan TB, Wickens CD (2000) A model for types and levels of human interaction with automation. IEEE Trans Syst Man Cybern-Part A: Syst Hum 30(3):286–297
59. Rashid A, Joosen W, Foley S (2016) Security and resilience of cyber-physical infrastructures. In: Proceedings of the 1st international workshop held on 06 April 2016 in conjunction with the international symposium on engineering secure software and systems, London, UK
60. Rosen R, Von Wichert G, Lo G, Bettenhausen KD (2015) About the importance of autonomy and digital twins for the future of manufacturing. IFAC-PapersOnLine 48(3):567–572
61. Rosenberg L, Hammer T, Shaw J (1998) Software metrics and reliability. In: 9th international symposium on software reliability engineering
62. Ruijters E, Stoelinga M (2015) Fault tree analysis: A survey of the state-of-the-art in modeling, analysis and tools. Comput Sci Rev 15:29–62
63. Sanislav T, Zeadally S, Mois GD, Fouchal H (2018) Reliability, failure detection and prevention in cyber-physical systems (CPSs) with agents. Concurrency Comput Pract Experience e4481
64. Schroeder B, Gibson GA (2007) Disk failures in the real world: what does an mttf of 1,000,000 hours mean to you? In: FAST, pp 1–16
65. Sha L, Meseguer J (2008) Design of complex cyber physical systems with formalized architectural patterns. In: Software-intensive systems and new computing paradigms. Springer, pp 92–100
66. Shi Y, Li M, Arndt S, Smidts C (2017) Metric-based software reliability prediction approach and its application. Empirical Software Eng 22(4):1579–1633
67. Singh C, Sprintson A (2010) Reliability assurance of cyber-physical power systems. In: IEEE PES general meeting. IEEE, pp 1–6
68. Singh C, Sprintson A (2010) Reliability assurance of cyber-physical power systems. In: Power and Energy society general meeting. IEEE, pp 1–6
69. Steger M, Karner M, Hillebrand J, Rom W, Rýmer K (2016) A security metric for structured security analysis of cyber-physical systems supporting SAE J3061. In: Modelling, analysis, and control of complex CPS (CPS Data), 2016 2nd International Workshop on IEEE, pp 1–6
70. Sun H, Han JJ (2002) The failure of MTTF in availability evaluation. In: Annual reliability and maintainability symposium. 2002 proceedings. IEEE, pp 279–284
71. Theorin A, Bengtsson K, Provost J, Lieder M, Johnsson C, Lundholm T, Lennartson B (2017) An event-driven manufacturing information system architecture for Industry 4.0. 55(5):1297–1311
72. Vesely WE, Goldberg FF, Roberts NH, Haasl DF (1981) Fault tree handbook. DTIC document
73. Wu L, Kaiser G (2013) FARE: a framework for benchmarking reliability of cyber-physical systems. In: 2013 IEEE long island systems, applications and technology conference (LISAT). IEEE, pp 1–6

74. Wu LL (2011) Improving system reliability for cyber-physical systems. http://academiccomm ons.columbia.edu/catalog/ac:146625
75. Yan R, Dunnett SJ, Jackson LM (2016) Reliability modelling of automated guided vehicles by fault tree analysis. In: 5th student conference on operational research (SCOR 2016)
76. Yang B, Li X, Xie M, Tan F (2010) A generic data-driven software reliability model with model mining technique. Reliab Eng Syst Saf 95(6):671–678

Sanja Lazarova-Molnar is an associate professor with the Faculty of Engineering at the University of Southern Denmark. Her current research interests include modeling and simulation of stochastic systems, reliability modeling and analysis, and data analytics for decision support. Since 2019, she is serving on the Board of Directors of The Society for Modeling & Simulation International (SCS). Sanja Lazarova-Molnar obtained her Ph.D. in Computer Science from the University in Magdeburg, Germany in 2005, specializing in the area of Modeling and Simulation, where she was also a member of the Simulation Group at the Institute of Simulation and Graphics. She has successfully collaborated with industry and academia, authoring more than 80 peer-reviewed research publications. Email: slmo@mmmi.sdu.dk.

Nader Mohamed is an associate professor in the Department of Mathematics, Computer Science and Information Systems at California University of Pennsylvania, PA, USA. He teaches courses in cybersecurity, computer science, and information systems. He was a faculty member with Stevens Institute of Technology, Hoboken, NJ, USA and UAE University, Al Ain, UAE. He received the Ph.D. in computer science from the University of Nebraska–Lincoln, Lincoln, NE, USA. He also has several years of industrial experience in information technology. His current research interests include cybersecurity, middleware, Industry 4.0, cloud and fog computing, networking, and cyber-physical systems.

Chapter 16
Dimensions of Trust in Cyber Physical Systems

Margaret L. Loper

Abstract The urban environment is becoming increasingly more connected and complex. In the coming decades, we will be surrounded by billions of sensors, devices, and machines, the Internet of Things (IoT). As the world becomes more connected, we will become dependent on machines to make decisions on our behalf. When machines use data from sensors, devices, and other machines (i.e., things) to make decisions, they need to learn how to trust that data, as well as the things they are interacting with. As cyber physical systems become more commonplace in IoT and smart city applications, it is essential that decision makers are able to trust the machines making decisions on their behalf. This chapter defines trust from a multidimensional perspective, which includes reliability. It describes a set of research projects conducted that span the multiple dimensions of trust. While the research described spans a range of trust topics, little has been done on the relevance of trust to simulation. Simulations that use and interact with real-world systems is growing, which means understanding their trustworthiness is of growing importance. The chapter concludes with a few ideas on a research agenda for the way forward.

16.1 Introduction

16.1.1 Internet of Things and Cyber Physical Systems

In the coming decade, we will be surrounded by billions of connected sensors, devices, and machines. This will lead to a pervasive presence of things (e.g., RFID tags, sensors, actuators, cell phones, vehicles), which have the ability to communicate and cooperate to achieve common goals. These things will be uniquely identifiable and addressable, and many will be smart and can capture, store, process, and communicate data about themselves, their physical environment, and their human

---

M. L. Loper (✉)
Georgia Tech Research Institute, Atlanta, USA
e-mail: margaret.loper@gtri.gatech.edu

J. L. Risco Martín et al. (eds.), *Simulation for Cyber-Physical Systems Engineering*,
Simulation Foundations, Methods and Applications,
https://doi.org/10.1007/978-3-030-51909-4_16

407

owners. Since there is not an "internet" exclusively dedicated to "things", the expression Internet of Things (IoT) is best understood as a metaphor that encapsulates the immersion of almost anything and everything into the communications space [5]. As the European Research Cluster on the Internet of Things (IERC) puts it, IoT is "A dynamic global network infrastructure with self-configuring capabilities based on standards and interoperable communication protocols where physical and virtual things have identities, physical attributes and virtual personalities; use intelligent interfaces; and are seamlessly integrated into the information network" [12].

A unique characteristic of the IoT is the presence of different modes of communication, including interaction between people (Human to Human or H2H), people and things (Human to Machine or H2M and M2H), and things (Machine to Machine or M2M). H2H communications are carried out in multiple forms and continue to innovate with social media and crowdsourcing. H2M or M2H communications assume human intervention and control. In contrast, the M2M communications have no explicit human intervention or very limited intervention.

The term Cyber Physical Systems (CPS) was coined in 2006, by the National Science Foundation, but was based on earlier concepts from mechatronics, embedded systems, pervasive computing, and cybernetics. According to the National Academies of Sciences, Engineering, and Medicine (NASEM), CPS are "engineered systems that are built from, and depend upon, the seamless integration of computational algorithms and physical components" [23]. According to their report, CPS can be small and closed, such as an artificial pancreas, or very large, complex, and interconnected, such as a regional energy grid. It also states that CPS bridges "engineering and physical world applications and the computer engineering hardware and computer science cyber worlds."

The National Institute for Standards and Technology (NIST) conducted an analysis of CPS trends by reviewing 31 published CPS definitions. They found that CPS definitions are largely consistent and highlight a set of six common characteristics: hybrid physical and logical systems, hybrid analytical and measurement methods, control, component classes, time, and trustworthiness [8]. NIST goes on to recognize the relationship between CPS and IoT.

- CPS emerged from a systems engineering and control perspective.
- IoT emerged from a networking and information technology perspective.

Despite their distinct origins, "CPS and IoT refer to a related set of trends in integrating digital capabilities (i.e., network connectivity and computational capability) with physical devices and engineered systems to enhance performance and functionality" [8].

The NIST report goes on to analyze 11 publications that compare and contrast CPS and IoT. Connecting the physical and logical worlds was a central characteristic attributed to both CPS and IoT, which provided the basis for unified components and interaction models based on four issues: control, platform, internet, and human interactions. The conclusion of the report is "the lack of consistent distinguishing metrics and the convergence of definitions, indicate an emerging consensus around the equivalence of CPS and IoT concepts". Based on this conclusion, the remainder of this chapter will discuss IoT and CPS interchangeably, as it relates to research and applications.

16.1.2   Internet of Things and Smart Cities

According to the United Nations (UN), the world's urban population is projected to grow by 2.5 billion from 2014 to 2050, and will account for 66% of the total global population by then [34]. The growing population in cities increases the demand for the fundamental needs of people living there, such as housing, utilities, medical care, welfare, education, and employment [32]. To deal with challenges faced during the growth of cities, the concept of Smart City has been envisioned, which denotes "the effective integration of physical, digital, and human systems in the built environment to deliver a sustainable, prosperous, and inclusive future for its citizens" [14].

With the growing economic and environmental problems in urban areas, the benefit of IoT technologies in a city are vast. A smart electrical grid will make cities more efficient by optimizing how energy is used and distributed. Device data will help inform and protect city residents by improving city service monitoring capabilities. Consumers will have better insights into the consumption of resources (energy, water, and gas). Application of IoT in cities, shown in Fig. 16.1, include the following industries [6].

- Retail: Autonomous point of sale systems, inventory/SKU management, store traffic monitoring
- Home and Business: Smart homes and buildings, building automation, energy management, access control

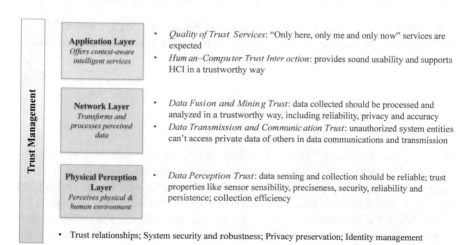

Fig. 16.1 IoT trust framework

- Transportation: Fleet management, routing planning, autonomous vehicles
- Healthcare: Home health, medical devices, telemedicine, aging in place
- Agriculture: Smart irrigation, farming/seeding systems, livestock monitoring
- Education: Remote education and digital education tools
- Entertainment/Hospitality: Venue management, safety and security, staff scheduling and routing, room automation and access control
- Manufacturing: Equipment monitoring and maintenance, product visibility
- Utilities: Smart grids and smart meters, grid performance, demand planning
- Construction: Asset tracking and worker safety monitoring.

Cities may be the first to benefit from the IoT, but being surrounded by billions of sensors, devices, and machines has profound implications for security, trust, and privacy. The more technology a city uses, the more vulnerable it is, so the smartest cities face the highest risks.

16.2 Trust

A review of the recently published book *Digital Transformation: Survive and Thrive in an Era of Mass Extinction* [30], succinctly captures IoT's central role in digital transformation, stating it is possibly the most important defining feature of the twenty-first century economy:

> We will have 50 billion small computers connected to a network. Fifty billion squared is equivalent to the number of stars in our universe. Siebel states: The Internet of Things may be the single most important defining feature of the 21st century economy. A powerful global network becomes a new computing platform. And much of the computing will take place within the sensors at the periphery of the network rather than at the core of the network. [31]

With that, it is incumbent on those looking to leverage IoT to have a clear view of the potential value and possible hurdles they will face as part of this transformative change. What are the risks, and how do we trust, IoT?

16.2.1 Definitions

With the grand vision of billions and trillions of cellphones, physical devices, vehicles, and other things embedded with electronics, software, sensors, actuators—things will outnumber people. It will be impossible for humans to monitor and control all these things; therefore, some decision-making will be delegated to things in the system. While the interconnection of things is understood in IoT, the intelligence of things is what makes the IoT paradigm "game-changing" [5]. There is an increasing desire to use things in lieu of humans in dangerous or routine situations, and also to make things more intelligent such that they can deliver personalized services. These trends of increasing complexity and scale raise questions about the trustworthiness of this emerging technology.

The connection between people and things is complex, and creates a set of trust concerns. Trust should be considered at two levels: (1) whether a thing trusts the data it receives or trusts the other things it interacts with (M2M) and (2) whether a human trusts the things, services, data, or IoT offerings that it uses (H2M or M2H). This leads to the idea that trust is multidimensional. Ahn et al. [1] described the concept of multidimensional trust by different agent characteristics, such as quality, reliability, and availability. For Matei et al. [22], trust refers to the trustworthiness of a sensor, whether it has been compromised, the quality of data from the sensor, and the network connection. Grandison and Sloman [10] define trust as the belief in the competence of an entity to act dependably, securely, and reliably within a specified context. To address behavior uncertainty in agent communities, Pinyol and Sabater-Mir [26] define three levels of trust based on human society: security, institutional, and social. Leisterm and Schultz [15] identify technical, computational, and behavioral trust, but focus primarily on a behavioral trust indicator. Lastly is the idea that trust is a level of confidence, the probability that the intended behavior and the actual behavior are equivalent given a fixed context, fixed environment, and fixed point in time [35].

For our work, we adopted the definition of trust that NIST uses in their report on trustworthiness of cyber physical systems. Trust is defined as "… the demonstrable likelihood that the system performs according to designed behavior under any set of conditions as evidenced by characteristics including, … security, privacy, reliability, safety and resilience" [24].

16.2.2 Types of Trust in IoT Systems

Work on trust management is often divided into two areas: security-oriented and non-security-oriented. The descriptions below are summarized from [33].

Security-oriented trust adopts a restricted view, where trustworthiness is equated to the degree to which an entity or object is considered secure. This traditional view sees trustworthiness as an absolute property that an entity either has or doesn't have. This is often accomplished by determining the credentials an entity possesses, and then iteratively negotiating how to disclose the certified digital credentials that verify properties of trust. This view of trust is also related to trusted computing, which is the expectation that a secure operating environment can be created by enforcing certain hardware and software behaviors with a unique encryption key inaccessible to the rest of the system. In software engineering, this view of trust is determined through formal verification. Managing trust in this context includes specifying and interpreting security policies, credentials, and relationships.

Non-security-oriented trust adopts a wider view similar to the social sciences. This includes a view of trust as a mechanism for achieving, maintaining, and reasoning about the quality of service and interactions. In this view, trust is a measurable property that different entities have in various degrees. Trust is determined on the basis of evidence (personal experiences, observations, recommendations, and overall reputation) and is situational, meaning an entity's trustworthiness differs depending on the context of the interaction. A goal of trust management is managing the risks of interactions between entities. This is also the basis of trust management in multiagent systems, which includes the notion of malicious and selfish behavior. Since non-security-oriented trust is similar to the human notion of trust, work related to computer-mediated trust between users, building human trust in computer systems, and human-computer interaction has led to sophisticated models of trust and reputation research.

To tie this together in a system model for IoT, we adopt a layered trust framework defined by Yan et al. [36]. These layers work together to create an environment in which things and humans can interact and make trustworthy decisions. The layers in the framework include (i) physical perception, which perceives physical environments and human social life; (ii) a network layer that transforms and processes perceived environment data; and (iii) an application layer that offers context-aware intelligent services in a pervasive manner. The fourth layer represents the cyber physical social relationships that connect layers. Figure 16.1 depicts these layers, with trust objectives. A trustworthy IoT system relies on the cooperation among layers. "Ensuring the trustworthiness of one IoT layer (e.g., network layer) does not imply that the trust of the whole system can be achieved" [36].

16.2.3 Trust Architecture

The next step we took is to translate the layers of the trust framework into an architecture, on which a research strategy was developed. The trust architecture and its system components are shown in Fig. 16.2a. The lowest component called the sensor

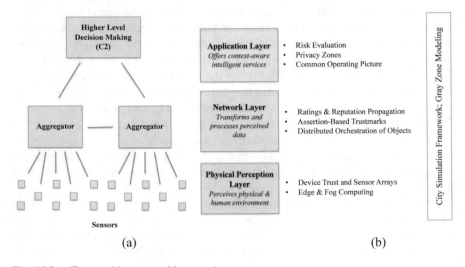

(a)   (b)

Fig. 16.2 a Trust architecture and **b** research strategy

is analogous to the physical perception layer, where sensors, devices, and machines are individually serving as a source for data. The data being generated by these sensors are assimilated and elevated through context into information by an aggregator.

An aggregator is an intelligent machine that collects data from sensors and uses that data to create knowledge for decision-making. In order for an aggregator to determine if the data communicated to it are worthy of being used, the notion of trust becomes an issue. When there are a number of different sensors, each presenting data, and these data have conflicts from sensor to sensor, aggregators must select which among them to trust, how much to trust, as well as some criteria, for establishing that trust. Reasons for competing sources of data to be in conflict with each other, at the Aggregator or Command and Control (C2) components, could be many: malfunction, bad actor, tampering, environmental conditions, context conditions, and so on. Finally, the C2 component is responsible for looking across aggregators to synthesize data, as well as provide an interface, to humans interacting with the system

16.3   Trust Research Strategy

Our research in IoT started with an ideation event that engaged a large number of researchers to discuss technologies, technical challenges, and application areas. This was followed by a number of internally-funded research projects, which spanned the trust architecture. Our initial research was on how aggregators trust data they receive from sensors, as well as how they trust other aggregators. That research was followed by trust and security of sensor arrays, and a trust negotiation language for aggregators. Our research broadened into smart cities, where we looked at the reliability of Fog Computing in communication constrained environments, in situ privacy algorithms, and a situational awareness system for the Georgia Tech police department. More recently, we have looked at the question of risk factors for smart cities, as well as modeling the danger, associated with manipulating technology in smart cities.

Many of these projects are shown in Fig. 16.2b, and described in the sections that follow, to give a better understanding of our trust research strategy. Note that these projects span the multiple dimensions of trust (defined earlier), as well as security- and non-security-oriented trust approaches.

16.3.1   Physical and Perception Layer

16.3.1.1   Device Trust and Sensor Arrays

With a growing demand for data collection, there is also a demand for privacy and safety. This project developed a sensor testbed and algorithms to collect scientific research data, while preserving privacy, in public environments. The technical

approach interfaced a reconfigurable System-on-Chip with multiple sensing periph-
erals and wireless network feedback. The sensor arrays, called Community Array
Nodes, contained ten different sensors that gave insight into urban environment monitoring and smart city behavior. Hardware- and software-based security and trust
mechanisms are allowed for operating in remote environments, providing protection to both deployed hardware and the back-end infrastructure. There were three
successful hardware deployments across the city of Atlanta's North Avenue smart
city test bed, each with the capability of collecting sensor data, processing it in situ,
and sending it to a central server over a secure connection. This approach minimized
the difficulty in deploying a distributed sensor network, while addressing privacy
and security concerns.

16.3.1.2   Fog Computing

Despite its widespread availability, high degree of scalability, and inexpensive cost,
cloud computing cannot address all possible computing needs for modern-day applications. For example, if the network connection to the Internet is severed, the uplink to
the server too slow, or the distance between the user and the server too great, the user's
cloud experience will be significantly degraded. Fog Computing [3], ameliorates
these issues by creating a continuum of compute, storage, and networking resources
from the cloud to the end device. Fog Computing primarily targets applications
which have high bandwidth requirements, low latency restrictions, or that operate in
environments where internet and cloud connectivity are restricted, denied, or inter-
mittent. We explored Fog Computing's ability to operate under degraded networking
conditions, which addresses the reliability aspects of trust. We used Georgia Tech's
MobileFog platform [11, 29], to develop a pedestrian statistics application that
located human figures in frames, tracked their movement between frames, and generated ground position estimates with decimeter-level accuracy. The experiments tested
the limitations of Fog Computing in three bandwidth configurations: benign, hostile,
and denied. Our results represent the first systematic exploration of a real-world Fog
Computing application's response to degraded networking conditions.

16.3.2   Network Layer

16.3.2.1   Trust Negotiation Language

A trust framework is any structure that builds trust among autonomous actors for the
purpose of sharing and reusing identities. The goal of the Trustmark framework [9], is
to facilitate federated identity and attribute management (i.e., the reuse of digital identities and associated attributes) in enterprise systems. This project revolved around
the use of Trustmarks as a secure and robust framework for exchanging trusted,
third-party-attested attributes in support of autonomous peer-to-peer trust decisions.

A Trustmark is a machine-readable, cryptographically signed digital artifact, issued by a Trustmark provider to a Trustmark recipient, and relied upon by one or more Trustmark-relying parties. Such a process is valuable in IoT, as devices come from manufacturers that are themselves Trustmark recipient organizations. These IoT-enabled devices can then present preloaded Trustmarks in order to establish trust. For example, if the device manufacturer Samsung created smartphones with a particular Android capability, it would make sense for Android to grant a Trustmark to Samsung which would then be preloaded and presentable on all Samsung-made devices, to prove that they adhere to a particular set of requirements laid forth by Google on the Android platform.

To securely exchange attribute information in support of IoT trust and trust negotiation, we developed a set of extensions to a preexisting Trustmark framework specification, to include parameter definitions and values within Trust marks. These parameters contain data-typing information that allows for conveying most modern attribute information, as well as human-readable names and descriptive text, for helping Trustmark assessors fill in the appropriate values before issuing Trusmarks. Within Trustmarks, these parameters provide the necessary third-party-attested values that are required for rich trust decisions.

16.3.2.2 M2M Trust

Connecting the physical world with the digital world not only creates new opportunities for innovation and discovery, but also opens doors for misuse and abuse. This work argues that reputation-based trust can be an effective countermeasure for securing M2M communications. We established M2M trust by taking into account both transaction/interaction service behaviors and feedback-rating behaviors in the presence of bogus transactions and dishonest feedback. Our trust model, called M2M Trust [16], introduces two novel trust metrics: pairwise-similarity-based feedback credibility and threshold-controlled trust propagation. We compute the direct trust from machine A to machine B by utilizing their pairwise rating similarity as the weight to the normalized aggregate of ratings that A has given to B.

We conducted extensive experiments using simulation and real datasets for a scenario of self-driving cars on road networks. Specifically, can self-driving cars trust one another to provide a safe driving experience, and can M2MTrust help alleviate traffic jams, whether accidental or malicious? Our direct trust computation model effectively constrained malicious nodes to gain direct trust from dishonest feedback ratings by leveraging feedback credibility. Furthermore, our threshold-controlled trust propagation mechanism successfully blocked the trust propagation from good nodes to malicious nodes. Experimental results showed that M2MTrust significantly outperformed other trust metrics in terms of both attack resilience and performance in the presence of dishonest feedback, and sparse feedback ratings against four representative attack models.

16.3.3 Application Layer

16.3.3.1 Trusting Smart Cities

The benefits of making cities smart must be considered against the potential harm that could come from being massively interconnected. To understand how trust applies to smart cities, we developed a set of risk factors that capture a range of issues that cities should consider when deploying smart city technologies [20]. Risk emerges when the value at stake in a transaction is high, or when this transaction has a critical role in the security or the safety of a system. "In most trust systems considering risk, the user must explicitly handle the relationship between risk and trust by acknowledging that the two notions are in an inverse relationship" [25]. Or put simply—the more risk associated with a transaction, the less we trust it.

To apply this inverse relationship to smart cities, we defined three key risk factors [17]: nontechnical, technical, and complexity. Nontechnical risk includes aspects of a smart city where humans are involved, such as management, training and education, governance, and security practices. Technical risk factors focus on the technology aspects of a smart city, including both hardware and software systems. This also includes the concept of cyber physical systems, which are systems of collaborating computational elements controlling physical entities. The last risk factor is complexity. A smart city is not a discrete thing; it is the complex multidimensional interconnection of diverse systems (human and technology) that deliver services and promote optimum performance to its users. There is risk in the complexity of these systems, especially as the scale becomes very large. Building on these risk factors, a threat analysis matrix for capturing how well smart cities address these risks was proposed.

16.3.3.2 Privacy Zones

While location-based services and applications are increasing in popularity, there are growing concerns over users' location privacy. Although there exist general-purpose mobile permission systems and cloaking techniques, they are often rigid, coarse-grained, not sufficiently personalizable, and unaware of road network semantics. For example, permission decisions are static and follow a one-size-fits-all principle—once a user allows or denies GPS access to an app, the setting is applied to future location requests, unless the setting is manually changed. In Yigitoglu et al. [37], we proposed Privacy Zone, a novel system for constructing personalized fine-grained privacy regions . Privacy Zone allows users to seamlessly enter their privacy specifications under spatial, temporal, and semantic customization. For example, a user can allow location access when at the park, but deny access when at the hospital.

To efficiently compute privacy zones without excessive energy consumption, we developed advanced processing techniques based on the concept of safe hibernation—a time period or a geographic region, within which the client is guaranteed to not enter a privacy zone. We empirically evaluated our techniques to demonstrate their trade-offs with respect to hibernation time, computation effort, and network bandwidth usage. Our results show that Privacy Zone is efficient, scalable, and flexible, while preserving the users' location privacy.

16.4   Simulation Trust

Among the projections concerning the IoT, the pervasive thread in all of them is the sheer size of the expected associated market. While there is still uncertainty and uneasiness on how the whole process is going to unfold, there is common agreement that the number of communications links and smart endpoints (sensors, actuators, etc.) and the induced complexity will be of a significantly different order of magnitude than we have ever experienced.

By its very nature, IoT is entering completely new territory. While IoT technologies aim at integrating (everyday) objects into the communications space, they also bring about a new set of challenges. For example, ergonomics, health implications, as well as risk of cyberattacks needs to be rigorously researched. As a result, the way IoT solutions are going to be designed, developed, and deployed is bound to go through a radical transformation. Related methods and procedures will have to incorporate sound and thorough analysis, and technological and financial considerations will dictate that IoT innovations be modeled and simulated before going live. Modeling and simulation (M&S) is already core to IoT [18]:

> Modeling and simulation is a vital ingredient in creating the connected products at the heart of the Internet of Things. It can support early evaluation and optimization of designs and ongoing verification as changes occur—to make sure the right product is developed and delivered with the required speed and quality.

However, in a universe fraught with nonlinearities shaped by a fast-growing pool of intelligent objects interacting with each other and submitted to a variety of contingent disturbances, extrapolations are no longer clear-cut. Given the IoT complexity and scale, M&S is no longer a luxury, it is a necessity.

M&S is all the more needed since it takes place among the continuous and unstable interaction between enterprises and their environment. IoT enterprises can be complex adaptive socio-technical systems. They consist of many independent agents, whose behavior can be described by social, psychological, and physical rules, rather than dictated by the dynamics of the entire system. The overall enterprise system adapts and learns often resulting in emergent patterns and behaviors. Given that no single agent is in control, complex enterprise system behaviors are often unpredictable and uncontrollable. It follows that creating a model, executing a simulation, and performing experimental runs should become prerequisite steps in any IoT-related project or undertaking. Do the concepts of trust, outlined earlier, apply to M&S?

16.4.1 LVC and IoT

As mentioned previously, IoT is characterized by a wide variety of tags, sensors, actuators, analytics, and embedded systems that are uniquely identifiable and addressable, and cooperate over networks. Also discussed is that IoT has multiple modes of interaction, which include people and things (e.g., H2H, H2M, M2H, and M2M). Characterizing IoT interactions between people and things resembles the framework developed for how people and simulation models interact, known as Live, Virtual, and Constructive (LVC). The similarities between these two areas are quite interesting, and lead us to the observation that IoT is another type of LVC system. The LVC taxonomy, shown in Fig. 16.3, is defined as

- Live simulation refers to M&S involving real people operating real systems (e.g., a pilot flying a jet).
- Virtual simulation is one that involves real people operating simulated systems (e.g., a pilot flying a simulated jet).
- Constructive simulation applications are those that involve simulated (or no) people operating simulated systems (e.g., a simulated pilot flying a simulated jet).

There is no name for simulated people operating real equipment. In the late 1980s, when the LVC taxonomy was created, there were no examples of this type of interaction. However, technology has advanced to the point where simulated humans are operating real systems. For example, driverless cars have proved that the interaction between the real and simulated worlds is possible. Even though that quadrant of the

Fig. 16.3 Categorizing simulation models by the way humans interact with them [13]

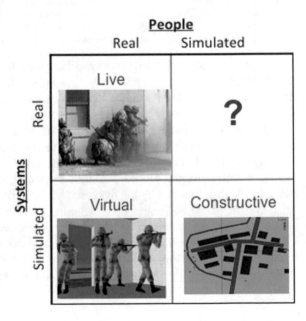

	Simulation	IoT
Live	Real people operating real systems	Sensors, Devices, Smart Phones, Security Cameras
Virtual	Real people operating simulated systems	Mobile Apps, Social Media, Driving Directions
Constructive	Simulated (or no) people operating simulated systems	Embedded Simulation, Machine Learning, Analytics
Autonomy	Simulated (or no) people operating real systems	Driverless Vehicles, Robots, Home/Building Automation, Embedded Systems

Fig. 16.4 Using LVC taxonomy for IoT

matrix has not been officially named, it bears resemblance to artificial intelligence and autonomy.

First presented in [19], we can use the LVC taxonomy to describe IoT, as shown in Fig. 16.4.

- Live refers to real people operating real IoT systems (e.g., a smartphone).
- Virtual refers to real people operating simulated IoT systems (e.g., social media).
- Constructive refers to simulated (or no) people operating simulated IoT systems (e.g., analytics).
- Autonomy refers to simulated (or no) people operating real IoT systems (e.g., driverless vehicles).

The map between IoT and LVC in Fig. 16.4, helps us think about how trust applies to M&S. Using the projects discussed in Sect. 16.3, as a roadmap, we can posit whether trust research applies to live, virtual, or constructive simulation. Let's explore several examples: First, the device trust for sensor arrays and privacy zone work could be applied to live simulation. Context is often necessary for representing things in simulation models (e.g., location), however, knowing the data are different than knowing the person(s)/vehicle(s) that generated the data. This highlights the need to protect the privacy of the data collected from sensors. A second area applies to one of the fundamental elements of LVC simulation: communication mechanisms (exchanging data). The work on M2M Trust and Trustmarks could easily apply to the messages exchanged by simulation systems, as well as determining which simulation model to trust in a message exchange. Third, the reliability of Fog Computing has direct application to live simulation, and constructive in regards to simulation in the cloud or simulation as a service. When simulation computations are pushed closer to the edge to improve response time, the trustworthiness of these computing platforms is critical. Lastly, the risk framework created for looking at smart cities as a whole is directly relevant to looking at the development and execution of LVC + A systems, when one or more of the simulation components are driven by live sensor data. From this quick analysis, it appears that the portfolio of trust research we have explored has relevance to the continuum of LVC simulation.

16.4.2 Internet of Simulation Things

IoT enables distributed control and computational architectures: one can trust (or distrust) abstract concepts, abstract entities, or physical things; including persons, organizations, information, systems, etc. Since simulation models can be used to control or give commands to sensors and actuators, or provide faster-than-real-time prediction to systems, we need to enhance trust relationships when the simulation is part of the IoT system. Expanding IoT's modes of interaction, we have

> Machine to Machine (M2M) → Simulation to Simulation (S2S), Machine to Simulation (M2S) or Simulation to Machine (S2M)
>
> Human to Machine (H2M) → Human to Simulation or Simulation to Human (H2S or S2H)
>
> Human to Human (H2H) → Human to Human (H2H)

An example of where M2S and S2M are already happening is a data-driven online simulation. The Dynamic Data-Driven Application Systems (DDDAS) concept is a unique paradigm for exploiting maturing computational and sensor networking technologies to compensate for model deficiencies and unforeseen system evolution and stimulus conditions, mitigate the effect of design imperfections on long-term, as well as short-term system safety, and enable informed decision for maintenance planning and crisis management [7]. This paradigm utilizes online data to drive simulation computations, and the results are then used to optimize the system or adapt the measurement process. For example, live sensor data and analytics can be used to construct or infer the current state of a system and faster-than-real-time simulation can then be used to project the system's future state. Also, simulation can be used to control an operational system, e.g., data from a real system are fed directly into the simulation model which analyzes alternate options and produces recommended courses of action. With the availability of data from IoT and smart city instrumentation, paradigms such as DDDAS can be expected to grow in importance.

As discussed by Carothers et al. [4], when simulation uses data from things in the network to make decisions, users need to learn how to trust these data, as well as the things (sensors) they are interacting with. Currently securing sensors and devices is accomplished through information security technologies, including cryptography, digital signatures, and electronic certificates. This approach establishes and evaluates a trust chain between devices, but it does not tell us anything about the quality of the information being exchanged over time. Data from sensors or aggregators may be in conflict with each other due to malfunction, bad actors, tampering, environmental conditions, context conditions, and so on. Thus, whether or not the simulation should trust these data must be established by an agent that is capable of a trust evaluation prior to them being deemed useful as information. Further, if the simulation has a role in controlling or giving commands to a sensor or actuator in the IoT system, then any data the simulation uses from an external source to make those decisions must be trustworthy. In other words, the simulation should not purposely be misled into issuing malicious commands.

To illustrate the emerging importance of trust for simulation, let's first look at where work is already underway to use simulation to monitor, control, and predict aspects of cities. Related to the built domain (e.g., a dwelling, building, block, neighborhood, district, city), Farhat et al. [7], are using a DDDAS to monitor the health of large-scale structural systems. Their work is focused on composite materials of aircraft, but we can envision it being applied to city structures like stadiums, bridges, or dams. The overall goal for their work is to enable and promote active health monitoring, failure prediction, aging assessment, informed crisis management, and decision support for complex and degrading structural engineering systems.

Related to the infrastructure (e.g., communications, water and energy, movement of goods and food, mobility networks), simulation and optimization can be used to monitor a city's water supply. Mahinthakumar et al. [21] recognize that urban water distribution systems are vulnerable to accidental and intentional contamination incidents that could result in adverse human health and safety impacts. When a contamination event is detected, e g , data from a water quality surveillance sensor network and reports from consumers, they use a DDDAS approach to answer critical questions like what response action (e.g., shut down portions of the network, implement hydraulic control strategies, introduce decontaminants) should be taken to minimize the impact of the contamination event. Real-time answers to complex questions can be addressed through the dynamic integration of computational components (including models and simulation) and real-time sensor data. The last example is also related to infrastructure, focused on transportation. In Saroj et al. [28], a real-time data-driven transportation simulation model was used to evaluate and visualize network performance, and provide dynamic operational feedback. The study used a hybrid traffic simulation model to represent 17 consecutive intersections on a traffic corridor partially equipped with smart devices. The architecture would enable control of the signals and the vehicle volumes using real-time data from in-field detectors.

As more data-driven simulation is used in smart cities, a concern is that the sensors that provide data to the simulation systems can be hacked, resulting in fake data being sent to simulations. This could be used for all manner of mischief, like causing signal failures that shut down subways or allowing contaminants into the water supply. For example, what if the data driving transportation simulation systems made traffic signals stay red or green, tweak electronic speed limit signs, or messed with ramp meters to send cars onto the freeway all at once? What would commuting look like if erroneous sensor data sent to simulation changes the routes of public transportation or changes subway schedules? How would cities respond to an inadequate supply of electricity or water, or worse yet, not be notified that drinking water was contaminated? What if the waste collection was interrupted during the summertime, and garbage piled up in the streets because the data from smart trash cans that feed a simulation to optimize trash routes was misrepresented? Many systems in cities are interconnected, so erroneous data, driving one simulation could cause a cascade effect, impacting other systems in the city.

Many of these issues get at data integrity, and how to detect misbehavior in the sensor system. Too many false positives may remove valuable sensor resources from the network, while too many false negatives may pollute the data generated and veer the simulation off track [7]. Research that looks at the sensor networks that drive simulation models, and how to discover and correct node misbehavior is critical for simulation trust.

16.5   Internet of Trusted Simulations

As the number of sensors and simulation applications connected to the network grows, we will see different patterns of communication and trust emerge. Data from the sensors and aggregators will be fed into models and simulations that make predictions and decisions which impact our lives. Creating, understanding, and managing large-scale distributed simulation systems interacting with each other to manage operational systems present major challenges and risks.

The IoT can be viewed as a "system of systems" or even "systems of systems". Similar to any other system engineering endeavor, M&S is a critical foundational building block. It can be used early in the life cycle to determine the efficacy of a proposed product, it is an effective means of defining product requirements, and can be used to test and confirm the viability of meeting requirements, as well as to verify the performance of a product. Therefore, looking at the Internet of Trusted Simulations from a systems engineering life cycle perspective can help frame a research agenda.

To construct a set of research issues to consider, we can look to a recently published NIST report which identifies 17 technical concerns that negatively affect the ability to trust IoT products and services [35]

1. Scalability
2. Heterogeneity
3. Ownership and Control
4. Composability, Interoperability, Integration, and Compatibility
5. "Ilities" (availability, compatibility, ...)
6. Synchronization
7. Lack of Measurement
8. Predictability
9. Testing and Assurance
10. Certification
11. Security
12. Reliability
13. Data Integrity
14. Excessive Data
15. Speed and Performance
16. Usability
17. Visibility and Discovery.

Framing the problem in this construct may be unique, but research into this space is underway. Some of these concerns—testing and assurance, certification, heterogeneity, interoperability, composability—are areas where the simulation and LVC community has spent considerable time developing solutions. In other areas—reliability, data integrity—we have spent less time. Several groups are already looking at parts of the trust and risk problem for simulation, as described below.

The Interactive Model-Centric Systems Engineering (IMCSE) research program [27], is investigating the various aspects of humans interacting with models and model generated data, in the context of systems engineering practice. The areas of research they are investigating include: how individuals interact with models; how multiple stakeholders interact using models and model generated information; facets of human interaction with visualizations and large data sets; how trust in models is attained; and what human roles are needed for model-centric enterprises of the future. Their work has developed seven guiding principles for model trust.

1. Transparency should always be possible, but tailorable.
2. Model-context appropriateness is a key determinant of trust.
3. Real-time interaction with models has upsides and downsides.
4. Trust may be implicitly on the models, but explicitly on people.
5. Trust emerges from the interaction between human actors, through models.
6. Availability of model pedigree engenders trust.
7. Trust is influenced by the entangled technological and social factors.

Engineering practice is becoming increasingly model-centric, where models are valuable assets for designing and evolving systems. Thus, human effectiveness in digital engineering and human acceptance of model-centric practice will be essential for future acquisition programs. The objective of the IMCSE research program is to generate knowledge impacting human effectiveness in model-centric environments.

Another group that is tackling an area related to trusted simulations from a systems engineering perspective is the Analysis, Simulation & Systems Engineering Software Strategies (ASSESS) initiative. Their mission is to guide and influence software tool strategies for performing model-based analysis, simulation, and systems engineering, by expanding the use and business benefit of tools. Their most recent strategic insight paper is on Engineering Simulation Risk Models (ESRM). The premise of this work is that engineering simulation is being used more broadly to make technical and business decisions, especially during the early stages of developing a new product. An ESRM is needed to improve credibility by providing a clearer understanding of the predictive capabilities and "appropriateness" of the simulation(s), which will increase confidence in engineering decisions that result from the simulation [2]. The proposed credibility reviews for each simulation phase include

- Previous Phase (if applicable)
- Usage
- Pedigree (as appropriate to that phase)
- Verification

- Fidelity
- Validation
- Uncertainty
- Robustness.

The ESRM includes a set of recommended best practices and associated metrics to understand and manage the appropriateness and risk associated with the simulation.

16.6   Conclusions

This chapter has covered the topic of trust in IoT and smart cities, and posed an argument for why this work is directly relevant to simulation. The definition of trust has many dimensions, which means that there is a rich landscape of problems to address. As pervasive simulation becomes more commonplace in IoT applications, it is essential that they are secure or at least tolerant of cyber threats. Privacy and trust issues must also be adequately addressed to realize widespread adoption. Simulation trust is not an area that research has traditionally focused, and future research should include fundamental principles concerning how trust is established, maintained, and used in simulation, and the theory behind their operations. All of these factors, plus insurability and risk measurement, represent new areas of research that we should pursue to ensure simulation trust in untrusted environments.

Acknowledgements I would like to thank the Georgia Tech Research Institute and the Georgia Tech Institute for Information Security and Privacy for supporting this research. This work could not have been accomplished without my collaborators in GTRI and Dr. Ling Liu in the Georgia Tech College of Computing. I would also like to thank Alain Louchez, the Managing Director for the Center for Development and Application of IoT Technologies (CDAIT) for his collaboration and conversations on Trust and IoT. Lastly, I would like to acknowledge that portions of this chapter were published in the 2019, Winter Simulation Conference, and were presented there as a Titan Talk.

References

1. Ahn J, DeAngelis D, Barber S (2007) Attitude driven team formation using multi-dimensional trust. In: Proceedings of the IEEE/WIC/ACM international conference on intelligent agent technology (IAT '07), 2nd–5th November, Fremont, CA, pp 229–235
2. ASSESS (2018) Understanding and engineering simulation risk model. Credibility theme strategic insight research paper
3. Bonomi F, Milito R, Zhu J, Addepalli S (2012) Fog computing and its role in the internet of things. In: Proceedings of the first MCC workshop on mobile cloud computing, 13th–17th August, New York, NY, pp 13–16
4. Carothers C, Ferscha A, Fujimoto R, Jefferson D, Loper M, Marathe M, Taylor S, Vakilzadian H (2017) Computational challenges in modeling and simulation. In: Fujimoto R, Bock C, Chen

W, Page E, Panchal JH (eds) Re-search challenges in modeling & simulation for engineering complex systems. Springer Nature, Heidelberg, pp 45–74

5. CDAIT (2018) Driving new modes of IoT-facilitated citizen/user engagement. Center for the Development and Application of Internet of Things Technologies. Technical report, July 2018. https://cdait.gatech.edu/sites/default/files/georgia_tech_cdait_thought_leadership_working_group_white_paper_july_9_2018_final.pdf

6. CDAIT (2019) Digital transformation and the internet of things. Center for the Development and Application of Internet of Things Technologies. Technical report, October 2019. https://www.news.gatech.edu/2019/11/25/georgia-tech-internet-things-research-center-releases-white-paper-digital-transformation

7. Farhat C, Michopoulos JG, Chang FK, Guibas LJ, Lew AJ (2006) Towards a Dynamic data driven system for structural and material health monitoring. In: Alexandrov VN, van Albada GD, Sloot PMA, Dongarra J (eds) International conference on computational science. Springer, Berlin, Heidelberg, pp 456–464

8. Greer C, Burns M, Wollman D, Griffor E (2019) Cyber-physical systems and internet of things. NIST Spec Publ 202:52

9. GTRI (2013) NSTIC trustmark pilot. Georgia Tech Research Institute https://trustmark.gtri.gatech.edu

10. Grandison T, Sloman M (2000) A survey of trust in internet applications. IEEE Commun Surv Tutor 3(4):2–16

11. Hong K, Lillethun D, Ramachandran U, Ottenwälder B, Koldehofe B (2013) Mobile fog. In: Proceedings of the second ACM SIGCOMM workshop on mobile cloud computing (MCC'13), 12th–16th August, Hong Kong, China. https://doi.org/10.1145/2491266.2491270

12. IERC (2014) IoT European research cluster website. http://www.internet-of-things-research.eu/about_iot.htm. Accessed 17 June 2014

13. IITSEC (2018) Fundamentals of modeling and simulation. Interservice/industry training, simulation and education conference (I/ITSEC), 26th November, Orlando, FL, Tutorial Number 1819

14. ISO/IEC (2014) ISO/IEC_JTC_1, Smart cities preliminary report

15. Leisterm W, Schultz T (2012) Ideas for a trust indicator in the internet of things. In: Proceedings of the first international conference on smart systems, devices and technologies (SMART 2012), 27th May–1st June, Stuttgart, Germany, pp 31–34

16. Liu L, Loper M, Ozkaya Y, Yasar YA, Yigitoglu E (2016) Machine to machine trust in the IoT era. In: The 18th international workshop on trust in agent societies (Trust 2016), 9th–13th May, Singapore, pp 18–29

17. Loper M (2015) Trusting smart cities. Technical report, Sam Nunn Security Fellows Program, Georgia Institute of Technology, Atlanta, Georgia

18. Loper M, Louchez A (2015) The internet of things and the importance of modeling and simulation: a look at why modeling and simulation capabilities are becoming an indispensable element of the internet of things toolbox. Automation World, 3 August. http://www.automationworld.com/all/internet-things-and-importance-modeling-and-simulation

19. Loper M (2017) Trust as a service in LVC simulations. Invited panel on research challenges in M&S in the era of big data and the internet of things. In: Interservice/industry training simulation & education conference (I/ITSEC), 27th November–1st December, Orlando, FL

20. Loper M (2018) Trusting smart cities: risk factors and implications. Small Wars J. In: Presented at the TRADOC mad scientist conference on installations of the future, June 19th, Atlanta, GA. http://smallwarsjournal.com/jrnl/art/trusting-smart-cities-risk-factors-and-implications

21. Mahinthakumar K, von Laszewski G, Ranjithan R, Brill D, Uber J, Harrison K, Sreepathi S, Zechman E (2006) An adaptive cyberinfrastructure for threat management in urban water distribution systems. In: International conference on computational science. Springer, Berlin, Heidelberg, pp 401–408

22. Matei I, Baras J, Jiang T (2009) A composite trust model and its application to collaborative distributed information fusion. In: Proceedings of the 12th international conference on information fusion (FUSION 2009), 6th–9th July, Chicago, IL, pp 1950–1957

23. National Academies of Sciences, Engineering, and Medicine (NASEM) (2017) A 21st century cyber-physical systems education. National Academies Press
24. NIST (2017) National Institute for Standards and Technology, NIST special publication 1500-202 framework for cyber-physical systems: volume 2, Working group reports, version 1.0, June 2017. http://nvlpubs.nist.gov/nistpubs/SpecialPublications/NIST.SP.1500-202.pdf
25. Patrick A (2002) Building trustworthy software agents. IEEE Internet Comput 6(6):46–53
26. Pinyol I, Sabater-Mir L (2013) Computational trust and reputation models for open multi-agent systems: a review. Artif Intell Rev 40(1):1–25
27. Rhodes DH, Ross AM, Reid J (2019) Interactive model-centric systems engineering (IMCSE) (No. SERC-2019-TR-003). Massachusetts Institute of Technology, Cambridge, United States
28. Saroj A, Roy S, Guin A, Hunter M, Fujimoto R (2018) Smart city real-time data-driven transportation simulation. In: Rabe M, Juan AA, Mustafee N, Skoogh A, Jain S, Johansson B (eds) Proceedings of the 2018 winter simulation conference. Institute of Electrical and Electronics Engineers, Inc, Piscataway, New Jersey, pp 857–868
29. Saurez E, Hong K, Lillethun D, Ramachandran U, Ottenwälder B (2016) Incremental deployment and migration of geo-distributed situation awareness applications in the fog. In: Proceedings of the 10th ACM international conference on distributed and event-based systems (DEBS'16). ACM Press, New York, NY, pp 258–269
30. Siebel TM (2019) Digital transformation: survive and thrive in an era of mass extinction. Rosetta Books, New York
31. Stybel L, Peabody M (2019) Board members: ask this one question every year—'Change comes gradually' is an assumption. Psychol Today. https://www.psychologytoday.com/us/blog/pla tform-success/201908/board-members-ask-one-question-every-year?amp
32. Tascikaraoglu A (2018) Evaluation of spatio-temporal forecasting methods in various smart city applications. Renew Sustain Energy Rev 82(2018):424–435
33. Terzis S (2009) Trust management. IEEE Comput Soc Comput Now 8(9):1296
34. United Nations (UN) (2015) UN DESA, world urbanization prospects: the 2014 revision. United Nations Department of Economics and Social Affairs, Population Division, New York, NY, USA
35. Voas J, Kuhn R, Laplante P, Applebaum S (2018) Internet of things (IoT) trust concerns. NIST cybersecurity white paper, 17th October. https://csrc.nist.gov/CSRC/media/Publications/white-paper/2018/10/17/iot-trust-concerns/draft/documents/iot-trust-concerns-draft.pdf
36. Yan Z, Zhang P, Vasilakos AV (2014) A survey on trust management for internet of things. J Netw Comput Appl 42:120–134
37. Yigitoglu E, Gursoy ME, Liu L, Loper M, Bamba B, Lee K (2018) PrivacyZone: a novel approach to protecting location privacy of mobile users. In: 2018 IEEE international conference on big data, 10th–13th December, Seattle, WA, pp 1238–1247

Margaret L. Loper received a B.S. degree in electrical engineering from Clemson University, Clemson, SC in 1985, a M.S. degree in computer engineering from the University of Central Florida, Orlando, FL in 1991, and a Ph.D. in computer science for the Georgia Institute of Technology, Atlanta, GA in 2002. Dr. Loper is the Associate Director (interim) and Chief Scientist for the Information & Communications Laboratory at the Georgia Tech Research Institute. She has been involved in modeling and simulation research for more than thirty years, specifically focused on parallel and distributed systems. She teaches simulation courses for both academic and professional education, and edited a book on how modeling and simulation is used in the systems engineering life cycle based on her teaching. Dr. Loper also serves as Chief Technologist for the Georgia Tech Center for the Development and Application of Internet of Things Technologies, and she is an associate director for the Georgia Tech Institute for Information Security and Privacy. Dr. Loper was recently named a Regents' Researcher by the University System of Georgia (USG) Board of Regents. This title represents the highest academic and research recognition bestowed by the USG. Dr. Loper's awards and honors include three service awards for her contributions to the development of distributed simulation standards, including Distributed Interactive Simulation and High Level Architecture. She has won the best tutorial award at the Interservice/Industry Training, Simulation & Education Conference, and was named a Mad Scientist by the U.S. Army TRADOC. Most recently she earned the title of Titan of Simulation by the Winter Simulation Conference.

Chapter 17
Ethical and Other Highly Desirable Reliability Requirements for Cyber-Physical Systems Engineering

Tuncer Ören

Abstract In this chapter, first, Cyber-Physical Systems Engineering approach to reliability issue of CPS is cited. Accordingly, following dimensions of reliability of CPS are outlined and relevant references are given: Categories of reliability issues; Reliability and security aspects of computation; Reliability and failure avoidance in simulation studies, including: Validity and verification issues of modeling and simulation, A frame of reference for the assessment of the acceptability of simulation studies, Failure avoidance in artificial intelligence, and Failure avoidance in or due to simulation studies; Aspects of sources of errors. The last one covers the following aspects: Ethics and value systems, including: Ethical requirements from humans and Ethical requirements from AI in simulation and system studies; Decision-making biases, including: Human decision-making biases and AI biases; Improper use of information, such as misinformation, disinformation, and mal-information; Attacks by humans and by autonomous AI systems; Flaws; Accidents; and Natural disasters.

17.1 Introduction

Advancements and ubiquity of Cyber-Physical Systems (CPS) make them Achilles heel of technologically advanced countries. Advancements and increased usage of CPS accelerate their vulnerability. Even though proper use of simulation can enhance the reliability of CPS, other precautions are also in order. Simulation-based system studies are important for almost all disciplines [35, 61]. Hence, cloud simulation, with its ubiquitous aspect, is an important possibility for these disciplines. Furthermore, Simulation-as-a-Service increases the accessibility and utility of simulation [40]. Simulation of Cyber-Physical Systems and their reliability are gaining momentum since a longtime [49, 50]. Their reliability issues cover a large spectrum of activities such as validation and verification, as well as failure avoidance in simulation [62]. From another perspective, reliability issues of systems become a primordial problem

T. Ören (✉)
University of Ottawa, Ottawa, ON, Canada
e-mail: oren.tuncer@sympatico.ca

© Springer Nature Switzerland AG 2020
J. L. Risco Martín et al. (eds.), *Simulation for Cyber-Physical Systems Engineering*,
Simulation Foundations, Methods and Applications,
https://doi.org/10.1007/978-3-030-51909-4_17

429

with the increase of complexity and interdependence of component systems, as well as autonomous smart systems. In this chapter, the emphasis is on the avoidance of failure of complex systems such as cyber-physical systems, as well as their simulation studies. Therefore, the main categories of issues to take into account to assure the reliability of CPS are outlined and basic references are given. Especially many chapters of this book cover in-depth several related issues, therefore, this chapter is deliberately kept as short as possible.

Ethics is a vital link in the reliability of human systems, as well as simulation studies [48, 64]. For CPS, the importance of ethics has already been underlined [49, 50].

For advanced CPS, simulation can be used for several purposes, such as experimentation and gaining experience for training. CPS rely on computation and communication by definition and especially also on artificial intelligence including learning, understanding, and computational awareness.

In *machine learning* (or computational learning), the knowledge base of the computational system is updated with the learned knowledge. However, in a learning system, the learned knowledge should be verified that it is appropriate. Otherwise, the knowledge base of the learning system may deteriorate. Simulation can provide a significant possibility to test the efficacy of the new knowledge base.

Machine understanding (or computational understanding) is very important in computational (artificial) intelligence [65]. In machine understanding systems, an important reliability issue is the avoidance of *misunderstanding* [60, 63]. Other desirable features include *multi-understanding* and *switchable understanding* [66], to assure proper coverage of the machine understanding ability.

Computational awareness is becoming a prerequisite of any "intelligent" system and a priori of advanced artificial intelligence systems. A CPS would be deficient for proper functioning, if its computation and communication abilities are not context aware [68] and/or self-aware [29].

Furthermore, CPS can also be vulnerable especially under *cyberattacks* [72], as well as flaws and malfunctions.

17.2 Cyber-Physical Systems and Cyber-Physical Systems Engineering

"A recent book which elaborates on cyber-physical systems adds: '*In this definition, 'cyber' refers to the computers, software, data structures, and networks that support decision-making within the system, and "physical" denotes not only the parts of the physical systems (e.g., the mechanical and electrical components of an automated vehicle) but also the physical world in which the system interacts (e.g., roads and pedestrians). CPS is closely related to terms in common use today, such as Internet of Things (IoT), the Industrial Internet, and smart cities, and to the fields of robotics and systems engineering.*' [36, p. 14]" [57]. Cyber-Physical Systems are special cases of connected entities; and their development within the realm of the stages of tool development was clarified by [58, 64].

A recent publication gives a good account of the following application areas of Cyber-Physical Systems: Agriculture, education, energy management, environmental monitoring, intelligent transportation, medical devices and systems, process control, security, smart city and smart home, and smart manufacturing [10]. The fact that CPSs are manifesting in many aspects of our existence, redefines the priority of engineering trustworthy systems [71]. A recent article clarifies 7 types of cybersecurity measures to prevent malware-based attacks. They are: (1) Antiviruses and anti-malware, (2) Firewalls, (3) Sandboxing, (4) Content disarm and reconstruction, (5) E-mail scanning and spam filtering, (6) Phishing training, and (7) Browsing and download protection [21]. As it is the case in reliability engineering, the aim is to build reliable systems based on components some of which can be unreliable.

A good survey of academic aspects of cybersecurity, as well as review of basic concepts of cybersecurity, is developed by Abu-Taieh et al. [1]. Complexity challenges in Cyber-Physical Systems are elaborated in a recent book by Mittal and Tolk [34]. A Cybersecurity Body of Knowledge is being developed [12, 13]. A general guide for directors and officers about cybersecurity is available [32, 67], and provides a psychological and behavioral examination in cybersecurity. A recent dictionary clarifies several aspects of cybersecurity [2].

The emergence of cyber-physical-social systems (CPSS) as a novel paradigm has revolutionized the relationship between humans, computers and the physical environment. [86].

17.3   Systems Engineering Approach for Reliability Issues

Systems Engineering approach provides a comprehensive paradigm to cover many aspects of systems studies. Due to many types of systems, there are accordingly many types of systems engineering as clarified by Ören [58].

Systems Engineering is defined by The International Council on Systems Engineering (INCOSE) as "a transdisciplinary and integrative approach to enable the successful realization, use, and retirement of engineered systems, using systems principles and concepts, and scientific, technological, and management methods." [22]. The emphasis is on "successful." Hence, while evidence-based decision-making and success aspects of simulation studies are very important, success of the real system is of primordial importance. In this chapter, in addition to reliability of simulation studies, we also concentrate on the assurance of success of real systems.

By considering on how systems can fail, systematic elaborations of avoidance of failure can be carried out for systems, as well as for their simulation studies. Tables 17.1 and 17.2 list categories of reliability issues and their definitions.

Table 17.1 Categories of reliability issues and their definitions (Definitions are from Merriam-Webster, Unabridged [33])

Error	"an act involving an unintentional deviation from truth or accuracy: a mistake in perception, reasoning, recollection, or expression" "an act that through ignorance, deficiency, or accident departs from or fails to achieve what should be done"
Failure	"omission of performance of an action or task; especially, neglect of an assigned, expected, or appropriate action" "inability to perform a vital function" "a collapsing, fracturing, or giving way under stress: inability of a material or structure to fulfill an intended purpose"
Malfunction	"to function badly or imperfectly: fail to operate in the normal or usual manner"
Accident	"an event or condition occurring by chance or arising from unknown or remote cause" "an unexpected happening causing loss or injury which is not due to any fault or misconduct on the part of the person injured but for which legal relief may be sought"
Defect	"an irregularity in a surface or a structure that spoils the appearance or causes weakness or failure: FAULT, FLAW" "want or absence of something necessary for completeness, perfection, or adequacy in form or function: DEFICIENCY, WEAKNESS"
Fault	"a physical or intellectual imperfection or impairment"
Flaw	"a faulty part" "a fault or defect especially in a character or a piece of work"
Glitch	"malfunction" "a minor problem that causes a temporary setback"
Blunder	"to make a mistake or commit an error usually as a result of stupidity, ignorance, mental confusion, or carelessness"
Bug	"an unexpected defect, fault, flaw, or imperfection"
Mistake	"to choose wrongly" "to be wrong in the estimation or understanding of"
Shortcoming	"the condition or fact of failing to reach an expected or required standard of character or performance"
Inaccuracy	condition of "containing a mistake or error: INCORRECT, ERRONEOUS" condition of "not functioning with accuracy or precision: FAULTY, DEFECTIVE"
Misconception	"a wrong or inaccurate conception"
Misinterpretation	"incorrect interpretation"
Misunderstanding	"a failure to understand" "disagreement"
Falsehood	"absence of truth or accuracy: FALSITY" "an untrue assertion especially when intentional: LIE"

Table 17.2 Categories of fallacies that may create reliability issues and their definitions (Definitions are from Merriam-Webster, Unabridged [33])

Fallacy	"a false or erroneous idea" "a plausible reasoning that fails to satisfy the conditions of valid argument or correct inference"
• Formal fallacy	"a violation of any rule of formal inference — called also *paralogism*"
• Material fallacy	"a reasoning that is unsound because of an error concerning the subject matter of an argument"
• Sophism	"an argument that is correct in form or appearance but is actually invalid; *especially*: an argument used for deception, disputation, or the display of intellectual brilliance"
• Verbal fallacy	"unsound reasoning that uses words ambiguously or otherwise violates a condition for the proper use of language in the argument"
– Amphibology	"ambiguity in language" "a phrase or sentence susceptible of more than one interpretation by virtue of an ambiguous grammatical construction"
– Fallacy of composition	"the fallacy of arguing from premises in which a term is used distributively to a conclusion in which it is used collectively or of assuming that what is true of each member of a class or part of a whole will be true of all together (as in *if my money bought more goods I should be better off; therefore we should all benefit if prices were lower*)"
– Fallacy of division	"unsound reasoning that uses words ambiguously or otherwise violates a condition for the proper use of language in the argument"

17.4 Reliability of and Failure Avoidance in Computation and in Simulation Studies

It is known that software errors can have serious consequences including life hazards. For example, *"Between June 1985 and January 1987, a software-controlled radiation therapy machine called the Therac-25 massively overdosed six people, resulting in serious injury and deaths. A widely cited paper published in 1993 detailed the causes of these accidents [26, 28]"* [27]. Since CPS depends on software, their security also depends on the security of associated software [69, 74]. The software security issue of safety critical CFS is compounded with the many types of sources of failures as cited in Sect. 17.6. There are over 100 definitions of simulation [54]. However, two categories of its usage, namely, experimentation and experience are widespread and important. Accordingly, the following definitions are suggested [55]:

> From the point of view of experiment: *"Simulation* is performing goal-directed experiments with models of dynamic systems."
>
> *Simulation-based experience* may be for training or for entertainment purposes. Accordingly, from the point of view of experience: *"Simulation* is providing experience under controlled conditions for *training* or for *entertainment* [55]."

- Training aspect of experience: "*Simulation* is providing experience under controlled conditions for *training*, i.e., for gaining / enhancing competence in one of the three types of skills: (1) motor skills (virtual simulation), (2) communication and decision-making skills (constructive simulation), and (3) operational skills (live simulation)."
- Entertainment aspect of experience: "*Simulation* is providing experience for entertainment purpose (gaming simulation)."

Failure for not using simulation may have catastrophic consequences as it happened in the case of Boeing 737 Max 8 fatal crashes [19, 24, 84].

The following are outlined in the sequel: (1) Validity and verification issues of modeling and simulation, (2) A frame of reference for the assessment of the acceptability of simulation studies, (3) Failure avoidance in artificial intelligence, and (4) Failure avoidance in or due to Simulation Studies.

17.4.1   Validity and Verification Issues of Modeling and Simulation

A framework for validation and verification of simulation models for systems of systems which "addresses problems arising especially in recently emerging Systems of Systems such as cyber-physical autonomous cooperative systems" was developed by Zeigler and Nutaro [85]. Their definitions of some of the basic concepts are given in the sequel:

A **simuland** is the real-world system of interest. It is the object, process, or phenomenon to be simulated

A **model** is a representation of a *simuland*, broadly grouped into conceptual and executable types.

Abstraction is the omission or reduction of detail not considered necessary in a *model*.

Validation is the process of determining if a *model* behaves with satisfactory accuracy consistent with the study objectives within its domain of applicability to the *simuland* it represents.

Verification is the process of determining if an implemented *model* is consistent with its specification.

Validity and verification issues of modeling and simulation have been well examined since a longtime [3–5]. An early comprehensive bibliography was developed by longtime contributors to the field, Balci and Sargent [6]. Oberkampf [43] provides a bibliography for the period 1967–1998.

17.4.2 A Frame of Reference for the Assessment of the Acceptability of Simulation Studies

An early frame of reference for the assessment of the acceptability of simulation studies was given by Ören [44]. The essence is a matrix where the rows represent elements of simulation studies and the columns represent criteria with respect to which assessments need to be done.

The following elements of simulation studies were considered: simulation results, data (real-world data, simulated data), model (parametric model, values of parameters), experimentation specification (experimental frame, runs (number, length)), program (representation, execution), methodology/technique (modeling, experimentation, simulation, programming).

The following criteria were considered: Goal of the study, real system (structure, data), specific model (parametric model, model parameter set), another model, experimentation specification, norms of (modeling methodology, experimentation technique, simulation methodology, software engineering methodology).

This framework can be expanded in a systematic way. For example, the goal of the study can be evaluated with respect to ethical norms. A more elaborate version of the framework is published in 1993 [73].

17.4.3 Failure Avoidance: Artificial Intelligence (AI)

Synergies of simulation with some disciplines such as AI and software agents, raise other reliability issues for simulation [56]. Intelligence, natural (human or animal) or computational (aka artificial or machine) "is adaptive and goal-directed knowledge processing ability" [47]. A classification and a systematic glossary of 20 types of intelligence was given by Ghasem-Aghaee et al. [17]. An early framework for Artificial Intelligence in quality assurance of simulation studies is given by Ören [45, 46].

Depending on the scope and knowledge processing power, three types of AI are distinguished: Narrow, General, and Super AI.

Artificial Narrow Intelligence (ANI) is AI focused on some area(s). It is also called Narrow or Weak AI. With the inclusion of Artificial Intelligence in simulation, the following issues need to be considered for the reliability of simulation studies: adequacy, completeness, consistency, correctness, and integrity of knowledge bases of the rule base. However, adequacy, completeness, consistency, correctness, and integrity of model-base and parameter base are important with or without the involvement of AI.

Artificial General Intelligence (AGI) is AI having capabilities like humans. Synergy of software agents and simulation brings additional reliability issues [53] such as unconstraint autonomy and cooperation. Possible problems with *autonomy* and *cooperation*—to avoid dangerous coalition of artificial intelligence entities—need to be well studied [64]. The reference gives a list of 57 types of autonomy and clarifies some counterintuitive views of autonomy and cooperation [64].

Artificial Super Intelligence (ASI) is AI with capabilities above human beings. Some of the future dangers of computational intelligence are already elaborated by Barrat [7] and Bostrom [9], and should be considered seriously.

17.4.4 Failure Avoidance in or Due to Simulation Studies

Systems can still fail after all issues mentioned in the previous section are properly treated. For this reason, the concept of Failure Avoidance (FA) was proposed [30, 62]. FA is especially important to assure the reliability of systems. The Risk Digest has already 31 volumes in 2019, and carries many examples of possibilities of failures in systems [78]. Some examples taken from TDR for the possible contribution of simulation to failure avoidance in systems were given in the previously mentioned references [30, 62].

Like any powerful tool, simulation should be used properly. Some possible negative consequences of simulation if not used properly in training and education were cited in the following categories (details in the reference): Training to enhance motor and operational skills (and associated decision-making skills), training to enhance decision-making and operational skills, and education ([62, p. 214], from [51]).

Examples of negative consequences of simulation if not used properly in areas other than training and education were cited in the following categories: Evaluation of alternative courses of actions, engineering design, prototyping, diagnosis, and understanding ([62, pp. 214, 215], from [51]).

The following categories of sources of failures in modeling and simulation are also given: Common mistakes in modeling, in experimentation, in computerization, in project management, and in the expectations of users [62, pp. 216, 217].

17.5 Aspects of Sources of Failures

In this section, eight categories of sources of failures are elaborated on. They are (1) Ethics and value systems, (2) Decision-making biases, (3) Improper use of information: Misinformation, disinformation, and mal-information, (4) Attacks, (5) Flaws, (6) Accidents, and (7) Natural Disasters.

17.5.1 Ethics and Value Systems

Intelligence and knowledge are necessary but not sufficient to solve or to prevent problems. A value system where the metric of success is different from financial gain is needed to have the motivation to solve or to prevent some important problems. Hence, for some important cases, ethical behavior is needed. Ethical behavior requires respect for the rights (of mostly others). The sources of ethical behavior can be:

(i)   self-initiated (or genuine) as it is the case of personal belief or philosophical belief (e.g., altruism of Zen) or

(ii) imposed (or emulated) where the norms can be imposed by the state (legislation) or by the society (customs, peer pressures, regulations, codes of conduct, and codes of ethical conduct).

(iii) The case of the religious source of ethical behavior can be self-initiated or imposed.

Educating cybercitizens, namely cyber ethics in education [8], as well as cyber-physical systems education [37], are important issues. A rationale for a code of professional ethics for simulationists is given by Ören [48].

17.5.1.1 Ethical Requirements from Humans

The oldest simulation society (SCS) has a code of professional ethics for simulationists [59, 75]. Many professional societies have codes of professional ethics also. The behavior may be a responsible behavior by self-imposed restriction or by imposed restriction(s), accountable behavior [48, p. 429]. The joint ACM/IEEE-CS Software Engineering Code of ethics is applicable to any member of the software engineering profession [18]. Code of ethics for Systems Engineers is given by INCOSE [23].

Normally, ethical behavior would be expected from humans. However, this aspect of human behavior, as witnessed by ecological disasters, among many other problems, is unfortunately not yet well developed. Corruption Perceptions Index documents this aspect of 180 countries and territories [11].

17.5.1.2 Ethical Requirements from AI in Simulation and System Studies

CPS are already pervasive, and their domains of application are widening. These aspects make them fragile and create a weakness in the functioning of societies. Hence, reliability of CPS becomes a primordial issue. So far as the ethical aspect of possible solutions is concerned, this important problem is well recognized [49, 50, 80].

17.5.2   Decision-Making Biases

17.5.2.1   Human Decision-Making Biases

An important flaw in human decision-making, in addition to cultural biases and ignorance-induced biases, is dysrationalia, a term coined by Keith Stanovich, an emeritus professor of applied psychology and human development. Some fundamental concepts clarified by Stanovich are:

- Traditional IQ tests miss some of the most important aspects of real-world decision making. It is possible to test high in IQ yet to suffer from the logical-thought defect known as dysrationalia.
- One cause of dysrationalia is that people tend to be cognitive misers, meaning that they take the easy way out when trying to solve problems, often leading to solutions that are wrong.
- Another cause of dysrationalia is the mindware gap, which occurs when people lack the specific knowledge, rules and strategies needed to think rationally.
- Tests do exist that can measure dysrationalia, and they should be given more often to pick up the deficiencies that IQ tests miss [76].

Several very convincing cases of dysrationalia are given in the article [76]. Hence, in Cyber-Physical Systems Engineering studies, the cases of dysrationalia need to be avoided.

17.5.2.2   AI Biases

There is an overreliance on AI techniques regardless of which type of AI (namely, Artificial Narrow Intelligence, Artificial General Intelligence, or currently inexistent Artificial Super Intelligence) is used. For example, it is misleading to claim that machine learning, though with impressive abilities—with adequate training data— represents all aspects of computational intelligence. Therefore, equating machine learning with computational intelligence is deceptive. Computational understanding [60, 63, 65], as well as computational awareness abilities [87], are other important aspects of computational intelligence.

The types of AI bias and the difficulty of fixing it are identified by Hao [20]. Ho clarifies the biases in framing, collecting, and preparing data in deep learning. For the hardness to fix AI bias, Ho elaborates on the following issues: Unknown unknowns, imperfect processes, lack of social context, and the definitions of fairness.

17.5.3 Improper Use of Information: Misinformation, Disinformation, and Mal-information

Information can be true or false. *Mal-information* is the use of true information with the intention of causing harm to individuals, group of people, institutions, or countries. The distinction between misinformation and disinformation is important. Both are based on false information. *Misinformation* is not intentional. "The creation of misinformation might be the result of a variety of situations including honest mistakes, ignorance, or some unconscious bias [15, p. 333, 83]." However, in *disinformation*, false information is "intentionally communicated to mislead people about some aspect of reality [15, p. 333, 83]." Malicious signals may be part of disinformation aiming for autonomous and autonomic systems. Disinformation is a part of the attacks of systems. Regardless of their origins, both misinformation and disinformation may lead to serious consequences in system studies. Strategies to protect systems from disinformation and misinformation need to be developed.

17.5.4 Attacks (by Humans and Autonomous AI Systems)

Attacks of CPS can be made by humans and by autonomous AI systems. Some attacks, such as disinformation are at the information level. However, systems need to be protected from physical, as well as cyberattacks, which are becoming more effective and pervasive than physical assaults [38]; hence cybersecurity and cyber resilience are of prime importance [39, 79]. The vision of the radar of an antimissile system can easily become blurry, once a prespecified signal is received to activate a preloaded software component. Hence, the counterattack cannot be as effective as anticipated. A different aspect of the security problem is the avoidance of planting of spy chips [81]; early detection of their existence becomes very important.

17.5.5 Flaws

Flaws of some components, for example, of sensors or chips—if they are used alone—can have very serious influences on the security of systems; therefore, such components of important CPS should be tested to assure that they are free from flaws. Redundant components may be useful. Sources of flaws on the components of autonomous and especially autonomic systems need to be systematically checked and eliminated. Unfortunately, even early in 2019, autopilot crash did happen [31], and the number of flaws even on government websites is very high [77].

17.5.6 Accidents

Accidents with serious consequences happen, as for example, Three Mile Island nuclear accident of 1979 [82], Chernobyl disaster of 1986 [25], and Fukushima Daiichi nuclear accident of 2011 [14]. Accident prevention techniques [41, 42], need to be applied especially for the infrastructure of CPS.

17.5.7 Natural Disasters

Some types of potential "accidents" can be avoided based on common sense; by not building a nuclear central or a dam on the natural fault line, for example. However, some unforeseen disasters may be expected especially with the current drastic climate change. One source of prevention or protection is to turn to scientific knowledge and scientific thinking [16, 70].

17.6 Conclusion

In the connected world we live in, the connected elements, as well as their connections, are increasing. Hence, the vulnerability of our environment is also increasing with an accelerated rate. A possible additional reliability problem is chained reactions in connected elements, namely, failure of a connected element causing failure in other connected elements.

Based on the maturing process of the simulation discipline [35, 52], a possibility for the future advance artificial intelligence systems can be a *simulation-based evidential decision-making*. Such autonomous or quasi-autonomous advanced artificial intelligence systems can formulate with the human decision maker(s), relevant hypotheses, and accordingly, can access models from cloud-based simulation infrastructures, if necessary, modify the models, or develop appropriate models, and to formulate relevant experimentation scenarios to perform simulation studies. Based on the outcomes of the simulation studies, the artificial intelligence system can make evidence-based decisions. Such a scenario that currently appears to be futuristic may be one of the several possibilities to advance computational intelligence systems. However, another approach can be human decision-making augmented with computational intelligence. In such a case, simulation-based decision-making can be the essence of evidence-based rational decision-making, provided several aspects of failure avoidance, as outlined in this chapter, can be taken into consideration.

References

1. Abu-Taieh EM, Al Faries AA, Alotaibi ST, Aldehim G (2018) Cyber security body of knowledge and curricula development. https://doi.org/10.5772/intechopen.77975, https://www.intechopen.com/books/reimagining-new-approaches-in-teacher-professional-development/cyber-security-body-of-knowledge-and-curricula-development. Accessed 29 Nov 2020
2. Ayala L (2016) Cybersecurity lexicon. Apress, New York, NY
3. Balci O (1998) Verification, validation, and accreditation. In: Proceedings of the winter simulation conference, pp 41–48
4. Balci O (1998) Verification, validation, and testing. In: Banks J (ed) The handbook of simulation, Chap. 10. Wiley, New York, NY, pp 335–393
5. Balci O, Sargent RG (1981) A methodology for cost–risk analysis in the statistical validation of simulation models. Commun ACM 24(4):190–197
6. Balci O, Sargent RG (1984) Bibliography on the credibility assessment and validation of simulation and mathematical models. ACM Simuletter 15(3):15–27
7. Barrat J (2013) Our final invention—artificial intelligence and the end of the human era. Thomas Dunne Books, St. Martin's Press, New York, NY
8. Blackburn A, Chen IL, Pfeffer R (2018) Emerging trends in cyber ethics and education. IGI Global, Hershey, PA
9. Bostrom N (2014) Superintelligence—paths, dangers, strategies. Oxford University Press, Oxford, UK
10. Chen H (2017) Applications of cyber-physical systems: a literature review. J Indus Integra Rev 2(3):1750012-1–1750012-1-28
11. CPI (2018) Corruption perceptions index 2018. https://www.transparency.org/files/content/pages/2018_CPI_Executive_Summary.pdf. Accessed 23 Nov 2019
12. CyBOK-IEEE. IEEE—scoping the cyber security body of knowledge. https://www.computer.org/csdl/magazine/sp/2018/03/msp2018030096/13rRUytF47N. Accessed 23 Nov 2019
13. CyBOK-UK. Cyber security body of knowledge. UK National Cyber Security Centre. https://www.ncsc.gov.uk/section/education-skills/cybok. Accessed 23 Nov 2019
14. EB-Fukushima (2011) Encyclopedia Britannica. Fukushima accident. https://www.britannica.com/topic/accident-safety. Accessed 23 Nov 2019
15. Fallis D (2016) Mis- and dis-information (lying, propaganda etc.). In: Floridi L (ed) The Routledge handbook of philosophy of information. Routledge, New York, NY
16. Foster J (ed) (2019) Facing up to climate reality: honesty, disaster and hope. London Publishing Partnership, London, UK
17. Ghasem-Aghaee N, Ören T, Yilmaz L (2017) Simulation, intelligence and agents: exploring the synergy. In: Mofakham FN (ed) Frontiers in artificial intelligence—intelligent computational systems. Bentham Science Publishers, Sharjah, UAE, pp 1–58
18. Gotterbarn D, Miller K, Rogerson S (1997) Software engineering code of ethics. Commun ACM 40(11):110–118
19. Guerra P, Gilbert D (2019) US airlines still don't have flight simulators for Boeing 737 Max 8 pilot training. Vice News. https://www.vice.com/en_us/article/vbwj9m/us-airlines-still-dont-have-flight-simulators-for-boeing-737-max-8-pilot-training. Accessed 23 Nov 2019
20. Hao K (2019) This is how AI bias really happens—and why it's so hard to fix. MIT Technology Review, Feb 4, 2019. https://www.technologyreview.com/s/612876/this-is-how-ai-bias-really-happensand-why-its-so-hard-to-fix/. Accessed 23 Nov 2019
21. IE-Cybersecurity (2019) Interesting engineering. 7 cybersecurity measures that can prevent malware-based attacks. https://interestingengineering.com/7-cybersecurity-measures-that-can-prevent-malware-based-attacks. Accessed 23 Nov 2019
22. INCOSE-def—definition of systems engineering. https://www.incose.org/systems-engineering. Accessed 23 Nov 2019
23. Incose-ethics. https://www.incose.org/about-incose/Leadership-Organization/code-of-ethics. Accessed 23 Nov 2019

24. Kriel R, Mukhtar I, Hu C (2019) This is the flight simulator and manual used to train pilots of doomed Ethiopian airlines flight. CNN. https://www.cnn.com/2019/03/28/africa/ethiopian-air lines-boeing-max-8-simulator-manual-intl/index.html. Accessed 23 Nov 2019
25. Leatherbarrow A (2016) Chernobyl 01:23:40: the incredible true story of the world's worst nuclear disaster. Published by Andrew Leatherbarrow
26. Leveson N (1995) Safeware: system safety and computers. Addison Wesley, Boston, MA
27. Leveson NG (2017) The Therac-25: 30 years later. Computer, Nov. 2017, pp 8–11. https:// www.computer.org/csdl/magazine/co/2017/11/mco2017110008/13rRUxAStVR. Accessed 23 Nov 2019
28. Leveson NG, Turner CS (1993) An investigation of the Therac-25 accidents. Computer 26(7):18–41
29. Lewis PR et al (2016) Self-aware computing systems: an engineering approach. Springer, New York
30. Longo F, Ören TI (2010) Enhancing quality of supply chain nodes simulation studies by failure avoidance. In: Proceedings of the EMSS 2010—22nd European modeling and simulation symposium (within the IMMM-7—The 7th international Mediterranean and Latin American modeling multi-conference), October 13–15, Fes, Morocco, pp 201–208
31. Marshall A (2019) Another fatal Tesla autopilot crash and more car news this week. WIRED. https://www.wired.com/story/another-fatal-tesla-autopilot-crash-more-car-news/. Accessed 23 Nov 2019
32. McAlaney J, Frumkin LA, Benson V (2019) Psychological and behavioral examinations in cyber security. IGI Global, Hershey, PA
33. Merriam-Webster, Unabridged. Accessed 23 Nov 1999
34. Mittal S, Tolk A (eds) (2019) Complexity challenges in cyber-physical systems: using modeling and simulation (M&S) to support intelligence, adaptation and autonomy. Wiley, Hoboken, NJ
35. Mittal S, Durak U, Ören T (eds) (2017) Guide to simulation-based disciplines: advancing our computational future. Springer, New York, NY
36. NAP-CPSE (2016) A 21st century cyber-physical systems education. The National Academic Press, Washington, DC
37. NAP-education (2016) A 21st century cyber-physical systems education. The National Academic Press, Washington, DC. https://www.nap.edu/catalog/23686/a-21st-century-cyber-physical-systems-education. Accessed 23 Nov 2019
38. NAP-lethality (2019) The growing threat to air force mission-critical electronics—lethality at risk. The National Academic Press, Washington, DC. https://www.nap.edu/catalog/25475/the-growing-threat-to-air-force-mission-critical-electronics-lethality. Accessed 23 Nov 2019
39. NAP-resilience (2019) Forum on cyber resilience—beyond spectre: confronting new technical and policy challenges. The National Academic Press, Washington, DC. https://www.nap.edu/catalog/25418/beyond-spectre-confronting-new-technical-and-policy-challenges-proceedin gs-of. Accessed 23 Nov 2019
40. NATO/STO (2015) Modelling and simulation as a service: new concepts and service-oriented architectures. NATO—Science and Technology Organization, Brussels, Belgium. TR-MSG-131
41. NIST (2016) Systems security engineering considerations for a multidisciplinary approach in the engineering of trustworthy secure systems. Commerce Dept. National Institute of Standards and Technology, Gaithersburg, MD
42. NSC (2015) Accident prevention for business & industry: engineering & technology, 14th edn. National Security Council, Washington, DC
43. Oberkampf WL (1998) Bibliography for verification and validation in computational simulation. SANDIA Report, SAND98-2041, Albuquerque, NM and Livermore, CA
44. Ören TI (1981) Concepts and criteria to assess acceptability of simulation studies: a frame of reference. CACM 24(4):180–189
45. Ören TI (1986) Artificial intelligence in quality assurance of simulation studies. In: Elzas MS et al (eds) Modelling and simulation methodology in the artificial intelligence era. North-Holland, Amsterdam, The Netherlands, pp 267–278

46. Ören TI (1987) Quality assurance paradigms for artificial intelligence in modelling and simulation. Simulation 48(4):149–151
47. Ören TI (1994) Artificial intelligence in simulation. Ann Oper Res 53(1):287–319
48. Ören TI (2002) Rationale for a code of professional ethics for simulationists. In: Proceedings of the 2002 summer computer simulation conference, pp 428–433
49. Ören TI (2002, Invited Plenary Paper) Ethics as a basis for sustainable civilized behavior for humans and software agents. Acta Systemica 2(1):1–5. Also published in the Proceedings of the InterSymp 2002—The 14th international conference on systems research, informatics and cybernetics of the IIAS, July 29–August 3, Baden-Baden, Germany
50. Ören TI (2002, Invited Paper) In search of a value system for sustainable civilization: ethics and quality. In: Special focus symposium (within InterSymp 2002) on: cognitive, emotive and ethical aspects of decision making and human actions, August 1–2, Baden-Baden, Germany, pp 69–72
51. Ören TI (2002, Invited Plenary Paper) Growing importance of modelling and simulation: professional and ethical implications. In: Cheng Z et al (eds) Proceedings of the Asian simulation conference/the 5th international conference on system simulation and scientific computing, Nov. 3–6, 2002, vol 1, International Academic Publishers/Beijing World Publishing Corp, Shanghai, China, pp 22–26
52. Ören TI (2005, Keynote Article) Maturing phase of the modeling and simulation discipline. In: Proceedings of ASC—Asian simulation conference 2005 (The sixth international conference on system simulation and scientific computing (ICSC '2005), 2005 October 24–27, International Academic Publishers—World Publishing Corporation, Beijing, P.R. China, pp 72–85
53. Ören TI (2007) Reliability in agent-directed simulation. In: Proceedings of the EMSS 2007— 19th European modelling and simulation symposium. Bergeggi, Italy, October 4–6, 2007, pp 78–86
54. Ören TI (2011) The many facets of simulation through a collection of about 100 definitions. SCS M&S Mag 2(2):82–92
55. Ören TI (2011) A critical review of definitions and about 400 types of modeling and simulation. SCS M&S Mag 2(3):142–151
56. Ören T (2018, Keynote) Powerful higher-order synergies of cybernetics, systems thinking, and agent-directed simulation for cyber-physical systems. In: Proceedings of InterSymp 2018— The 30th international conference on systems research, informatics and cybernetics, and the 38th annual meeting of the international institute for advanced studies, July 30–August 3, 2018, Baden-Baden, Germany
57. Ören T (2018) Simulation of cyber-physical systems of systems: some research areas—computational understanding, awareness and wisdom. J Chinese Simul Syst (2)
58. Ören T (2020) Agent-directed simulation and nature-inspired modeling for cyber-physical systems engineering. In: Risco-Martin J-L, Mittal S, Ören T (eds) Simulation for cyber-physical systems engineering: a cloud-based context. Springer, New York
59. Ören TI, Elzas MS, Smit I, Birta LG (2002) A code of professional ethics for simulationists. In: Proceedings of the 2002 summer computer simulation conference, pp 434–435
60. Ören TI, Kazemifard M, Yılmaz L (2013) Machine understanding and avoidance of misunderstanding in agent-directed simulation and in emotional intelligence. In: Proceedings of SIMULTECH 2013, 3rd international conference on simulation and modeling methodologies, technologies and applications, July 29–31, Iceland, Reykjavik, pp 318–327
61. Ören T, Mittal S, Durak U (2018, Invited Paper) A shift from model-based to simulation-based paradigm: timeliness and usefulness for many disciplines. Internat J Comput Softw Eng 3(1)
62. Ören TI, Yilmaz L (2009) Failure avoidance in agent-directed simulation: beyond conventional V&V and QA. In: Yilmaz L, Ören TI (eds) Agent-directed simulation and systems engineering. Wiley, Systems Engineering Series, Berlin, Germany, pp 189–217
63. Ören TI, Yilmaz L (2011) Semantic agents with understanding abilities and factors affecting misunderstanding. In: Elci A, Traore MT, Orgun MA (eds) Semantic agent systems: foundations and applications. Springer, New York, pp 295–313

64. Ören T, Yilmaz L (2017) The age of the connected world of intelligent computational entities: reliability issues including ethics, autonomy and cooperation of agents (invited ebook chapter). In: Mofakham FN (ed) Frontiers in artificial intelligence—intelligent computational systems. Bentham Science Publishers, UAE, pp 184–213

65. Ören T, Yilmaz L, Ghasem-Aghaee N, Kazemifard M, Noori F (2016) Machine understanding in agent-directed simulation: state-of-the-art and research directions (Keynote article of the 2016 modeling and simulation of complexity in intelligent, adaptive and autonomous systems (MSCIAAS) symposium of the SpringSim'16). In: Durak U, Günal M (eds) Proceedings of the SpringSim'16, April, 3–6, 2016, Pasadena, CA. SCS—The Society for Modeling and Simulation International, San Diego, CA, pp 848–855

66. Ören TI, Yilmaz L, Kazemifard M, Ghasem-Aghaee N (2009) Multi-understanding: a basis for switchable understanding for agents. In: Proceedings of the summer computer simulation conference, July 13–16, 2009, Istanbul, Turkey. SCS, San Diego, CA, pp 395–402

67. Paloalto-NYSE (2018) The definitive cybersecurity guide for directors and officers, 2nd edn. https://www.securityroundtable.org/navigating-the-digital-age-2nd-edition/. Accessed 23 Nov 2019

68. Pathan KT (2013) Activity awareness in context-aware systems using software sensors. PhD dissertation. University of Leicester, Leicester, UK

69. Pfleeger CP, Pfleeger SL, Margulies J (2015) Security in computing, 5th edn. Prentice-Hall, Englewood Cliffs, NJ

70. Pielke R Jr (2018) The rightful place of science: disasters & climate change, 2nd edn. Revised and updated. Consortium for Science, Policy & Outcomes, Tempe, AZ: Arizona State University

71. Saydjari OS (2018) Engineering trustworthy systems: get cyber security design right for the first time. McGraw-Hill, New York, NY

72. Selby N, Vescent H (2017) The cyber attack survival manual: tools for surviving everything from identity theft to the digital apocalypse. Weldon Owen, Richmond, CA

73. Sheng G, Elzas MS, Ören TI, Cronhjort BT (1993) Model validation: a systemic and systematic approach. Reliab Eng Syst Safe 42:247–259

74. Sikorski M, Honig A (2012) Practical malware analysis: the hands-on guide to dissecting malicious software. No Starch Press, San Francisco, CA

75. SimEthics. https://scs.org/ethics/. Accessed 23 Nov 2019

76. Stanovitch KE (2015) Rational and irrational thought: the thinking that IQ tests miss. Scientific American mind, January 2015. https://www.scientificamerican.com/article/rational-and-irrational-thought-the-thinking-that-iq-tests-miss/. Accessed 23 Nov 2019

77. Stokel-Walker C (2019) Exclusive: thousands of security flaws found on UK government websites. New Scientist, 2019 March 23. https://www.newscientist.com/article/2197453-exclusive-thousands-of-security-flaws-found-on-uk-government-websites/#ixzz62C0upHm9. Accessed 23 Nov 2019

78. TRD—The risk digest (2019). https://catless.ncl.ac.uk/Risks/. Accessed 23 Nov 2019)

79. USC, USHoRs, CoFAs (2013) Cyber attacks: an unprecedented threat to U.S. national security. United States Congress, United States House of Representatives, and Committee on Foreign Affairs. Second Session, March 2013. https://www.govinfo.gov/content/pkg/CHRG-113hhrg80123/pdf/CHRG-113hhrg80123.pdf. Accessed 23 Nov 2019

80. Winfield AF, Jirotka M (2018) Ethical governance is essential to building trust in robotics and AI systems. Philosophi Trans R Soc A Mathemat Phys Eng Sci. https://royalsocietypublishing.org/doi/full/10.1098/rsta.2018.0085 Accessed 23 Nov 2019

81. WIRED-chips (2019) Planting tiny chips in hardware can cost as little as $200, 2019-10-10. https://www.wired.com/story/plant-spy-chips-hardware-supermicro-cheap-proof-of-concept/. Accessed Feb 5 2020

82. WNO (2012) World nuclear org. Three Mile island accident. https://www.world-nuclear.org/information-library/safety-and-security/safety-of-plants/three-mile-island-accident.aspx. Accessed 23 Nov 2019

83. Wolterman J (2019) Psychological warfare: a brief history. Information and politics. https://informationandpolitics.com/psychological-warfare/psychological-warfare-a-brief-overview/?gclid=CjwKCAjwldHsBRAoEiwAd0Jybe-1DVNkpgelqXUCvV-aaqt-1FoXcTImErVhC-D1bJTuC7gQLeRx3xoCKGwQAvD_BwE. Accessed 23 Nov 2019
84. Young C (2019) Boeing reportedly scrapped 737 Max safety system as it was too expensive. https://interestingengineering.com/boeing-reportedly-scrapped-737-max-safety-system-as-it-was-too-expensive. Accessed 23 Nov 2019
85. Zeigler BP, Nutaro J (2015) Towards a framework for more robust validation and verification of simulation models for systems of systems. J Defence Model Simul Appl Method Technol 13(1):2–16
86. Zeng J, Yang LT, Lin M, Ning H, Ma J (2016) A survey: cyber-physical-social systems and their system-level design methodology. Future Generat Comput Syst 105. https://doi.org/10.1016/j.future.2016.06.034
87. Zhao Q (2011) Computer awareness: ways towards intelligence. In: Madani K, Dourado A, Rosa A, Felipe J (eds) Computational intelligence. Revised and Selected papers of the international joint conference (IJCCI) 2011, Paris, France, October 24–16, 2011, pp 3–14

Tuncer Ören is a professor emeritus of computer science at the School of Electrical Engineering and Computer Science of the University of Ottawa, Canada. He has been involved with simulation since 1965. His Ph.D. is in Systems Engineering from the University of Arizona, Tucson, AZ (1971). His basic education is from Galatasaray Lisesi, a high school founded in his native Istanbul in 1481 and in Mechanical Engineering at the Technical University of Istanbul (1960). His research interests include: advancing methodologies for modeling and simulation; agent-directed simulation; agents for cognitive and emotive simulations (including representations of human personality, understanding, misunderstanding, emotions, and anger mechanisms); computational awareness; reliability, QA, failure avoidance, ethics; as well as body of knowledge and terminology of modelling and simulation. He has over 540 publications, including 54 books and proceedings (+3 in press and in preparation). He has contributed to over 500 conferences and seminars held in 40 countries. Dr. Ören has been honored in several countries: USA: He is a Fellow of SCS (2016), an inductee to SCS Modeling and Simulation Hall of Fame–Lifetime Achievement Award (2011), and received SCS McLeod Founder's Award for Distinguished Service to the Profession (2017). Canada: Dr. Ören has been recognized, by IBM Canada (2005), as a pioneer of computing in Canada. He received the Golden Award of Excellence from the International Institute for Advanced Studies in Systems Research and Cybernetics (2018). Turkey: He received "Information Age Award" from the Turkish Ministry of Culture (1991), an Honor Award from the Language Association of Turkey (2012), and Lifetime service award from the Turkish Informatics Society and Turkish Association of Information Technology (2019). A book was edited by Prof. Levent Yilmaz: Concepts and Methodologies for Modeling and Simulation: A Tribute to Tuncer Ören. Springer (2015).

Index

© Springer Nature Switzerland AG 2020
J. L. Risco Martín et al. (eds.), *Simulation for Cyber-Physical Systems Engineering*,
Simulation Foundations, Methods and Applications,
https://doi.org/10.1007/978-3-030-51909-4

Printed in the United States
by Baker & Taylor Publisher Services